1080/ND

ENZYME BIOCHEMISTRY OF THE ARTERIAL WALL
AS RELATED TO ATHEROSCLEROSIS

To the memory of
my parents

IRENE
and
DAVID ZEMPLÉNYI, M.D.

ENZYME BIOCHEMISTRY OF THE ARTERIAL WALL

AS RELATED TO
ATHEROSCLEROSIS

TIBOR ZEMPLÉNYI
M.D., D.Sc.

Head of Atherosclerosis Research, Institute for Cardiovascular Research, Prague-Krč and Asst. Professor of Medicine, Charles University, Prague

LLOYD-LUKE (MEDICAL BOOKS) LTD
49 NEWMAN STREET
LONDON
1968

© LLOYD-LUKE (MEDICAL BOOKS) LTD., 1968

This book is protected under the Berne Convention and may not be reproduced by any means in whole or in part. Application with regard to reproduction should be addressed to the Publisher

PRINTED AND BOUND IN ENGLAND BY
HAZELL WATSON AND VINEY LTD
AYLESBURY, BUCKS
SBN 85324 048 5

PREFACE

It has been my purpose in this book to present an account of the biochemistry of vascular enzymes, particularly in relation to atherosclerosis and to our own work at the Institute for Cardiovascular Research in Prague. The author felt that this monograph would be useful to those interested in problems of vascular disease and should be directed especially towards two types of investigator, the biochemist and the "medical biologist". However, the former may not be too aware of the medical problems, whereas the latter, to be candid—often has forgotten most of the biochemistry that he learnt as a student. For these reasons the author decided to attempt the difficult task of addressing himself both to "chemists" and "doctors".

Part one of this text is a review of vascular enzyme biochemistry with background material that will obviously be common knowledge among biochemists; it may help to clarify this field for the biologist. My task has been simplified in many ways by Hoffman-Ostenhof's *Enzymologie* and in particular Dixon and Webb's *Enzymes*.

Parts two and three are particularly aimed at exploring some of the relationships between enzyme function and vascular disease. In these sections the medical biologist's indulgence is sought for the presentation in some chapters of elementary material well known to the medical graduate.

Prague, February, 1968   TIBOR ZEMPLÉNYI

ACKNOWLEDGMENTS

It is a great pleasure to express my thanks and appreciation to my associates Ing. O. Mrhová, C.Sc. and Dr. D. Urbanová, C.Sc. Without their close co-operation the programme of investigations discussed in Part Two of this book could not have been accomplished. I should also like to express my gratitude to my former associates Ing. D. Grafnetter, C.Sc. and Dr. Z. Lajda, D.Sc., for co-operation in many aspects of our studies; to Ing. B. Buriánová for statistical calculations; to Dr. K. Ośancová for reading the English text; to Mrs. E. Baburková, Mrs. J. Hájková, Mrs. L. Hášová, Mrs. I. Knížková, Miss A. Krimláková, Mrs. M. Krabcová, Mrs. E. Procházková and Mrs. J. Vacková for indispensable technical assistance. I wish to acknowledge the excellent assistance of Miss M. Dobíhalová, Mrs. L. Gréová, Mrs. I. Knížková and Miss A. Krimláková in preparing the manuscript and the illustrations of this monograph.

I would like also to thank Academician Klement Weber, D.Sc. and Professor Jan Brod, D.Sc., head of the Institute for Cardiovascular Research, for their continued interest and encouragement.

Several of the illustrations in this book first appeared in articles by the author and his associates in the following journals or books: *Arch. Mal. du Coeur* (Figs. 30, 32, 34, 35, 39, 40); *Ann. N.Y. Acad. Sci.* (Figs. 44–49 and 62); *Circulation Research* (Fig. 74); *Brit. J. Exp. Path.* (Figs. 75 and 76); *Enzymes of Lipid Metabolism* (1961), Pergamon Press, Oxford (Fig. 77); *Cor et Vasa* (Fig. 78); *J. Atheroscler. Res.* (Plate II, microphotographs 3 and 4); *Atherosclerosis and Its Origin* (1963), Academic Press, New York (Plate II, microphotographs 1, 2, 5, 6, 7 and 8); *Bull. Soc. Roy. Antwerp* (Fig. 79) and *Acta Zool. Pathol. Antwerp* (Plate VII, microphotographs 4, 5 and 6). I wish to record my gratitude to the editors and/or publishers who have so readily given their permission for the inclusion of the illustrations in this book.

I wish to express my hearty thanks to Professor C. W. M. Adams, Guy's Hospital Medical School, University of London, for invaluable advice throughout the preparation of this monograph. He also undertook the "literary adjustment" of the manuscript.

Finally my thanks are due to the publishers, Lloyd-Luke (Medical Books) of London, for their patience, consideration and co-operation at every stage of the preparation and production of this book.

TIBOR ZEMPLÉNYI

CONTENTS

PREFACE   v

ACKNOWLEDGMENTS   vi

LIST OF ABBREVIATIONS   xi

PART ONE
GENERAL PROPERTIES OF ENZYMES SO FAR INVESTIGATED IN THE VASCULAR WALL

I   INTRODUCTION   3

II   ENZYMES OF GLYCOLYSIS AND GLYCOGEN BREAKDOWN   15
    Glycogen-phosphorylase, E.C. 2.4.1.1.
    Hexokinase, E.C. 2.7.1.1.
    Phosphoglucomutase, E.C. 2.7.5.1.
    Glucosephosphate isomerase, E.C. 5.3.1.9.
    Fructosediphosphate aldolase, E.C. 4.1.2.13.
    Phosphoglycerate kinase, E.C. 2.7.2.3.
    Phosphoglyceromutase, E.C. 2.7.5.3.
    Phosphopyruvate hydratase, E.C. 4.2.1.11.
    Lacate dehydrogenase, E.C. 1.1.1.27.

III   ENZYMES OF THE TRICARBOXYLIC ACID (TCA) CYCLE   27
    Citrate synthase, E.C. 4.1.3.7.
    Aconitate hydratase, E.C. 4.2.1.3.
    Fumarate hydratase, E.C. 4.2.1.2.
    Isocitrate dehydrogenase, E.C. 1.1.1.42.
    Succinate dehydrogenase, E.C. 1.3.99.1.
    Malate dehydrogenase, E.C. 1.1.1.37.
    Malate dehydrogenase (decarboxylating), E.C. 1.1.1.40.

IV   ENZYMES OF THE GLUCOSE 6-PHOSPHATE OXIDATION SYSTEM (THE PENTOSE PHOSPHATE PATHWAY)   35
    Glucose 6-phosphate dehydrogenase, E.C. 1.1.1.49 and phosphogluconate dehydrogenase (decarboxylating), E.C. 1.1.1.44.
    Ribosephosphate isomerase, E.C. 5.3.1.6.

V   ENZYMES OF THE RESPIRATORY CHAIN   39
    $NADH_2$ cytochrome c reductase, E.C. 1.6.2.1. and lipoamide dehydrogenase, "diaphorase", E.C. 1.6.4.3.
    Cytochrome oxidase, E.C. 1.9.3.1.

VI	ENZYMES LINKING ENERGY-RICH BONDS WITH MUSCULAR CONTRACTION	45

    Creatine kinase, E.C. 2.7.3.2.
    Adenosinetriphosphatase, ATPase, E.C. 3.6.1.3. and E.C. 3.6.1.8.
    Adenylate kinase, E.C. 2.7.4.3.

VII	THE GLYOXALASE SYSTEM	51
VIII	OTHER ENZYMES: LISTED ACCORDING TO THE NATURE OF THE CATALYSED REACTION	52

    Oxidoreductases
    3-Hydroxyacyl-CoA dehydrogenase, E.C. 1.1.1.35.
    Glutamate dehydrogenase, E.C. 1.4.1.2.-4.
    Monoamine oxidase, E.C. 1.4.3.4.
    Glutathione reductase, E.C. 1.6.4.2.
    Transaminases or Aminotransferases
    Aspartate aminotransferase, E.C. 2.6.1.1.
    Alanine aminotransferase, E.C. 2.6.1.2.
    Glutamine-fructose-6-phosphate aminotransferase or hexosephosphate aminotransferase, E.C. 2.6.1.16.
    Pentosyltransferases
    Purine nucleoside phosphorylase, E.C. 2.4.2.1.
    Hydrolases
    Acetylcholinesterase and cholinesterase, E.C. 3.1.1.7. and E.C. 3.1.1.8.
    Alkaline phosphatase and acid phosphatase, E.C. 3.1.3.1. and E.C. 3.1.3.2.
    5′-Nucleotidase, E.C. 3.1.3.5.
    Arylsulphatase, E.C. 3.1.6.1.
    β-Glucuronidase, E.C. 3.2.1.31.
    Leucine aminopeptidase, E.C. 3.4.1.1.
    Cathepsins
    Lyases
    Carbonate dehydratase, carbonic anhydrase, E.C. 4.2.1.1.
    Isomerases
    Mannosephosphate isomerase, E.C. 5.3.1.8.

IX	ENZYME COFACTORS	72

    Nicotinamide nucleotide coenzymes
    Flavin nucleotides
    Glutathione
    Coenzyme A
    Purine and pyrimidine nucleotides as phosphate carriers
    Biotin

PART TWO

SPECIAL PROBLEMS OF VASCULAR ENZYMES AND ATHEROSCLEROSIS

- X INTRODUCTION — 87
- XI NOTES ON METHODS USED IN THE AUTHOR'S LABORATORY FOR STUDIES ON VASCULAR WALL ENZYMES — 95
  - General principles of "Optical Tests"
  - Notes on methods using tetrazolium salts for determination of dehydrogenase activities
  - Alkaline phosphatases, alkaline phosphomonoesterases
  - Acid phosphatase, acid phosphomonoesterase
  - 5′-Nucleotidase
  - Adenosinetriphosphatase, ATPase
  - "Non-specific" carboxylesterase
  - Lactate dehydrogenase (Optical Test)
  - Malate dehydrogenase (Optical Test)
  - Tetrazolium methods for determination of dehydrogenase activities
  - Lactate dehydrogenase isozymes
  - β-Glucuronidase
  - Aminotransferases
  - Other methods used
- XII SEX DIFFERENCES — 108
- XIII ENZYMES OF THE VASCULAR WALL IN EARLY STAGES OF EXPERIMENTAL RABBIT ATHEROSCLEROSIS — 119
- XIV PROBLEMS OF EXPERIMENTAL ATHEROSCLEROSIS IN THE RAT — 135
- XV THE EFFECT OF INJURY ON VASCULAR ENZYMES — 145
  - Feeding rats with excess vitamin D
  - Arterial wall injury caused by experimental hypertension
- XVI ARTERIAL HYPOXIA AND LACTATE DEHYDROGENASE ISOZYMES AS RELATED TO ATHEROSCLEROSIS — 161
- XVII COMPARISON OF ENZYME ACTIVITIES IN HUMAN VESSELS OR VASCULAR SEGMENTS DIFFERING IN SUSCEPTIBILITY TO ATHEROSCLEROSIS — 168
- XVIII PROBLEMS OF COMPARATIVE ATHEROSCLEROSIS AS RELATED TO THE DIFFERENT SUSCEPTIBILITY OF ARTERIES AND ARTERIAL SEGMENTS TO THE DISEASE — 181
  - *Enzymes of the vascular wall in the chicken and the duck*
  - *Enzymes of the vascular wall in the rhesus macaque*
  - *Enzymes of the vascular wall in the pig*
  - *Enzymes of the vascular wall in cattle*

XIX	CONCLUDING REMARKS TO PART TWO: AN ATTEMPT TO EVALUATE THE SIGNIFICANCE OF LOCAL VASCULAR FACTORS IN ATHEROGENESIS	195

PART THREE
THE LIPOLYTIC ACTIVITY OF THE VASCULAR WALL

XX	INTRODUCTION	203
XXI	MAST CELLS AND ATHEROSCLEROSIS	207
XXII	THE RELATIONSHIP BETWEEN ELASTASE, ATHEROSCLEROSIS AND LIPOPROTEIN LIPASE	209
XXIII	THE LIPOLYTIC ACTIVITY OF THE VESSEL WALL	214

XXIV	THE CHOLESTEROL ESTERASE AND PHOSPHOLIPASE A ACTIVITY OF ARTERIAL TISSUE
	    Cholesterol esterase, E.C. 3.1.1.13.
	    Phospholipase A, E.C. 3.1.1.4.
		226

XXV	CONCLUDING REMARKS TO PART THREE	231
	REFERENCES	232
	INDEX	255

LIST OF ABBREVIATIONS

AcP	Acid phosphatase (phosphomonoesterase II)
ADP	Adenosine 5'-diphosphate
AGPDH	Glycerol-3-phosphate dehydrogenase*
AMP	Adenosine 5'-phosphate
AP	Alkaline phosphatase (phosphomonoesterase I)
APP	ATPase
ATP	Adenosine 5'-triphosphate
CDP	Cytidine 5'-diphosphate
CoA	Coenzyme A
DEAE cellulose	Diethylaminoethyl cellulose
DFP	Di-isopropyl phosphorofluoridate
DNA	Deoxyribonucleic acid
FAD	Flavin-adenine dinucleotide
FFA	Free fatty acid
FMN	Flavin mononucleotide
GDP	Guanosine 5'-diphosphate
GSH	Reduced glutathione
GSSG	Oxidized glutathione
LDH	Lactate dehydrogenase
MDH	Malate dehydrogenase
NAD	Nicotinamide-adenine dinucleotide
$NADH_2$	Reduced NAD
NADP	Nicotinamide-adenine dinucleotide phosphate
$NADPH_2$	Reduced NADP
NBMT	Nitroblue monotetrazolium
NT	Neotetrazolium
5-Nu	5'-nucleotidase
PMS	Phenazine methosulphate
RNA	Ribonucleic acid
SDH	Succinate dehydrogenase
TCA cycle	Tricarboxylic acid cycle
UDP	Uridine 5'-diphosphate
UDPG	Uridinediphosphoglucose

Formerly known as α-*glycerophosphate dehydrogenase.*

PART ONE

GENERAL PROPERTIES OF ENZYMES
SO FAR INVESTIGATED IN THE
VASCULAR WALL

Chapter I

INTRODUCTION

THE arterial wall is a metabolically-active living organ that has its own equipment for the numerous anabolic and catabolic processes that are fundamental to all living matter and vital for synthesizing normal and abnormal tissue-components.

All such processes depend on the catalytic effects of enzymes and enzyme systems that link energy-producing reactions with energy consuming processes required for mechanical or osmotic work, for active transport through membranes, for biosynthesis and various other functions. Enzymes, which are specialized protein molecules, are the instruments for all the energy conversions in the cell.

For the proper understanding of changes which enzymes may undergo during life, it will be useful briefly to discuss the biosynthesis of enzymes, the factors regulating the amount of enzymes produced and the factors regulating their activity.

Enzymes are essentially proteins and many aspects of their formation have been elucidated by one of the most exciting chapters of discovery in modern biology. (See for example the excellent chapter on enzyme formation in Dixon and Webb's *Enzymes*, and the reviews by Riley and Pardee, 1962 and Bennett and Dreyer, 1964.)

It must be mentioned—without going into detail—that all the genetic information required for the exact synthesis of proteins is embodied in the deoxyribonucleic acid (DNA) of the genes contained within chromosomes of the cell nucleus. Work of recent years has thrown considerable light upon the way in which this genetic information is transmitted to the protein-synthesizing equipment of the cell.

The information specifying the amino acid sequence of the polypeptide chain in the protein (enzyme) to be synthesized is stored in the form of a specific nucleotide sequence in the DNA chain. (The four purine and pyrimidine bases of the DNA chain are adenine, thymine, guanine and cytosine.) The DNA molecule contains a coded message, the genetic code, with "instructions" for the synthesis of specific proteins.

The transcription of the information carried in the DNA molecule is brought about in the cell nucleus by a specific enzyme RNA nucleotidyltransferase (E.C. 2.7.7.6.). This enzyme catalyses the synthesis of polynucleotide chains from ribonucleoside triphosphates, and for this reaction DNA is required as a template. In this way a ribonucleic acid (RNA) chain is built up which is complementary to the DNA template, because of specific pairing of bases of the DNA template and those of the newly formed RNA chain. The base sequence of the latter is thus the same as in the DNA molecule, with the exception that uracil is produced instead of thymine.

The RNA leaves the nucleus, diffuses in the cytoplasm and becomes bound to ribosomes. (Submicroscopic particles attached to the membranes of the "rough" endoplasmic reticulum.) Usually several ribosomes form complexes (polysomes) with one molecule of RNA. Since this type of RNA molecule carries the genetic information from the gene to the ribosome, it is named messenger RNA (mRNA). The synthesis of the polypeptide chain of the protein takes place at the ribosomes and the mRNA acts as a template.

It is now generally accepted that the first stage in protein biosynthesis is the "activation" of the individual amino acid molecules which is brought about by a group of enzymes each being specific for one of 20 amino acids (amino-acyl-t-RNA synthetases, E.C. 6.1.1.). They catalyse reactions where energy-rich linkages with specific amino acid carrier molecules are produced. These amino acid carriers are RNA molecules of a short chain length containing 70–80 mononucleotide residues. They have a molecular weight of only about 25,000, while the molecular weight of mRNA is up to 1,000,000. They are preferably called transfer RNA (tRNA) instead of the former term soluble RNA (sRNA). There exists at least one tRNA for each amino acid and the activation and transfer of amino acids take place with a high degree of specificity.

In the second stage the amino acid carried by its tRNA becomes attached to the mRNA—ribosome complex.

It is extraordinarily interesting that the base sequence of three consecutive mononucleotide residues along the template mRNA chain has the ability to code a single amino acid and these "coding triplets", the so-called "codons" of mRNA are recognized by the tRNA molecules *via* corresponding "adaptor triplets" or "anticodons". Therefore, the tRNA molecule must possess two recognition sites, one which is specific for the binding of the correct amino acid in the first stage (see above) and another, which recognizes the codon of the mRNA. The attachment of tRNA to the mRNA-ribosome complex takes place by an enzymatic process which requires guanosine triphosphate (GTP), one molecule of which is broken down for each amino acid incorporated. During the reaction the amino acid is attached to the end of the growing polypeptide chain and a molecule of tRNA is released in an active form and can be used again as an amino acid carrier.

It appears most likely that the ribosomes are initially attached to the beginning of the mRNA template and subsequently move along the polynucleotide chain with the amino acid correctly "labelled" by the tRNA. Thus, step by step the polypeptide chain grows and when the ribosome reaches the end of the template the completed polypeptide is released by a mechanism so far not entirely understood (perhaps by an enzymatic process requiring ATP or GTP).

The formation of hydrogen-bonds and correct folding of the polypeptide chain may take place either during its synthesis on the ribosome or after the release of the completed chain. It appears that these events depend on the amino acid sequence and no additional genetic information is needed.

Remarkable progress has been made in deciphering the coding system whereby the four-character "language" of nucleic acids (embodied in the coding triplets) is translated into the twenty-character language of proteins. This deciphering is made possible by the use of synthetic polynucleotides and especially trinucleotides which can replace natural mRNA and which are added to cell-free systems for protein synthesis (containing suspensions of ribosomes, enzymes, tRNAs, ATP, GTP, K and Mg ions, amino acids etc.). Since the nature of the amino acids thus incorporated depends on the nucleotide composition of the synthetic polynucleotides, it was possible to identify the nucleotide composition of a number of coding triplets corresponding to individual aminoacids and to construct a "dictionary" of code words. The genetic code seems to be universal for all living matter.

It is superfluous to say that such enormous advances in knowledge, accumulating with phenomenal speed in the last few years, are fundamental to our understanding of how cells function and how basic events of life are being carried out.

It is of great interest to obtain further insight into the factors regulating the rate of synthesis of proteins, particularly enzymes.

First of all it must be realized that the formation of an enzyme is impossible without the presence of the corresponding gene, i.e. without the corresponding genetic information in the cell. However, this does not ensure the actual production of the enzyme. The presence of an "inducer" or the absence of a "repressor" may also be needed for the enzyme to be synthesized in appreciable quantity.

These phenomena have been investigated particularly in micro-organisms. A classical and much investigated example is the induction of β-galactosidase in *Escherichia coli* by growth on lactose. But there is little doubt that induction and repression also function in higher organisms. An important point is that the *amount* of enzymes actually synthesized *de novo* from amino acids is affected and not the *activity* of the enzyme.

The inducer is usually the substrate of the enzyme but sometimes a substance closely related to the substrate can also work as inducer, and the natural substrate must not necessarily be the best inducer. In some cases one metabolite may induce the synthesis of a group of related enzymes for a pathway by which the inducing substance can be metabolized.

In contrast to the inducer, the repressor is a product of the enzyme reaction or is closely related to it. Theoretically, if there is excess production of an enzyme, the concentration of its product will be high and this will tend to repress the formation of the enzyme. The opposite is true in the case of deficiency of an enzyme, where the concentration of the substrate will be increased. This would provide a self-regulatory mechanism to adjust the amounts of the enzyme synthesized in the cell to the needs of its metabolism. However, the available evidence seems to suggest that in the majority of cases those enzymes, whose formation is controlled by induction, are not subject to repression and *vice versa* (see Dixon and Webb, 1964).

The biochemical basis of induction and repression is unknown. It is not clear whether inducers act on nucleic acids, or accelerate activation of amino acids, or interfere with the formation or availability of some repressor (derepression).

Some theories postulate the existence of units of consecutive genes in the chromosomes ("operons") and explain induction and repression as action at some sensitive control points (operators) of these units. Other theories postulate the existence of cytoplasmic repressors, the production of which is directed by regulator genes, quite different from the other structural genes. An inducer could combine with such cytoplasmic repressor and inactivate it, thereby allowing the enzyme-synthesizing system to function.

As mentioned above, the phenomena of induction and repression have been studied mostly in micro-organisms, but there can be little doubt that they also play an important role in the adaptive changes in enzymatic activities of higher organisms. Here, however, more complicated control systems, in particular those of hormonal and neurohumoral character, are superimposed on the more primitive regulatory mechanisms. Nevertheless, these differences must not cause us to lose sight of the fundamental process of enzyme synthesis as the main factor on which the amount of the enzyme in the cell primarily depends.

The enzymes synthesized in the cell are confined to cytoplasmic structures (mitochondria, endoplasmic reticulum, ribosomes, lysosomes, cytoplasmic ground substance etc.) and to the nucleus. The localization of some enzymes is very specific, e.g. that of the enzymes of terminal oxidation in the mitochondria. The amount of enzymes in the cell may be affected by the tightness of their attachment to these intracellular structures, as well as by permeability changes in the cell membrane and in the membranes of the particulate structures (e.g. of mitochondria and of lysosomes). These changes can be part of a physiological mechanism (e.g. permeability changes of the mitochondrial membrane during the swelling-contraction cycle of the mitochondrion—see p. 159). However, more often they accompany pathological alterations of tissues and cause leakage of enzymes out of the cells. This mechanism of changes in the tissue concentration of enzymes appears to be of considerable importance in connection with atherogenesis and will be discussed in several chapters of this book.

It is obvious that one of the important means of regulating metabolism involves changes in the *amounts* of certain enzymes in tissues. There exist other important mechanisms that modify enzymatic *activity*. When working with pure enzymes, the term specific activity denotes units of enzyme per milligram of protein, whereby one unit is defined as the amount that will catalyse the transformation of 1 micromole of substrate per minute, under defined conditions. However, in many biological investigations one is working with impure preparations, or very often with blood plasma, homogenates or crude extracts. Under these circumstances, specific activity cannot be determined and the term "activity" is used only to express the amount of substrate (or product) per unit time and unit weight of protein, nitrogen, total solids, wet weight or other parameters of the preparation (extract, homogenate, tissue fluid etc.). Although less exact from the enzymological point of view, such arbitrary activity units of impure enzymatic sources clearly depend on the specific activity of the enzyme: work with such material has provided much information of basic importance in biology and medicine. Of course, if the specific activity of the enzyme is known, the "purity" of the biological preparation can be calculated.

There are many factors that may alter the activity of an enzyme *in vitro* (concentration of substrate, pH, temperature, specific inhibitors and activators —such as coenzymes, metal ions etc.). There is fairly good evidence to believe that many of these factors also exert a regulatory function *in vivo*, and, in particular, the availability of certain "vitamin cofactors" and metal ions appears to be essential under physiological conditions for enzyme activity.

An important regulatory mechanism is so-called *feedback inhibition*; it is characterized by inhibition of a biosynthetic pathway by its end-product. This mechanism has been identified in synthetic pathways for amino acids, nucleotides and other substances and, in all cases, the first enzyme of the pathway has been shown to be the site of inhibition. In contrast to repression, feedback inhibition abolishes the activity of an enzyme already synthesized. While repression is a relatively slow controlling mechanism, feedback inhibition is almost instantaneous and is able to adjust enzyme activity to momentary metabolic fluctuations (Dixon and Webb, 1964).

This introduction would be incomplete if mention were not made of the existence of multiple forms of an enzyme. Various techniques have shown that enzymes previously considered homogeneous are in fact heterogeneous, and the term isozymes (isoenzymes) was coined to refer to multiple molecular forms of an enzyme existing within a single cell or tissue.

The techniques used for the demonstration and isolation of isozymes are based on their physical, chemical, immunological or kinetic properties. Usually electrophoretic methods (e.g. starch gel, agar gel, paper) or chromatographic methods (e.g. DEAE-cellulose chromatography) have been used, but other methods—such as differential salting out—have also proved effective.

The isozymes most intensively investigated are those of lactate dehydrogenase, but more than a hundred other enzymes have been demonstrated in the last few years to exist in multiple molecular forms.

Isozymes may differ in primary structure and, in addition, the structure may be modified by attachment of small molecules to allosteric sites of the enzyme (i.e. to positions other than the active site). Other alterations consist in the removal of a portion of the polypeptide chain. It appears that the commonest molecular basis for the production of isozymes is the existence of polymers composed of different subunits.

For example, some dehydrogenases known to exist in multiple forms, such as alcohol dehydrogenase and glutamate dehydrogenase, appear to be polymers composed of only one type of subunit. On the other hand, a large body of evidence indicates that the structure of lactate dehydrogenase isozymes (see Chapter XVI) comprises two types of subunits (A and B), which form five types of tetramers, i.e. BBBB or LDH_1, BBBA or LDH_2, BBAA or LDH_3, BAAA or LDH_4 and AAAA or LDH_5 (Appella and Markert, 1961). Since the different kinds of subunits have different amino acid compositions, it is postulated that their synthesis is encoded in different genes that control the subunit synthesis independently. The genetic approach has been especially applied in the study of the regulation of lactate dehydrogenase isozyme formation under physiological and pathological conditions. (For a review of such genetic control see Vesell, 1966a.)

Of special interest is the question of the biological significance of isozymes. Their widespread occurrence and localization in different subcellular regions suggest that they have important regulatory functions in cell metabolism. For example, it is known that some intermediate metabolites are common to more than one biochemical process or that they lie at branching steps of more complicated metabolic cycles. It has been concluded that the existence of multiple forms of enzymes exhibiting similar functions enables these divergent metabolic pathways to be precisely regulated (Stadtman, 1963). We shall see in Chapter XVI that a function of this kind is ascribed to lactate dehydrogenase isozymes. Another aspect of the metabolic role of enzymes with similar functions is indicated, for example, by the existence of a soluble and mitochondrial form of an enzyme. In the case of glycerolphosphate dehydrogenase and, perhaps, also malate dehydrogenase this property provides a shuttle system for reduced NAD between the cytoplasm and mitochondria (Kaplan, 1963, see also p. 41).

These are only a few examples of the biological significance of isozymes. Many problems in this field are still unresolved but it can be expected that further advances in this new area of biochemical research will throw more light on several aspects of the precise regulation of cell metabolism.

In addition to the foregoing discussion of enzyme formation and regulation, some further introductory remarks must be made on the general problems of arterial metabolism.

The respiration of arterial tissue has been investigated by many authors, usually by means of the Warburg respirometer or other manometric methods. In 1943 Lazovskaya demonstrated a slight oxygen consumption by rat-aorta homogenates. Briggs et al. (1949) found that the aortas of rats can oxidize various substances of the glycolytic and tricarboxylic acid cycle. The oxygen uptake of intact rat aortas is about one-tenth of the uptake exhibited by the liver, and the Q_{O_2} (i.e. μl. of oxygen consumed per 1 mg. of dry tissue per hour) of the aorta is 1·09. For comparison, it is interesting that Kirk et al. (1954a) showed that the Q_{O_2} of the dog aorta is 0·63 and that of the human aorta is 0·26 (0·22–0·36), while Beaconsfield (1962) observed a Q_{O_2} value of only 0·14 for dog arterial tissue.

Wertheimer and Ben-Tor (1960, 1961) investigated the effect of age, diet and some hormones on rat aortic respiration. They demonstrated lower oxygen uptake in aortas of old rats and depression of the oxygen and glucose uptake by a diet containing 55 per cent butter-fat. They observed increased oxygen consumption under the influence of norepinephrine and insulin and a fall in alloxan-diabetic animals. In old rats heparin increased the oxygen uptake.

Hilz (1962) found increased oxygen uptake in vitamin-A deficient rats. Mineralocorticoids inhibited aortic O_2 consumption whereas glucocorticoids exhibited a biphasic effect; at low concentrations increased O_2 uptake occurred but high doses inhibited aortic respiration.

Since different aortic segments reveal a different susceptibility to atherosclerosis (see Chapter XVII), it seemed reasonable to investigate aortic oxygen uptake from this topochemical aspect. The results are, however, quite contradictory. In the few experiments that Briggs et al. (1949) carried out they did not observe any clearcut difference between rat aortic segments, but Christie and Dahl (1957) found lower oxygen uptake in the abdominal than in the thoracic segment of the rat aorta.

In rabbits, Dury et al. (1957) found that the oxygen consumption of the aortic arch was higher than in other portions of the vessel. Fischer and Geller (1960) observed a similar ratio and, moreover, cortisone administration and renal hypertension (with or without cholesterol feeding) were accompanied by a significant rise of oxygen consumption in all aortic segments.

In Munro et al.'s (1961) experiments the cockerel's aorta exhibited only an insignificant trend towards lower oxygen consumption in the thoracic part when compared with the arch and abdominal segment.

The detailed comparative studies of Maier and Haimovici (1957, 1958) dealing with human, rabbit and dog aortic segments are discussed on page 32 in connection with the succinoxidase system.

Sex differences have been reported by Munro and Rifkind (1964) who observed higher oxygen consumption in the aortas of older hens than cocks. However, hexoestrol implantation did not affect oxygen consumption. This is in contrast to what might be expected from the results of Malinow and Moguilevsky (1961), who reported increased oxygen consumption in the innominate trunk of similarly treated cockerels.

The results of investigations concerned with the effect of cholesterol feeding on aortic respiration are also very contradictory.

For example Fischer and Geller (1960), Wolleman and Kocsar (1964) and others found decreased aortic oxygen consumption in cholesterol-fed rabbits. On the other hand, Whereat (1961a and b; 1964) and Krčílek et al. (1962) observed increased respiration in such rabbit aortas. (The latter authors used the polarographic method described by Šerák et al., 1962, whereas all the other above-mentioned results were obtained by the conventional Warburg technique). Mandel et al. (1966) could not detect any difference in aortic oxygen uptake between cholesterol-fed and control rabbits when results were expressed on a wet weight basis, but on a surface basis atherosclerotic aortas exhibited increased oxygen uptake.

Loomeijer and Ostendorf (1959) and Munro et al. (1961) reported higher oxygen consumption in the cholesterol-fed rat aorta, but Wolleman and Kocsar (1964) found lower oxygen consumption in such aortas when compared with those of control rats.

The aortas of cholesterol-fed cockerels exhibited lower oxygen consumption than control aortas in Munro et al.'s (1961) experiments; these authors suggested—with regard to their inverse results in rats—that a relationship exists between interspecies susceptibility to experimental atherosclerosis and the metabolic response of the vessel wall to cholesterol feeding.

It can be seen that the results reported are extremely contradictory and confusing: the reason should perhaps be sought in the experimental conditions and methods used.

It may be useful, in this connection, to draw attention to some basic principles in the study of tissue respiration. Belicer demonstrated as early as 1939 that the respiration of a muscle homogenate is increased by about 300 per cent if a phosphate acceptor is added to the medium. Lardy and Wellman (1952) introduced the concept of "respiratory control" by showing that, in a tightly coupled system of oxidative phosphorylation, maximal respiration requires the presence of a phosphate acceptor, such as ADP.*

* Chance and collaborators (see Chance and Williams, 1956) showed that the respiratory rate of mitochondrial systems is highly dependent on several factors—especially ADP and substrate level—provided that sufficient oxygen is available. A slow respiration rate can result from a low level of both ADP and substrate in the system (state 1); from the absence of substrate in spite of a high level of ADP (state 2 or "starved" state); from a low level of ADP in spite of a high substrate level (state 4 or "resting" state). A fast respiration rate will result from the presence of a high level of ADP and substrate in the system (state 3 or "active" state). The

It is obvious that the presence of a sufficient concentration of a phosphate acceptor, especially ADP, is essential not only in work with mitochondria but also with homogenates. In the latter the concentration may vary, if it is not controlled, and this may introduce a bias into the results. It is not clear whether, in all the reported arterial respiration studies, the experimental conditions were controlled in this respect.**

Another possible interfering factor in studies on arterial respiration, especially that of atherosclerotic tissue, may be the presence of fatty acids. It is well known that cellular respiration and energy metabolism can be very significantly affected by the tightness of coupling of phosphorylation to oxidation. Thus, Pressman and Lardy demonstrated (1956) that very low concentrations of fatty acid anions activate mitochondrial ATPase (see also p. 143) and induce uncoupling of respiration from its dependence on phosphorylation.

Finally, it should be emphasized that most results dealing with arterial respiration have been expressed on a wet weight or dry weight or tissue nitrogen basis. In this respect we can quote the remarks made by Lehninger (1959) in connection with vascular respiratory studies: "It is regrettable that the measurements of respiratory rates were not also determined on the basis of cell counts or their equivalents, such as nucleic acid phosphorus etc., rather than dry weight, to avoid reckoning in the large and metabolically inert extracellular mass. Total nucleic acid, or better yet, DNA, represents an intracellular component which is *relatively* constant for different cell types, and presumably does not exist extracellularly". The introduction of such a *bias* appears to be especially important in atherosclerotic tissue with its masses of metabolically inert extracellular material.

Nevertheless, even if the quantitative data must be accepted often with some reservation, it seems to be well established that arterial tissue in man and all the above species has a low but distinct respiratory activity. Kirk *et al.* (1954a) found a mean RQ of 1·0 for the fresh dog aorta and 0·91 for the human aorta, indicating that carbohydrates are preferentially oxidized.

Obviously, the metabolism of the arterial wall must depend on a satisfactory oxygen supply. Unfortunately, as is well known, the nourishment of the artery through inward blood flow from the *vasa vasorum* and outward diffusion from the lumen is relatively poor: this shortcoming especially impairs the nourishment of the middle zone of the tunica media, i.e. the vascular "watershed"

---

rate limiting factor in states 1 and 4 is the level of ADP, in state 2 it is the level of substrate and only in state 3 is it the respiratory chain. The ratio of the respiration rates in the "active" and "resting" states is a measure of the degree of respiratory control.

States 3 and 4 are the only ones relevant to the above aortic respiration studies, if we suppose that in all of them an excess of an appropriate substrate (usually succinate) and oxygen were available. It is clear that to obtain results reflecting the activity of the respiratory chain, an excess of ADP in the system must be provided, because otherwise the amount of ADP becomes the rate limiting factor. This is achieved in work with mitochondria, for example, by adding a system that generates ADP, usually hexokinase with glucose and ATP (see p. 17). If this precaution is not taken ADP will be rapidly exhausted from the medium with consequent transition from state 3 to state 4.

** More recent evidence (Klinkenberg and Schollmeyer, 1961) indicates that—in addition to ADP and inorganic phosphate—respiratory control is effected by ATP that inhibits the rate of respiration. The concentration ratio of all three compounds seems to control the rate of respiration in intact cells; inhibition of respiration by ATP is regarded as perhaps the most sensitive criterion for intactness of mitochondrial structure (see Lehninger, 1965).

between the inner and outer zones (Adams, 1964a, 1967a). It might, therefore, be of considerable interest, to consider whether and to what extent the artery can derive its energy needs by other means than oxidation.

As far as glycolysis is concerned, the bovine aorta produces under anaerobic conditions 0·59–0·83 μl. of lactate per 1 mg. of tissue (Südhof, 1950). The corresponding value for the dog popliteal artery is 1·59 (Beaconsfield, 1962) and for the human aorta 0·90 (Kirk et al., 1954a, b). Experiments, under aerobic conditions using ^{14}C- labelled glucose, revealed that the glycolytic rate of arterial tissue is relatively high, because the amount of lactate produced corresponds to about 80 per cent of the glucose utilized (Beaconsfield, 1962). This is in agreement with data obtained previously in experiments without the use of labelled material. For example, Kirk et al. (1954a) observed that glycolytic activity in the presence of oxygen is only reduced by 30 per cent in human aorta and by 27 per cent in dog aorta, when compared with that under anaerobic conditions. Beaconsfield (1962) found an insignificant difference of only 17 per cent, while Fontaine et al. (1960) maintain that in the human, bovine and rabbit aorta approximately 75–80 per cent of glucose is utilized by glycolysis even under aerobic conditions. All these results seem to indicate, that in contrast to most other tissues, arterial wall respiration does not exert a "braking" action on glycolysis, i.e. there is no Pasteur effect and the arterial wall displays the phenomenon of "aerobic glycolysis".

However, as emphasized by Lehninger (1959), it is essential to be cautious in evaluating the significance of aerobic glycolysis of the arterial wall *in vitro*, because aerobic glycolysis can easily be induced in surviving slices of tissues, which ordinarily do not produce lactate aerobically, by means of periods of anaerobiosis or metabolic inhibitors.

Assuming that the arterial wall really exhibits aerobic glycolysis *in vivo*, this mechanism would be perhaps a suitable survival mechanism under hypoxic conditions, especially in view of the unfavourable blood supply to its inner-middle layers. However, the physiological significance of the absence of the Pasteur effect should not be overestimated, especially from the energetic point of view. This conclusion can be easily reached by very simple calculation.

The free energy change in glycolysis of 1 mole of glucose with the production of 2 moles of lactate is $-47,000$ calories but only a fraction of this energy, approximately 16,000 calories, is trapped in a form that can be converted into useful work.* This corresponds to the production of 2 moles of ATP (an efficiency of about 34 per cent—see Chapter II).

In contrast, the free energy change in the complete oxidation of 1 mole of glucose is $-688,000$ calories, of which about 304,000 calories are trapped in the form of 38 moles of ATP (an efficiency of about 42 per cent—Chapters III & V)**.

Returning now to the aortic investigations, arguing from the data mentioned, approximately 75 per cent of glucose would be utilized by glycolysis and 25 per cent by complete oxidation. This would mean, as would apply both to the aorta

* See footnote p. 47.
** According to this analysis each mg. of lactate produced corresponds to 0·0889 calories, and each ml. of oxygen consumed corresponds to 2·302 calories. If the efficiency were 100 per cent, the corresponding values would be 0·256 for lactate and 5·12 for oxygen. To use the latter two values for energy balance studies is obviously incorrect.

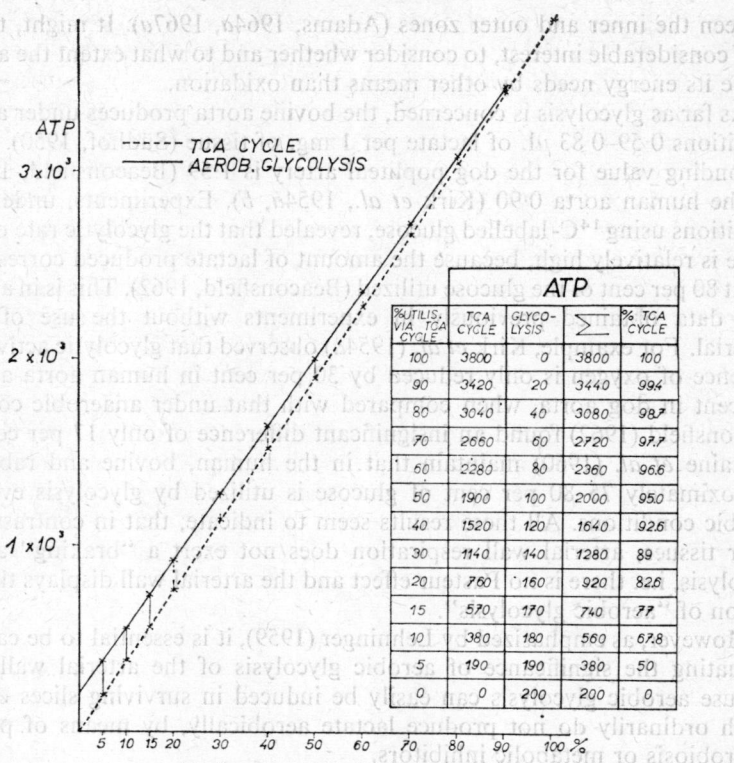

Fig. 1.—The production of theoretically available useful energy in aerobic glycolysis of 100 moles of glucose. X axis: Percentage of glucose utilization by the tricarboxylic acid (TCA) cycle. Y axis: ATP production in moles.

(Fontaine *et al.*, 1960) and the lens (Mandel, 1956), that out of 4 moles of glucose utilized, 3 give rise to lactic acid supplying only 2 moles of ATP per mole of glucose, instead of 38 moles of ATP provided by complete oxidation *via* the tricarboxylic acid cycle. Of the total amount of 44 moles of ATP produced (i.e. approximately 352,000 calories) 86·4 per cent are therefore supplied by the Krebs-Szent Györgyi tricarboxylic acid cycle and only 13·6 per cent by glycolysis. The main "producer" of useful energy remains, therefore, respiration, in spite of the high proportion of glycolysis. Even if the utilization of glucose by oxidation is only 10 per cent, nevertheless as much as 67·8 per cent of useful energy would be provided by this pathway (see Fig. 1).

The data in Fig. 1 also demonstrate that under aerobic conditions, a relative decrease in Krebs cycle activity causes the total production of useful energy to decline. This fall is less pronounced than it should be if the Krebs cycle were the only producer of energy, but the difference is impressive only at low levels of respiration. This is what may happen in poorly vascularized tissues where the ability to meet the energy needs from glycolysis may perhaps represent a mechanism for survival. As mentioned before, the inner-middle layers of the arterial tunica media might be considered as such a tissue (see Chapter XVI).

It has to be emphasized, however, that although the part played by glycolysis in the generation of energy in such tissues is relatively higher under hypoxic conditions, the amount of total energy supply is low and these tissues must necessarily be particularly vulnerable to further reduction in the oxygen supply.

It is also important to realize that tissues that rely on aerobic glycolysis require, under hypoxic conditions, relatively large amounts of glucose (or glycogen) to be able to compensate for their inefficient energy production. Therefore, the production of lactate is also increased. The possible consequences of the lowering of pH in the arterial wall are not clear. There is some evidence that "acidification" of the artery may induce destruction of the elastic membranes, and Baló (1963) considers it an important mechanism in the pathogenesis of atherosclerosis.

In the above discussion on the breakdown of glucose we have only considered the Krebs cycle and glycolysis. It is fair to add that, apart from these pathways, the glucose 6-phosphate oxidation system or pentose phosphate pathway (see p. 35) plays a certain role in the utilization of glucose in arterial tissue. This has been demonstrated not only by the presence of some enzymes of this pathway in the arterial wall, but also by experiments with $[1-{}^{14}C]$-glucose and $[6-{}^{14}C]$-glucose. The results obtained with dog aortas (Beaconsfield, 1962), guinea-pig aortas (Sbarra et al., 1961) and bovine aortas (Mandel and Kempf, 1963) reveal that the amount of ${}^{14}CO_2$ derived from $[1-{}^{14}C]$-glucose is 3 to 5 times higher than that derived from $[6-{}^{14}C]$-glucose. This indicates that the pentose phosphate pathway participates significantly in glucose utilization of arterial tissue. For comparison, it is of interest that the above ratio for organs such as the lung, kidney, heart, muscle and brain is 1 to 2 but for the lens approximately 15. Thus, according to Mandel and Kempf the participation of this pathway in glucose degradation seems to be relatively more important for aortic tissue than for many other organs. Because the pentose phosphate pathway produces reduced NADP (see p. 36, Fig. 4), which is essential for the reductive synthesis of lipids, one could speculate on the possible relationship between this feature of the vascular wall and its ability to synthesize lipids.

From what has been said it is evident that the arterial wall can meet its energy needs from carbohydrates. Until recently the coupling of phosphorylation to oxidation had been little investigated in arteries. Wolleman and Kocsar (1964) observed in a few rat aortic homogenates a mean P:O ratio of approximately 2, which is practically the same as that of liver and brain homogenates under identical conditions. Recently, Ritz and Kirk (1967) observed in guinea-pig aortas a P:O ratio of about 1·6. More information on this topic is much needed, especially concerning regulatory factors in arterial oxidative phosphorylation.

Similarly, there is very little information as to what extent the artery utilizes other energy sources apart from glucose, e.g. free fatty acids. In view of the connection between lipids and atherosclerosis, investigation of this problem is of basic importance.

There are many other important problems of vascular metabolism, such as the biosynthetic mechanisms for lipids, ground substance, mucopolysaccharides and mechanisms of calcification. Nevertheless, the above sketchy discussion of some metabolic properties of the vascular wall should provide sufficient introductory background for the presentation of the main topic of this volume, i.e. discussion of vascular enzymes as related to atherosclerosis.

In what follows we shall rely mainly on results obtained by biochemical methods and no attempt will be made to give a detailed picture of the data obtained by histochemical techniques. This has been discussed in a volume by Adams (1967a). Since, however, we obtained some useful complementary information by the histochemical approach, certain results obtained in this field will also be presented.

The first part of this volume (Chapters II to IX) comprises general information together with some methodological notes on those enzymes and cofactors which have been investigated by biochemical techniques in vascular tissue. Data on the specific properties of vascular enzymes and their changes under certain physiological and pathological conditions are also included in this part of the book: the reader will see that the backbone of these data is derived from the excellent pioneer work of Kirk and his co-workers.

The second part of the book (Chapters X to XIX) is confined almost exclusively to those special problems which have been studied by the author and his co-workers in the Institute for Cardiovascular Research in Prague in connection with the pathogenesis of atherosclerosis.

Finally the third part (Chapters XX to XXV) deals with the topic of lipolytic activity in the vessel wall as a possible arterial protective mechanism against lipid accumulation and development of atherosclerosis.

Chapter II

ENZYMES OF GLYCOLYSIS AND GLYCOGEN BREAKDOWN

THE energy for all activities of living cells is derived from carbohydrate, lipid and protein metabolism. Different tissues utilize these substances, particularly carbohydrates and fats, in varying proportions.

In the absence of oxygen the metabolism of carbohydrates in animal tissues is effected by glycolysis. The mechanism is essentially an oxidoreduction involving the transfer of electrons (hydrogen atoms) from the reducing to the oxidizing molecule *via* pyridine nucleotides (NAD and $NADH_2$). Starting from glycogen the overall reaction in muscular tissue can be summarized as follows:

$$\text{glycogen} + 3\text{ADP} + 3\text{PO}_4 \rightarrow 2 \text{ lactate} + 3\text{ATP}$$

When glucose serves as the starting substance only 2 moles of ATP (about 16 kcal) are produced per mole of glucose. This amount of calories represents the biologically utilizable energy of glycolysis. However, the free energy change in the transformation of one mole of glucose to two moles of lactate is 47 kcal, and the difference is dissipated as heat. This corresponds to an energy-capture efficiency of about 34 per cent. (The corresponding figure for aerobic utilization of carbohydrate is an efficiency of about 42 per cent, the free energy of hydrolysis of one mole ATP being taken as 8 kcal.)

An illustrative diagram of the glycolytic cycle is given in Fig. 2.

GLYCOGEN–PHOSPHORYLASE* E.C. 2.4.1.1.
(α-1, 4-Glucan: orthophosphate glucosyltransferase)

This enzyme is a glycosyltransferase catalysing the reversible reaction between the terminal non-reducing glucose residue in the outer chains of glycogen and orthophosphate; glucose 1-phosphate is an important product of this "phosphorolysis".

$$(\alpha\text{-1, 4-Glucosyl}) \text{ n} + \text{orthophosphate} = (\alpha\text{-1, 4-glucosyl}) \text{ n}-1 + \alpha\text{-D-glucose 1-phosphate}.$$

The enzymatically active enzyme (phosphorylase *a*) is converted in many animal tissues to an enzymatically inactive *b* form. Both forms exist in muscle and liver, but the molecular weight of the muscle enzyme is twice that of the

---

* The nomenclature recommended by the Commission on Enzymes of the International Union of Biochemistry (IUB) had been followed in this book. The *trivial name* is followed by the *code number* of the enzyme and the *systemic name* (where a satisfactory one exists) is given in brackets.

The trivial name of the enzyme is in most cases the one that is currently used; it is short but not necessarily very exact. The systemic name of the enzyme is formed according to

liver enzyme. In muscle the conversion into the inactive form is due to the hydrolytic action of phosphorylase phosphatase (E.C. 3.1.3.17.), formerly known as "PR enzyme" that splits the *a* form into two moles of the *b* form and 4 moles of orthophosphate. The reactivation of the enzymatically inactive *b* form is mediated by way of a kinase reaction (phosphorylase kinase, E.C. 2.7.1.38.) in

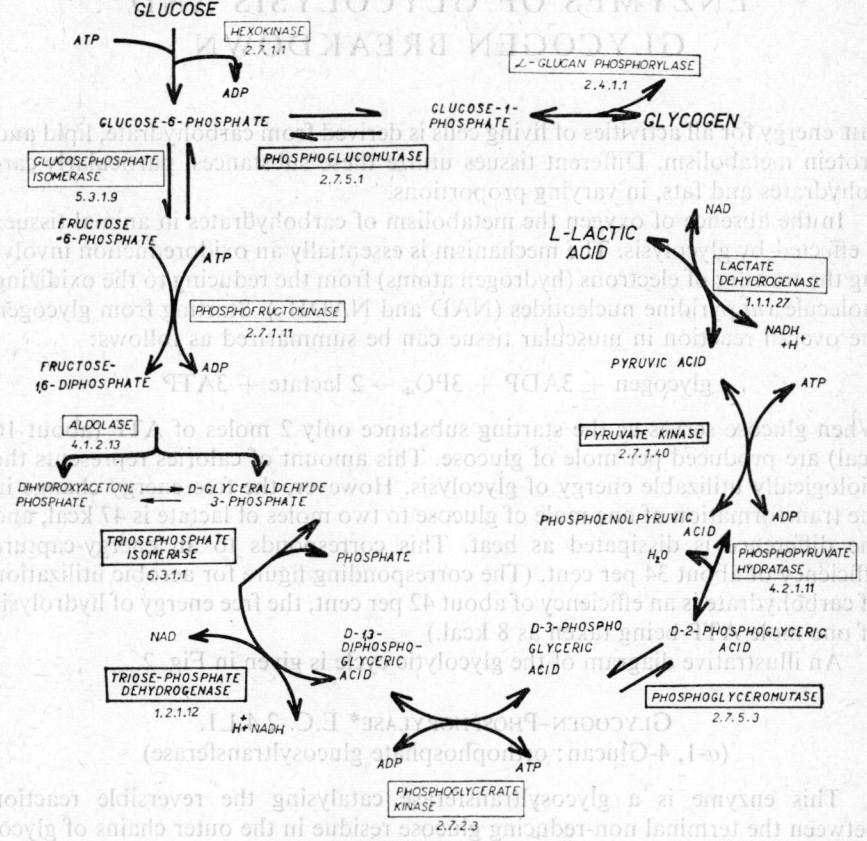

FIG. 2.—The glycolytic pathway.

definite rules; it identifies the enzyme precisely and shows the action of the enzyme as exactly as possible.

According to the recommendations of the Enzyme Commission of the IUB, enzymes can be divided into six main groups:

1. Oxidoreductases
2. Transferases
3. Hydrolases
4. Lyases
5. Isomerases
6. Ligases (synthetases)

The first figure of the code number indicates, therefore, to which of the above six main divisions the particular enzyme belongs. The further figures of the code number indicate respectively the sub-class, the sub-sub-class and the serial number of the enzyme in its sub-sub-class. (For details see *Comprehensive Biochemistry*, Eds. M. Florkin and E. H. Stotz. Elsevier Publ. Co., Amsterdam, Vol. 13, 1965.)

the presence of ATP and Mg. According to more recent evidence (see Sutherland and Rall, 1960) adenosine 3′, 5′-phosphate (cyclic adenylate) plays an important role in this reaction. The latter substance converts the inactive phosphorylase into the active form, and its tissue concentration is under the influence of catecholamines, ACTH, glucagon and serotonin. In addition, epinephrine also enhances the activity of phosphorylase kinase (Krebs et al., 1959). In the resting muscle only a small fraction of phosphorylase is in the active form, while in the working muscle practically all enzyme is rapidly converted into this form. In constantly working muscles, such as the heart or diaphragm, the relative proportion of the a form is much higher than in other muscle tissue.

In muscle the active (phosphorylated) form of the enzyme is made up of four presumably identical sub-units, whereas the inactive (dephosphorylated) form contains only two sub-units. Both forms of liver phosphorylase have the same molecular weight and seem to be made up of two sub-units (Krebs and Fischer, 1963).

It has been demonstrated that all forms of phosphorylase require pyridoxal phosphate as coenzyme (Baranovski et al., 1957; Yunis et al., 1962 and others); this compound seems to be important in stabilizing the structure of the enzyme.

In connection with the phosphorylase activity of the vascular wall it must be pointed out that this tissue is relatively rich in glycogen. According to Südhof (1950) the bovine carotid artery contains an average of 232 mg. of glycogen per 100 g. of tissue. In the presence of oxygen an intensive synthesis of glycogen from glucose takes place in arterial tissue; under anaerobic conditions glucose has an inhibitory effect on the spontaneous disappearance of glycogen (Südhof, 1950). In human peripheral arteries Schmidt and Hillenbrand (1953) found 70 mg. of glycogen per 100 g. of tissue but there was a considerable decrease in atherosclerotic vessels.

Phosphorylase activity is usually measured in the direction of synthesis, i.e. by colorimetric determination of the amount of orthophosphate formed from glucose 1-phosphate. Because cyclic adenylate reactivates the inactive form, the addition of this substance to the incubation medium enables the amount of both a and b forms of the enzyme to be estimated.

Kirk (1962b) determined the phosphorylase activity in homogenates of human aortas, as well as in pulmonary and coronary arteries, using the above mentioned procedure.

The enzymatic activity of the aorta and coronary arteries decreases significantly with age, when calculated on a wet weight or tissue nitrogen content basis. Atherosclerotic parts of the arteries display significantly lower activity when compared with intact segments. Kirk believes that the latter changes are related to atrophy of smooth muscle in the arterial wall. It is interesting that enzyme activity was highest in the pulmonary artery and no age-related changes were found in this vessel.

Hexokinase, E.C. 2.7.1.1.
(ATP: D-hexose 6-phosphotransferase)

This enzyme is found in yeast and animal tissues and catalyses the transfer of the terminal phosphatyl residue of ATP to a 6-carbon monosaccharide.

$$ATP + \text{D-hexose} = ADP + \text{D-hexose 6-phosphate}$$

Hexokinases are ubiquitous enzymes of animal tissues, moulds and yeasts. The enzyme requires the presence of magnesium ions for full activity. It is not endowed with very high substrate specificity and, for example, yeast and brain hexokinase are capable of phosphorylating D-glucose and D-fructose at approximately the same rate, while D-mannose and D-glucosamine also can act as acceptors.

In muscle and liver, however, somewhat more specific enzymes seem to be present and, hence, it is usual to refer to *gluco*kinase (E.C. 2.7.1.2.), *galacto*kinase (E.C. 2.7.1.6.), *keto*kinase (E.C. 2.7.1.3.) or to *fructo*kinase (E.C. 2.7.1.4.).

Unfortunately, there is a slight confusion in the nomenclature of hexokinases. The main reason seems to be the fact that enzymes of similar substrate specificity are not actually identical amongst different species (see Crane, 1962).

The "hexokinase reaction" is of fundamental importance because glucose can be neither metabolized nor stored unless it is phosphorylated. In this connection, it is interesting that phosphorylation of some compounds increases their selectivity for diffusion through cellular membranes and this is of great importance in metabolic reactions (see Neilands and Stumpf, 1958).

The substrate specificity of hexokinase has been investigated e.g. by Crane and Sols (1953). (See also Dixon and Webb, 1964.)

Several methods have been used to follow the hexokinase reaction, e.g. by determining the disappearance of hexose, glucose or "acid-labile phosphorus", or by estimating the formation of hexose 6-phosphate.

An elegant method is based on the determination of the rate of ADP production by way of two auxiliary reactions. In the first, ADP is phosphorylated with added phosphoenolpyruvate and pyruvate kinase. ATP and pyruvate is produced and the latter is reduced in the subsequent "indicator" reaction with $NADH_2$ and lactate dehydrogenase. The decrease in the concentration of $NADH_2$, a common step in many similar coupled reactions, is easily determined spectrophotometrically at 340 mμ. (see later "Optical Tests" in Chapter XI) and is a measure of the enzyme activity.

Brandstrup *et al.* (1957) determined the hexokinase activity of human aortic and pulmonary artery homogenates by measuring the rate of glucose disappearance from the medium. No significant variation with age was found. In some instances a decrease of 8–24 per cent in the enzyme activity of atherosclerotic tissue was observed, but only when results were calculated on a wet weight basis. No difference in activity was observed on a tissue nitrogen basis. It should be added that in these experiments on hexokinase in *homogenates* the utilization of glucose (1·80 mg. per g. fresh tissue per hour) was over four-fold higher than in previous studies by Kirk *et al.* (1954*a*), where respiration was determined in *intact* aortic tissue in the presence of glucose (see also Kirk, 1963*a*).

Phosphoglucomutase, E.C. 2.7.5.1.
(α-D-Glucose-1, 6-diphosphate: α-D-glucose-1-phosphate phosphotransferase)

This enzyme catalyses the conversion of glucose 1-phosphate, the initial product of phosphorolytic cleavage of glycogen (see above), to glucose 6-phosphate. Glucose 1, 6 diphosphate, which acts as a coenzyme, must be present and is needed in saturated concentration to achieve maximum activity of the enzyme. (For further details see section on phosphoglyceromutase, p. 21.)

The enzyme is present in the liver, heart, kidney, striated muscle and erythrocytes and probably in most living cells. In addition to its importance in the breakdown of glycogen, phosphoglucomutase has the crucial function of producing glucose 1-phosphate for the formation of uridine diphosphate glucose (UDPG), which is a common starting compound for many synthetic reactions (see also Chapter IX).

Magnesium ions are specific activators of the enzyme, whereas heavy metals and fluoride are potent inhibitors. Magnesium ions probably act by forming a magnesium-imidazole complex in the enzyme centre.

The determination of this enzyme activity is usually based on the different chemical properties of glucose 1-phosphate and glucose 6-phosphate, the former being a non-reducing sugar with acid-labile phosphate while the latter has the converse properties.

Kittinger *et al.* (1962) studied the activity of this enzyme in extracts of rat aortas. Activity was determined by a method in which the phosphoglucomutase reaction (whose end-product is glucose 6-phosphate) is coupled to glucose 6-phosphate dehydrogenase, a NADP-dependent enzyme. In this system spectrophotometrically assessed changes in NADP concentration (at 340 mμ.) are a measure of glucose 6-phosphate production. A significant decrease was found in enzyme activity of aortas with moderate and severe arteriosclerosis. (See also lactate dehydrogenase, page 24 and pentose phosphate pathway, page 38).

Kirk (1966*a*) studied the activity of the enzyme in specimens of human arteries using an assay in which the quantity of total hexose 6-phosphate produced is measured, essentially as fructose 6-phosphate, by the resorcinol colorimetric method at 490 mμ. The average enzyme activities of the intact parts of the arteries varied between 10·40 and 12·55 units per g. of tissue nitrogen. Fatty or fibrous atherosclerotic lesions of the aorta exhibited a significantly lower activity on a tissue nitrogen basis (80·4 per cent and 58·2 per cent respectively) than intact tissue from the same arteries. In the grossly normal aorta, pulmonary artery and coronary artery, enzyme activity tended to decline with age.

GLUCOSEPHOSPHATE ISOMERASE, E.C. 5.3.1.9.
(D-Glucose-6-phosphate ketol-isomerase)

This enzyme catalyses the attainment of a mobile equilibrium between D-glucose-6-phosphate and D-fructose 6-phosphate. It is present in yeast, plants and practically all animal tissues.

A colorimetric method for the assay of this enzyme is based on the colour-reaction of fructose 6-phosphate with resorcinol. Another "optical method" is based on the dehydrogenation of glucose 6-phosphate by glucose-6-phosphate dehydrogenase in the presence of its coenzyme NADP. The rate of $NADPH_2$ production, as measured by optical density at 340 mμ, is proportional to the enzyme activity (see page 96).

Brandstrup *et al.* (1957) determined the activity of glucosephosphate isomerase in homogenates prepared from human aortic and pulmonary arterial tissue by measuring the formation rate of fructose 6-phosphate from glucose-6-phosphate. The findings were similar to those for hexokinase (see p. 18).

In a later work from the same laboratory Kirk *et al.* (1958) studied the differences in glucose phosphate isomerase activity between normal and atherosclerotic parts of the same human coronary artery. They found a significant decrease (23 per cent) in activity of atherosclerotic samples, but only when calculated on a wet weight basis. The activity of this enzyme is rather high when compared with other glycolytic enzymes in human arteries, in particular the coronaries.

In rabbits with experimental atherosclerosis (see Chapter XIII), no significant changes of this enzyme in the aorta have been observed (Zemplényi et al., 1963c).

FRUCTOSEDIPHOSPHATE ALDOLASE, E.C. 4.1.2.13.
(Fructose-1, 6-diphosphate D-glyceraldehyde-3-phosphate-lyase)

"Aldolase" splits fructose 1, 6-diphosphate to equivalent amounts of D-glyceraldehyde 3-phosphate and dihydroxyacetone phosphate. It is specific only for dihydroxyacetone phosphate, while the other product can be replaced by other aldehydes. Such enzyme activity is highest in skeletal muscle; crystalline aldolase has been prepared from this tissue as well as from yeast. Brain, heart, liver and erythrocytes also display high aldolase activities. Under normal conditions, the serum activity is comparatively low, but in certain pathological conditions (e.g. certain cancers, liver diseases, muscular dystrophy etc.) the serum aldolase activity is markedly elevated.

Aldolase is usually found in the soluble fraction of the cytoplasm, but brain aldolase is confined to the mitochondrial fraction.

Aldolase is a key enzyme in glycolysis, and aldolase activity values are often considered a measure of the rate of glucose metabolism in a given tissue.

Several methods exist for its assay, for example a colorimetric method, in which the dinitrophenyl hydrazones of both free trioses are determined, or a spectrophotometric method based on Warburg and Christian's (1936) principle (see Chapter XI). The pH optimum of the enzyme depends on the assay system.

Kirk and Sørensen (1956) determined aldolase activity in homogenates of human aortas and pulmonary arteries, using the colorimetric method with 2,4-dinitrophenylhydrazine. Enzyme activity (referred to either wet weight or tissue nitrogen) decreased in the second decade but progressively rose in subsequent decades. The value for enzymatic activity in human arterial tissue was approximately 1 per cent of that reported for human striated muscle. Enzyme activities of atherosclerotic portions of the aorta were about 30 per cent lower than normal samples.

The age dependent increase of aldolase activity in human aortas discussed above does not accord with Mandel et al.'s (1959) findings in bovine aortas studied by essentially the same method. These authors found a definite decrease of aldolase activity with age, when results were calculated on a wet weight basis.

PHOSPHOGLYCERATE KINASE, E.C. 2.7.2.3.
(ATP: 3-phospho-D-glycerate 1-phosphotransferase.)

This enzyme catalyses the following reaction:

1, 3-diphospho-D-glycerate + ADP = 3-phospho-D-glycerate + ATP

It is important to realize that two triose phosphate molecules are produced from one molecule of glucose and, therefore, the above reaction should give rise to two moles of ATP per mole of glucose utilized. Thus, the regeneration of the two moles of ATP required in the preceding steps of glycolysis (i.e. in the hexokinase and phosphofructokinase reaction, see Fig. 2) is effected through the action of phosphoglycerate kinase.

The enzyme is widely distributed in animal tissues, plants and yeast. For full activity it requires the presence of magnesium or manganese ions. Heavy metals are inhibitors, but EDTA as well as pyrophosphate protect the enzyme.

An optical test for enzyme activity determination is based on the production of diphosphoglycerate and the conversion of this compound into glyceraldehyde-3-phosphate by the action of glyceraldehydephosphate dehydrogenase (E.C. 1.2.1.12.) in the presence of $NADH_2$. Changes in the concentration of $NADH_2$ measured at 340 or 366 mμ. (see Chapter XI) are proportional to the activity of phosphoglycerate kinase.

Kirk and Ritz (1966) used another method where the product of "reverse" reaction, 1, 3-diphosphoglycerate, was assessed colorimetrically. In human vascular tissue the average activity observed in the thoracic aorta, pulmonary artery and coronary artery was 7·15, 8·30 and 7·89 mM of 1, 3-diphosphoglycerate per g. of tissue nitrogen. The abdominal aorta displayed lower activity than the ascending aorta.

Samples of atherosclerotic coronary artery and fibrous aortic plaques exhibited activity that was 14 and 18 per cent higher, respectively, than normal segments. Activity in the inferior vena cava was only 61 per cent of that recorded for the thoracic aorta.

PHOSPHOGLYCEROMUTASE, E.C. 2.7.5.3.
(2, 3-Diphospho-D-glycerate: 2-phospho-D-glycerate phosphotransferase)

This enzyme catalyses the conversion of 3-phosphoglycerate to 2-phosphoglycerate; the reaction could be simply formulated as

2-phospho-D-glycerate $\rightleftarrows$ 3-phospho-D-glycerate.

However, it has been shown that the reaction takes place in the above direction only in certain plants: the enzyme catalysing such an intramolecular transfer reaction is termed phosphoglycerate phosphomutase (E.C. 5.4.2.1.). The enzyme, which is widely distributed in animal tissues, plants and yeast, requires a catalytic amount of 2, 3-diphospho-D-glycerate for its action. In this respect it is analogous to phosphoglucomutase (see p. 18). The enzyme functions as a phosphotransferase with the interesting feature that the phosphate donor is continuously regenerated and acts as a constant cofactor of the reaction:

2, 3-diphospho-D-glycerate + 2-phospho-D-glycerate = 3-phospho-D-glycerate + 2, 3-diphospho-D-glycerate.

The enzyme has been purified from rabbit muscle, yeasts and other sources. The yeast enzyme, when subjected to electrophoresis, yields five or six separate components that all exhibit phosphoglyceromutase activity.

Zinc, silver and cupric ions inhibit the activity of the enzyme in decreasing degree. Fluoride is also an inhibitor, the yeast enzyme being most sensitive to its action. The pH optimum of phosphoglyceromutase is at 5·9. For further details see Pizer's (1962) review.

Enzyme assays are based on "optical tests", where the mutase is coupled with other reactions that produce compounds with certain absorption characteristics in UV light. Other methods make use of the enhancing effect of molybdate ions on the optical rotation of 3-phosphoglycerate.

The assay described by Grisolia (1962) depends on spectrophotometric estimation of phosphoenolpyruvate (at 240 mμ.) that is formed by coupled action of the mutase and an excess of phosphopyruvate hydratase (E.C. 4.2.1.11).

Kirk (1966a) measured the activity of the enzyme in supernatants from homogenates of the same arterial samples where phosphoglucomutase was determined. The mean activity in normal aortic tissue (7·22 millimoles of substrate metabolized per gram of tissue nitrogen) was distinctly lower than the activity of the pulmonary and coronary artery. Atherosclerotic segments of the aorta and coronary artery exhibited lower activities than intact parts of the same vessels.

Phosphopyruvate Hydratase, E.C. 4.2.1.11.
(2-phospho-D-glycerate hydro-lyase)

This widely distributed enzyme (known as "enolase") catalyses a dehydration reaction leading from 2-phosphoglyceric acid to phosphoenolpyruvic acid. The free energy of hydrolysis of the latter compound is very high (-12 kcal) and the reaction is therefore important as it provides a high-energy bond.

$$\text{2-phospho-D-glycerate} = \text{phospho-enolpyruvate} + H_2O$$

Enolase is present in most living cells. It has been crystallized from yeast and from muscle, and is one of the most thoroughly investigated enzymes, particularly in regard to the physical and chemical properties of the protein (for a review, see Malmström, 1961).

The muscle enzyme is activated by some divalent metal ions (Mg^{++}, Zn^{++}, Mn^{++}); magnesium is probably its physiological activator. On the other hand, calcium (and strontium) are competitive inhibitors.

The spectrophotometric method for the determination of enzyme activity is based on the absorption of light by phosphoenolpyruvate at 240 mμ. Another spectrophotometric method is based on the conversion of phosphoenolpyruvate to pyruvate and lactate. The latter is then oxidized by lactate dehydrogenase and NAD; this last step is measured in the usual way (see "Optical Tests" in Chapter XI).

Wang and Kirk (1959) observed considerable enolase activity in supernatants of homogenates from human aortas, pulmonary and coronary arteries. The activity was approximately 50 per cent of that reported for the rabbit brain. No significant change in activity (calculated on a wet weight basis) with age was found in aortas from adult persons and no significant difference was observed between activities of normal and atherosclerotic tissue portions.

Kirk (1963a) emphasizes the possibility that the high-energy phosphate bonds, which are generated in the enolase-catalysed reaction (see above), may play an important role in the energetics of the arterial wall.

Lactate Dehydrogenase, E.C. 1.1.1.27.
(L-Lactate: NAD oxidoreductase)

This is an important glycolytic enzyme which catalyses the reaction

$$\text{L-lactate} + \text{NAD} = \text{pyruvate} + \text{NADH}_2$$

Although pyruvate and lactate are the substrates that are, respectively, most rapidly reduced or oxidized, the enzyme also reduces some other α-keto (and α, γ-diketo) acids and oxidizes other L-2-hydroxy-monocarboxylic acids (see Chapter XVI). In the case of C_4 to C_9 α-keto acids, the rate of reduction diminishes rapidly with increasing chain length.

The enzyme is ubiquitous in animal tissues and has been isolated from skeletal muscle, heart muscle, liver and erythrocytes of several animal species. In human tissues, so far studied, enzymatic activity is highest in the kidney and heart and lowest in the lung and serum.

Lactate dehydrogenase from heart, skeletal muscle, liver and serum of several species is said to be inhibited by high pyruvate concentrations. Oxamate and oxalate inhibition of the enzyme has also been investigated in great detail. It appears that only anionic substances function as inhibitors of the "substrate analogue type". Sodium sulphite is a strong inhibitor of the enzyme from rat liver, but not of that from bovine heart and rabbit muscle. (For references to inhibitors see Schwert and Winer, 1963.)

Animal lactate dehydrogenase is NAD-linked, but NADP can also act as coenzyme with less efficiency. The analogous enzymes from bacteria and yeast are flavoproteins and do not require nicotinamide-containing coenzymes.

In 1952 the electrophoretic experiments of Neilands revealed the presence of two catalytically active components in crystalline lactate dehydrogenase of bovine heart muscle, the major anodic component accounting for about 75 per cent of the total protein. Fractional precipitation with ammonium sulphate also indicated the presence or two components. The heterogeneity of lactate dehydrogenase from several animal tissues was later extensively investigated in several laboratories. (See Chapters I and XVI.)

The activity of lactate dehydrogenase can be measured very conveniently either spectrophotometrically or by colour reactions using tetrazolium salts as electron acceptors. For details see Chapter XI.

Kirk and Laursen (1955) devised a quantitative method, using triphenyl tetrazolium chloride as electron acceptor, for determinating the activities of various dehydrogenases in human aortic tissue. They used glucose, acetate, pyruvate, lactate and intermediate compounds of the Krebs cycle as substrates. Strong activity (i.e. formazan production) was observed with succinate as substrate, but there was only slight activity in the presence of lactate or pyruvate. It is, however, difficult to interpret the latter findings from the enzymological point of view, because the reactions were performed without adding phosphopyridine nucleotides to the medium and these coenzymes are rapidly destroyed after death. For similar reasons it is difficult to interpret some histochemical findings, e.g. those of Fried and Zweifach (1955), concerning endogenous enzymatic activity of peripheral vessels in the presence of "intermediary metabolites". It is surprising, however, that these authors did not observe increased formazan deposition even in the presence of NAD.

In another paper Matzke et al. (1957) determined lactate dehydrogenase activity in homogenates of human aortas and pulmonary arteries, using the classical method based on the spectrophotometric measurement of $NADH_2$ concentration changes at 340 mμ. Enzyme activity of both vessels was found to increase up to the sixth decade when expressed on a wet weight basis. A 16–28 per cent decrease in activity was observed in atherosclerotic portions of vessels from the sixth and seventh decade when compared with normal portions. This difference was apparent only when results were calculated on a wet weight basis; results referred to tissue nitrogen did not differ.

In a later paper Kirk et al. (1958) performed a similar study using homogenates of human coronary arteries. No significant changes in enzyme activity with age were observed (results were referred to wet weight). The activities of the atherosclerotic portions of the coronaries were found to be 17 per cent lower than normal portions.

Mandel and Kempf (1960) estimated lactate dehydrogenase activity by a similar technique in extracts of bovine aortas, and found a significant decrease in activity with age.

Kittinger et al. (1962) also used the spectrophotometric method for estimating activity in extracts of aortas from repeatedly-bred arteriosclerotic rats. The extracts of aortas with moderate or severe arteriosclerosis contained significantly less lactate dehydrogenase than extracts of normal or only slightly affected aortas. (The activity was calculated on a wet weight basis.)

Alekseeva and Ushkalov (1963) observed that in the aorta of atherosclerotic rabbits lactate dehydrogenas activity initially rises but subsequently decreases perhaps before the appearance of lipid deposits.

Using histochemical techniques, Adams et al. (1962) observed a zonal defect in the activity of lactate dehydrogenase and of other NAD-dependent dehydrogenases in the ageing human aorta. The enzymatic failure was attributed to ischaemic damage to the mid-zone of the tunica media (see Chapter XVI). In subsequent work (Adams et al., 1966; Saudek et al., 1966) lactate dehydrogenase activity was also estimated by a quantitative tetrazolium method in multiple consecutive layers of the human aortic wall, cut unfixed from the intimal to the adventitial surface. For determining enzyme activity, phenazine methosulphate and the tetrazolium salt INT were used as electron acceptors. In contrast to the histochemical results, this technique revealed unchanged or even increased lactate dehydrogenase activity in the middle and inner medial zones of the senescent aorta. Since phenazine methosulphate eliminates the need for endogenous tissue diaphorase as an electron carrier (see Chapter XI), it was concluded that in histochemical and biological determinations without phenazine methosulphate, $NADH_2$-tetrazolium reductase activity ("diaphorase" activity) becomes a rate-limiting factor. (See also p. 97.)

In our own experiments we studied lactate dehydrogenase activity in animal as well as in human vessels. Results will be presented in part two of this volume.

Before closing this review of vascular glycolytic enzymes a few additional comments may prove useful.

1. Phosphofructokinase Reaction

Perusal of Fig. 2 (p. 16) shows that only nine out of thirteen enzymes involved in the breakdown of glycogen to lactate so far have been investigated in vascular tissue. It must be pointed out that it is not feasible nor desirable exhaustively to study the large number of enzyme activities that may be involved in vascular metabolism. Moreover, useful data can be derived from seemingly fragmentary information, when based on the study of the more important rate limiting or key reactions in intermediary metabolism. Such an important key reaction that so far has not been investigated in the vascular wall is the phospho-fructokinase reaction, catalysed by phosphofructokinase (E.C. 2,7,1,11.).

This enzyme plays an important role in metabolic control, as a probable rate-limiting factor in glycolysis. It has been suggested that the enzyme is the mediator of the Pasteur effect (see Chapter I); the main argument being that phosphofructokinase could be inactivated by oxidation and reactivated by reduction. (For a review and references see Lardy, 1962.) Recent evidence indicates that in some tissues oxidation of free fatty acids inhibits glycolysis, induces intracellular accumulation of glucose 6-phosphate and enhances glycogen deposition. Newsholme and Randle (1962) presented evidence showing that these events are the result of inhibition of the phosphofructokinase reaction. The mechanism of inhibition is not clearly understood, but it could be due to inhibition of phosphofructokinase by the elevated concentration of intracellular ATP that results from oxidation of fatty acids. This explanation is strengthened by the observation that the enzyme is inhibited by ATP (Mansour et al., 1962; Passonneau and Lowry, 1962). The increased glycogen synthesis seems to be associated with the activation of UDPglucose-glycogen glucosyltransferase (see below) by glucose 6-phosphate.

It appears that many of the abnormalities of carbohydrate metabolism in diabetes and starvation (carbohydrate deprivation) are due to the increased release and oxidation of free fatty acids and to the interplay of various pathways in the "glucose fatty acid cycle" (Randle et al., 1963).

As disorders of lipid metabolism are doubtlessly important in the whole picture of atherogenesis, it would be very relevant to investigate the activity of phosphofructokinase in arterial tissue.

2. Glycogen Pathway

When discussing glycolytic enzymes we intentionally include enzymes that catalyse the conversion of glycogen to glucose 6-phosphate. They can be also considered, however, as parts of a separate pathway of glycogen synthesis and breakdown. With the exception of phosphoglucomutase, which acts in both synthetic and degradative directions, the arterial enzymes of glycogen synthesis have not so far been extensively investigated.

In a few preliminary experiments Kittinger et al. (1962) studied the activity of glucose-1-phosphate uridylyltransferase, E.C. 2.7.7.9. (UTP: α-D-glucose-1-

phosphate uridylyltransferase). This enzyme catalyses the formation of the strongly activated form of glucose, uridinediphosphoglucose (UDPglucose)

$$\alpha\text{-D-glucose-1-phosphate} + \text{UTP} = \text{pyrophosphate} + \text{UDPglucose}$$

(The glucose from UDPG is transferred to the end of the polysaccharide chain by the action of UDPG-glycogen glucosyltransferase, E.C. 2.4.1.11. The latter enzyme requires glucose 6-phosphate as a primer). The activity of the enzyme can be estimated by measuring the amount of glucose 1-phosphate formed. In an auxiliary reaction this compound is converted with excess phosphoglucomutase to glucose-6-phosphate and, in an indicator reaction with excess glucose 6-phosphate dehydrogenase, the latter is oxidized by NADP. This can be followed at 340 mμ. spectrophotometrically (Kalckar, 1953; Munch-Petersen, 1955). See also the principle of "Optical Tests", Chapter XI.

According to the above-mentioned preliminary studies by Kittinger et al. on the arteriosclerotic aortas of repeatedly-bred female rats the activity of glucose-1-phosphate uridylyltransferase significantly increases in all stages in the development of arteriosclerosis.

The enzyme is found in cell nuclei, especially in those of the liver. For activation it requires several divalent metal ions. It has been also detected in the mammary gland, in plants and can be isolated from dry baker's yeast.

It is of special interest that UDPglucose is a key compound, not only in the biosynthesis of glycogen but also in the biosynthesis of some mucopolysaccharides that contain glucuronic acid. According to Strominger et al. (1954), Dorfman (1962) and others, an important step in this process is the oxidation of UDPglucose by a specific NAD-dependent dehydrogenase (E.C. 1.1.1.22.) to UDPglucuronate (see also Hilz, 1962).

Because it is held by many investigators that mucopolysaccharides play an important part in atherogenesis, the study of glucose-1-phosphate uridylyltransferase may shed light on the synthesis of these compounds within the arterial wall.

Addendum in proof.

Recently Ritz and Kirk (*J. Geront.*, **22**, 433, 1967) reported on phosphofructokinase studies in human vascular tissue. The rather low activity of the enzyme suggests that it may be one of the rate-limiting steps in the vascular glycolytic pathway. No significant differences were recorded between activities in normal and atherosclerotic aortic segments. Concomitant sorbitol dehydrogenase studies revealed reduced activity in atherosclerotic aortic segments. The activities of both enzymes were lower in vascular samples from children than adults. The same authors (*J. Geront.*, **22**, 427, 1967) also observed decreased glycerol-3-phosphate dehydrogenase (E.C. 1.1.1.8.) and glyceraldehydephosphate dehydrogenase (E.C. 1.2.1.12.) activities in atherosclerotic aortic areas. The findings on enzyme E.C. 1.1.1.8. agree with data in experimental rabbit atherosclerosis but are the reverse of those observed in canine atherosclerosis (see p. 41). The reason of such contradictory results remains so far unresolved.

Chapter III

ENZYMES OF THE TRICARBOXYLIC ACID (TCA) CYCLE

THE basic reaction by which all living cells provide the free energy indispensable for life is essentially the reduction of molecular oxygen by hydrogen in the relevant substrates.

In the presence of oxygen, lactic acid is not produced in appreciable amounts by most tissues. Instead, pyruvic acid is oxidized *via* the tricarboxylic acid (citric acid) cycle. This occupies a central position in the oxidative metabolism of carbohydrates, lipids and aminoacids.

As seen from the accompanying diagram (Fig. 3), during the clock-wise operation of the citric acid cycle 3 moles of CO_2 are produced from one mole of pyruvic acid together with the net production of two moles of water *via* the cytochrome system.* This corresponds to the complete oxidation of pyruvic acid and is accompanied by the generation of "energy-rich" phosphate bonds. Under aerobic conditions the complete oxidation of a mole of glucose gives rise to 38 moles of ATP or about 304 kcal.

In the discussion that follows we shall see that only about half of the enzymes of the TCA cycle (six out of nine) have so far been extensively studied in vascular tissue. Furthermore, we lack information about the three enzymes involved in the conversion of pyruvate to acetyl-CoA. Nevertheless, a large body of valuable information has been accumulated on some enzymes especially in Kirk's laboratory, and it is to be hoped that in the future new techniques will extend and deepen our knowledge about the role of vascular enzymes in this extremely important metabolic pathway.

CITRATE SYNTHASE, E.C. 4.1.3.7.
(Citrate oxaloacetate-lyase [CoA-acetylating])

The mechanism of the tricarboxylic acid cycle became clear when Stern *et al.* (1951) succeeded in crystallizing the "condensing enzyme" from pig heart and demonstrated that it catalyses the reaction:

$$\text{Acetyl-CoA} + H_2O + \text{oxaloacetate} = \text{citrate} + \text{CoA}$$

The reaction is strongly exergonic in the direction of citrate synthesis and, thus, the conversion of a four-carbon acid into a six-carbon compound (tricarboxylic acid) is made possible.

The enzyme is widely distributed in animal and plant tissues and some

---

* Five moles of water are produced but three are consumed in the cycle.

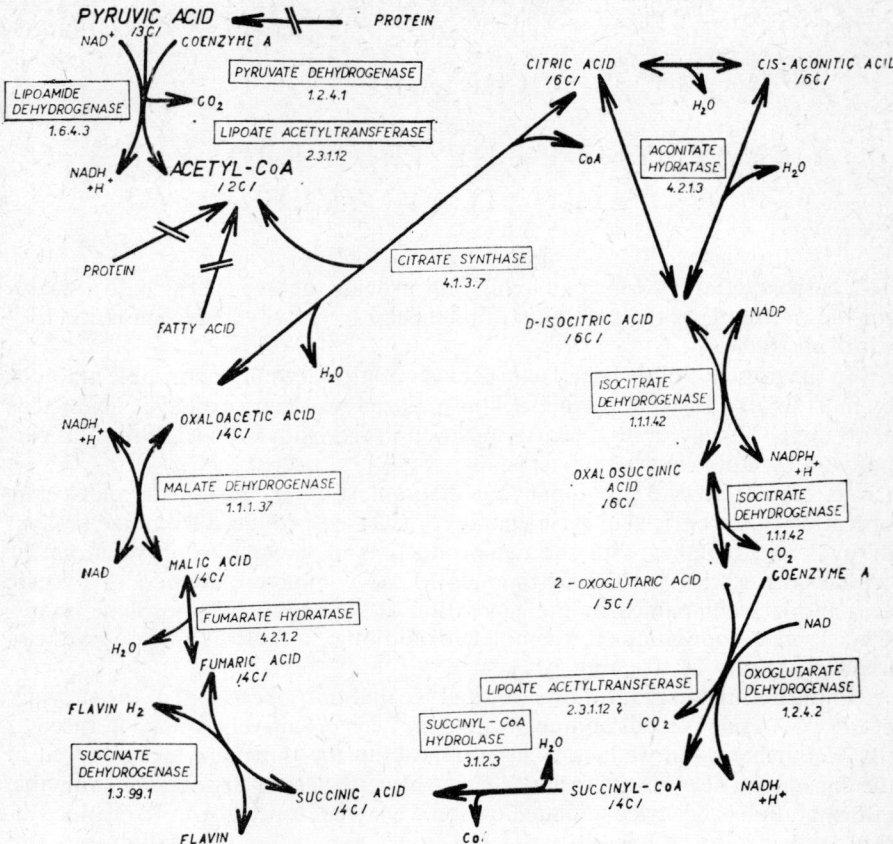

Fig. 3.—The tricarboxylic acid cycle (citric acid cycle, Krebs-Szent Györgyi cycle).

microorganisms. The highest levels are found in tissues with a high level of aerobic metabolism (e.g. heart, pigeon breast muscle).

An optical test for enzyme activity (see Chapter XI) is based on the production of oxaloacetate from malate in the presence of NAD and malate dehydrogenase. Subsequently acetyl-CoA reacts with oxaloacetate and the rate of $NADH_2$ production—measured at 340 mμ.—is proportional to the citrate synthase activity (Ochoa, 1955).

Kirk (1966c) measured the activity of the enzyme in human vessels using another optical method where the amount of CoA is measured spectrophotometrically at 412 mμ. after reacting it with 5, 5-dithio-bis (2-nitrobenzoic acid). Citrate synthase activity in the normal aorta, coronary artery and pulmonary artery is, respectively, 25·5, 47·4 and 29·2 μM of CoA formed per g. of wet tissue per hour. Aortic enzyme activity progressively decreases with advancing age from 20–84 years. Specimens of atherosclerotic aorta and coronary artery also showed a decline in enzyme activity.

Aconitate Hydratase, E.C. 4.2.1.3.
(Citrate [isocitrate] hydro-lyase)

This enzyme (known as "aconitase") catalyses the interconversion of citric and D-isocitric acids through the intermediate stage of *cis*-aconitic acid.

$$\text{Citric acid} \underset{}{\overset{+H_2O}{\rightleftarrows}} \textit{cis}\text{-Aconitic acid} \underset{}{\overset{+H_2O}{\rightleftarrows}} \text{D-isocitric acid}$$

The enzyme occurs in a wide variety of animal tissues and is abundant in liver, kidney, heart and prostate. It has been found also in plants, yeast and bacteria; its intracellular localization is mainly in the soluble cytoplasmic fraction.

Crude preparations of the enzyme rapidly lose activity on storage or dialysis but can be reactivated by Fe^{++} ions and a reducing agent.

According to Morrison (1954) one molecule of the metal and one molecule of the reducing agent combine with the protein to form active aconitase. Iron-binding agents inhibit the enzyme. Aconitase is also inhibited by copper and mercury ions and particularly by fluoroacetate or its metabolic product fluorocitrate. The latter compounds are specific inhibitors of aconitase and of the whole TCA cycle.

The methods most commonly employed for measuring this enzyme activity are based either on the chemical determination of citrate or on light absorption of *cis*-aconitate at 240 mμ. An "optical test" is based on the conversion of isocitrate to oxalosuccinate by isocitrate dehydrogenase (see below) and measurement of the resulting reduced NADP at 340 or 366 mμ.

Laursen and Kirk (1955a) demonstrated the presence of aconitase activity in homogenates of 17 human aortas by measuring the amount of citrate formed from *cis*-aconitate. In this relatively small series of determinations aortic aconitase activity did not appear to be age-dependent (with results expressed on a wet weight basis). In a further paper Kirk (1961a) reported significantly lower activity in atherosclerotic parts of arteries. The pulmonary artery displayed 36–40 per cent higher activity than the aorta.

In comparison with the other enzymes of the TCA cycle, arterial aconitase activity is low. Therefore, Kirk (1963a) has suggested that further study may disclose that it is a limiting step in the whole TCA cycle.

Fumarate Hydratase, E.C. 4.2.1.2.
(L-Malate hydro-lyase)

This enzyme ("fumarase") catalyses the interconversion of fumaric and malic acid:

$$\text{L-Malate} = \text{fumarate} + H_2O$$

It is widely distributed in tissues and, in particular, the liver, striated muscle and the heart reveal high activity. Some inorganic anions, especially phosphate ions, are activators, whereas thiocyanate and iodides are inhibitors. The enzyme is also competitively inhibited by some compounds structurally similar to its substrates, the most potent inhibitor being *meso*-tartrate.

Enzyme activity can be measured by determination of fumaric acid at 240 mμ. or by oxidation with permanganate.

Laursen and Kirk (1955a) assessed fumarase activity by measuring the amount of L-malic acid produced, using a fluorometric method devised by Lowry *et al.* (1954). In contrast to the findings with aconitase, a significant negative correlation (r = -0.62) was noted between age and fumarase activity in aortic homogenates, expressed on a wet weight basis.

In a later paper Sørensen and Kirk (1956) extended their investigations to include homogenates of pulmonary arteries, as well as aortic homogenates from the same subjects. In agreement with the previous study aortic enzyme activity decreased with advancing age, while a smaller decrease was noted in the pulmonary arteries. Analysis of the relationship between the degree of atherosclerosis and enzyme activity in the age group 55–79 years revealed decreased activity in aortic samples with moderate and severe atherosclerosis. Enzyme activity also correlated with the cholesterol content of the sample. (Calculations were based both on wet weight and tissue nitrogen bases.)

ISOCITRATE DEHYDROGENASE, E.C. 1.1.1.42.
(*threo*-D_s-Isocitrate: NADP oxidoreductase decarboxylating)

This enzyme catalyses the conversion of *threo*-D_s-isocitrate to 2-oxoglutarate in a NADP-linked reaction. Highly purified enzyme preparations from heart exhibit both isocitrate dehydrogenase and oxalosuccinate decarboxylase activity (see review by Plaut, 1963). The overall reaction proceeds as follows:

$$\textit{threo}\text{-}D_s\text{-isocitrate} + NADP = 2\text{-oxoglutarate} + CO_2 + NADPH_2$$

Manganese and magnesium ions are necessary for the decarboxylating action of the enzyme and, in the absence of manganese ions, oxalosuccinate accumulates during the reaction.

The enzyme is widely distributed in animal tissues, in particular in the liver, heart, striated muscle, erythrocytes and thrombocytes. In addition to this enzyme, some animal tissues, yeast and moulds contain another isocitrate dehydrogenase (E.C. 1.1.1.41.) that is *NAD*-dependent and does not decarboxylate added oxalosuccinate. It is claimed that the latter is a mitochondrial enzyme while the *NADP*-dependent enzyme is mainly cytoplasmic. Some controversy has centred around the relative importance of the two catalytic reactions in isocitrate oxidation (see Plaut, 1963).

The usual assay method is based on the measurement of NADP reduction in the presence of Mn^{++} at 340 mμ. (see "Optical Tests" in Chapter XI). Oxalosuccinate decarboxylase activity can be also determined manometrically by measuring the rate of CO_2 liberation from oxalosuccinate. In a further method the formation of an oxalosuccinate-Mn complex is measured at 240 mμ.

It must be stressed that reduced NADP is produced during the oxidative decarboxylation of isocitrate and the former compound is needed for the reductive synthesis of cholesterol and fatty acids. Therefore, investigation of isocitrate dehydrogenase activity (as well as the study of glucose-6-phosphate dehydrogenase activity—see later) are particularly relevant to arterial lipid metabolism.

Kirk and Kirk (1959) and Kirk (1960c) investigated the activity of this enzyme in supernatants of homogenates prepared from human aortas, and from pulmonary and coronary arteries; they measured the rate of reduction of NADP spectrophotometrically. The average activities in atherosclerotic aortas and coronary arteries were of the same magnitude as those observed in normal vessels.

SUCCINATE DEHYDROGENASE, E.C. 1.3.99.1.
(Succinate: [acceptor] oxidoreductase)

This enzyme of the TCA cycle has been intensely studied. For many years it has been recognized that the addition of succinate to tissue preparations considerably increases their oxygen consumption. The complete catalytic system responsible for the oxidation of succinate termed "succinoxidase" has been localized in discrete intracellular structures, particularly mitochondria. It was shown that the succinoxidase system comprises, in addition to other components of the terminal respiratory chain (see Fig. 5), a specific catalytic flavoprotein, i.e. succinate dehydrogenase. As pointed out by Singer and Kearney (1963) some confusion in the nomenclature has arisen from the loose application of the term "succinate dehydrogenase" to various respiratory chain preparations. It is clear that only the activity of the latter flavoprotein enzyme represents succinate dehydrogenase activity.

Singer and Kearney (1954) succeeded in separating this enzyme from the other components of the succinoxidase system. The solubilization of the enzyme from animal tissue involves pretreatment with an organic solvent (acetone or butanol) and extraction with alkaline buffers, to break the bonds between the dehydrogenase and the respiratory chain. The enzyme isolated from mammalian heart mitochondria was shown to contain 1 mole of flavin-adenine dinucleotide (FAD) in peptide bond with the apoenzyme and four atoms of nonhaem iron (Kearney, 1960). The enzyme is widely distributed in animal tissues, in bacteria and yeast.

In contrast to the other dehydrogenases of the glycolytic and TCA cycle, succinate dehydrogenase is not NAD or NADP dependent and directly catalyses the oxidation of succinic to fumaric acid:

Succinate + acceptor = fumarate + reduced acceptor.

The other components of the succinoxidase system facilitate electron transfer from succinate to molecular oxygen through cytochrome b, coenzyme Q, cytochrome c_1, cytochrome c and cytochrome oxidase. The exact nature of the junction between the flavoprotein (succinate dehydrogenase) and the other electron acceptors has so far not been unequivocally established. In artificial systems certain dyes may serve as acceptors and in particular phenazine methosulphate has been used with great success as an oxidation-reduction acceptor (see p. 98).

Succinate dehydrogenase is an SH-dependent enzyme and is, therefore, inhibited by p-chloromercuribenzoate and reactivated by glutathione. Well-known classical competitive inhibitors of the enzyme are malonate and oxaloacetate.

An unusual and characteristic property of the enzyme is the activation by fumarate, phosphate and in particular by its substrate, i.e. succinate. It is usual to activate the enzyme prior to assay, especially in experiments of short duration where activation by the substrate is slight (Singer and Kearney, 1963).

The activity of the enzyme can be measured by a manometric method based on oxidation of succinate with phenazine methosulphate (PMS) and determination of the initial rate of O_2 uptake in a Warburg apparatus. In another method, also based on oxidation with PMS, neotetrazolium serves as the final electron acceptor (see Chapter XI).

Briggs et al. (1949) observed that succinate increases oxygen consumption in rat thoracic aorta by about 110–150 per cent and that malonate depresses respiration of this tissue when previously stimulated by succinate.

Kirk and Laursen (1955) used the quantitative tetrazolium technique, mentioned in connection with lactate dehydrogenase (see page 23), for determining succinate dehydrogenase activity in the human aorta. In a succinate-containing medium about 0·160 μg. of formazan were produced per g. of wet tissue per hour.

Kirk et al. (1955) extended their observations to aortic respiration studies. The oxygen consumption of the intact human aorta in succinate-phosphate buffer media was found to be over four-fold higher than in glucose-phosphate buffer media; the same ratio was found in parallel tetrazolium-reduction studies. Such respiration studies clearly reflect the activity of the whole "succinoxidase" system. However, the marked depression of respiration following addition of malonate to the incubation medium confirms the presence of succinate dehydrogenase activity in the human aorta, as do the above-mentioned studies by Briggs et al. on rat vessels.

Maier and Haimovici (1957) investigated the succinoxidase system in *tissue sections* of intact vessels of man, rabbit and dog. They measured oxygen consumption in the presence of succinate and results were expressed on a dry tissue weight basis. The highest values were observed in the dog's aorta and the lowest values in the human aortic specimens. The values for rabbit aortic segments were intermediate between those of man and dog. In the aortas of male infants and young men (1 day to 17 years) the activity of the abdominal segments was significantly lower than that of the arch, ascending aorta and that of the thoracic aorta. A similar "gradient" of activity was observed in the dog aorta. However, in the aortas of older men (21 to 73 years) the gradient disappeared, chiefly due to some 50 per cent decrease in the activity of the arch and thoracic segments. Likewise the rabbit aorta displays no gradient in activity. Activity in the human vena cava is much higher than in the aorta, but the converse applies to the dog. Activity in the liver is about 24, 13 and 9-fold higher than in the human, dog or rabbit aorta, respectively.

In a subsequent paper, Maier and Haimovici (1958) investigated the same problems using vascular *homogenates* instead of tissue sections. Similar species differences were noted as before. A gradient in aortic succinoxidase activity was now observed not only in the dog but also in the rabbit.

In our own experiments we have been studying the activity of succinate dehydrogenase in the vessels of man and animals under different physiological and pathological conditions; the results will be presented in part two of this book.

Malate Dehydrogenase, E.C. 1.1.1.37
(L-Malate: NAD oxidoreductase)

This enzyme catalyses the conversion of L-malic acid into oxaloacetic acid and it is NAD-dependent.

$$\text{L-Malate} + \text{NAD} = \text{oxaloacetate} + \text{NADH}_2$$

It is widely distributed in animal tissues, plants and bacteria. It occurs in mitochondria and in cytoplasm but its cell distribution differs greatly in different organs. The liver, heart, striated muscle and the brain display the highest cytoplasmic activity. Following damage to the cell membrane malate dehydrogenase activity increases in the serum. As with lactate dehydrogenase (see p. 23), several workers have detected isozymes of malate dehydrogenase, i.e. different molecular types of this enzyme, in tissue and serum. It appears that at least two isozymic fractions of malate dehydrogenase can be isolated from one cell type in animal tissues.

The enzyme is competively inhibited by dicarboxylic α-hydroxy acids related to malate. The inhibition by traces of sulphite has been ascribed to the zinc content of the enzyme and the formation of a metal complex.

The activity of malate dehydrogenase can be measured colorimetrically, using different dyes as electron acceptors, or by a spectrophotometric method (see Chapter XI).

Matzke *et al.* (1957) determined spectrophotometrically malate dehydrogenase activity in homogenates of human aortas and pulmonary arteries and observed appreciable activity. They did not find any significant change in enzyme activity with age while there was a small 14–19 per cent decrease in atherosclerotic aortas, but only on a wet weight and not on a tissue nitrogen basis.

In a later paper Kirk *et al.* (1958) investigated malate dehydrogenase activity on a wet weight basis in human coronary arteries. No significant change with age could be detected, but the activities of the atherosclerotic parts of the vessels were reduced by some 33 per cent. It is interesting that the malate dehydrogenase activity in *coronary artery* tissue was higher than that of lactate dehydrogenase by an average ratio of about 3:1; the converse was found in aortic tissue.

Mandel and Kempf (1960) studied malate dehydrogenase activity in extracts of bovine aorta. A significant decrease in enzyme activity was found in the aortas of old animals. (Activity was calculated on a wet weight basis).

As with succinate dehydrogenase we also have investigated in our laboratory malate dehydrogenase activity in many animal and human vessels; the results will be presented in later chapters in this book.

Malate Dehydrogenase (Decarboxylating), E.C. 1.1.1.40.
(L-malate: NADP oxidoreductase [decarboxylating])

This enzyme is not part of the classical TCA cycle but is related to the previous enzyme. It was originally known as the "malic enzyme"; it is NADP-dependent and catalyses the reversible oxidative decarboxylation of malic and oxaloacetic acids.

$$\text{L-Malate} + \text{NADP} \rightleftarrows \text{pyruvate} + CO_2 + \text{NADPH}_2$$

The enzyme is activated by magnesium and manganese ions. "Malic enzyme" has been identified in animal tissue and plants, and it has been extensively purified from pigeon liver. It is mainly localized in the soluble cytoplasmic fraction.

The enzyme is sensitive to SH reagents and, for optimum activity, the presence of cysteine is required. Fumarate, malonate, oxaloacetate and those dicarboxylic α-hydroxy acids that are related to malate all exhibit inhibitory action. Bacteria, in particular *Lactobacillus arabinosus*, contain a similar adaptive enzyme (E.C. 1.1.1.38.) which is, however, NAD-dependent. Lyophilized *L. arabinosus* cells are used for preparing this "malic enzyme". It is used by biochemists to measure malate, formed from fumarate, in the Singer and Lusty (1963) method for fumarate estimation. From the biological point of view, the enzyme is important in that it fixes CO_2 and passes it through the components of the TCA cycle both in plants and animals.

Kirk studied spectrophotometrically the activity of the "malic enzyme" by measuring the increase of reduced NADP in supernatants from homogenates of human thoracic aortas, pulmonary and coronary arteries (Kirk and Kirk, 1959; Kirk, 1960*c*). The activity, calculated on a wet weight basis, was only about 1/50th of that of the NAD-dependent malate dehydrogenase. No difference in activity was observed between atherosclerotic and normal tissue.

Chapter IV

ENZYMES OF THE GLUCOSE 6-PHOSPHATE OXIDATION SYSTEM (THE PENTOSE PHOSPHATE PATHWAY)

THE presence of the pentose phosphate pathway has been repeatedly demonstrated in the arterial wall (See Chapter I). It comprises a series of enzyme reactions whereby glucose 6-phosphate is oxidized to CO_2 by means of NADP. In contrast to glycolysis this metabolic pathway does not depend on ATP breakdown, while the oxidation product is CO_2.

During one anticlockwise "rotation" of the reaction sequences indicated in Fig. 4, the first carbon of glucose 6-phosphate is decarboxylated. The other carbon atoms are converted through certain intermediary products (viz. pentoses) to newly synthesized glucose 6-phosphate on one side, and to glyceraldehyde phosphate on the other. Ribose 5-phosphate, which itself may be formed in these reactions, acts in a catalytic capacity in the reaction sequences. The net result of several "rotations" of the pentose phosphate pathway is the oxidation of a mole of glucose phosphate into three moles of CO_2 and one mole of glyceraldehyde phosphate. (The production of the latter compound, which can be further metabolized by the glycolytic pathway, gave rise to the well-known term "hexose monophosphate shunt" for this series of reactions). From two moles of glucose phosphate, two moles of glyceraldehyde phosphate are produced and these can be synthesized into one mole of glucose phosphate. Thus, starting with two moles of glucose phosphate one of them is completely oxidized to CO_2 and inorganic phosphate, while the other is regenerated. In addition, twelve moles of reduced NADP are produced per mole of glucose 6-phosphate and these can be utilized for the generation of high-energy bonds either immediately or by the action of transhydrogenases (E.C. 1.6.1.1.) converting $NADPH_2$ to $NADH_2$ (see Fig. 5). However, the more important biological function of the pentose phosphate pathway probably consists in the further utilization of the produced $NADPH_2$ for reductive fatty acid and steroid synthesis. Moreover, this pathway is important for synthesizing the pentoses of nucleic acids and nucleotides.

The pentose phosphate pathway operates in many animal tissues and, for example, has been extensively investigated in the mammary gland (see review by Glock and McLean, 1958). Possibly the pentose phosphate pathway becomes particularly active in tumours in order to meet the demands of malignant tissue for deoxyribonucleic acid. In some plant tissues, in particular leaves, the glucose 6-phosphate oxidation system seems to be the major metabolic pathway for glucose catabolism.

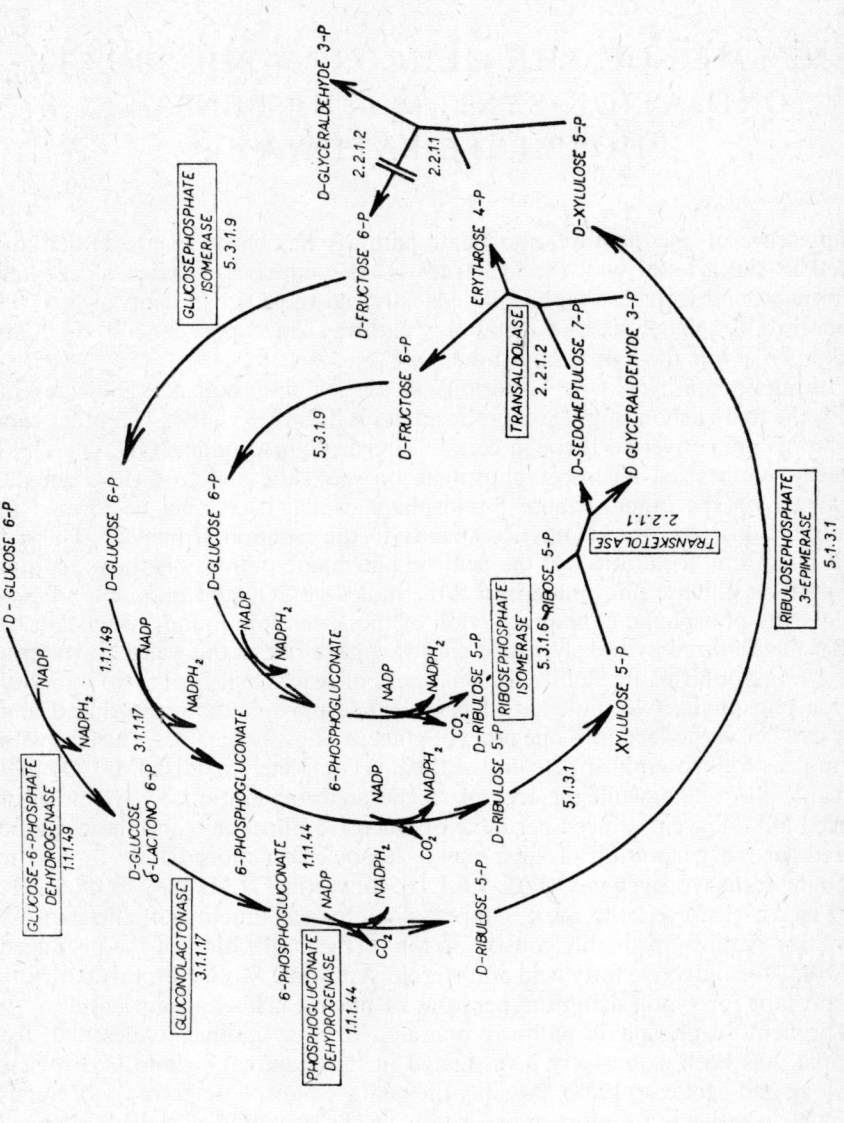

Fig. 4.—The pentose phosphate pathway or glucose 6-phosphate oxidation system. (Modified after Glock and McLean, 1958).

Glucose 6-phosphate Dehydrogenase, E.C. 1.1.1.49.

and

Phosphogluconate Dehydrogenase (Decarboxylating), E.C. 1.1.1.44.
(D-Glucose-6-phosphate: NADP oxidoreductase and 6-phospho-D-gluconate: NADP oxidoreductase [decarboxylating])

The first step of the pentose phosphate pathway is the oxidation of glucose 6-phosphate to the lactone of phosphogluconic acid:

$$\text{D-glucose 6-phosphate} + \text{NADP} = \text{D-glucono-}\delta\text{-lactone 6-phosphate} + \text{NADPH}_2$$

The dehydrogenase catalysing this reaction is found in plants, yeast and bacteria. In animal tissues the highest activity is observed in red blood cells, adipose tissue and in the lactating mammary gland.

Originally the enzyme was considered to have only an intermediary function and was termed "Zwischenferment" (Warburg and Christian, 1933). Although its function has been clarified, the term "Zwischenferment" is still in use due to its unique position in the history of modern biochemistry. As pointed out by Noltmann and Kuby (1963), the interest in the problems of "Zwischenferment" initiated studies which ultimately led to the discovery of NADP and the flavin-adenine nucleotides, and prompted advances linked to the discovery of alternate pathways of metabolism, dehydrogenases and oxidative enzymes in general.

The enzyme seems to exercise a key role in the regulation of carbohydrate metabolism, because it lies at the critical metabolic crossroad that determines whether glucose 6-phosphate is channelled through the glycolytic pathway or through the pentose phosphate pathway (Noltmann and Kuby, 1963).

The most convenient method for measurement of glucose 6-phosphate dehydrogenase activity is spectrophotometric assessment of the rate of reduced NADP formation. Moreover, due to the ease of spectrophotometric estimation and other advantageous features of the enzyme-catalysed reaction, glucose 6-phosphate dehydrogenase has become a valuable "indicator enzyme" in many "optical tests" (see Chapter XI). This principle is used for example in the assay of hexokinase, phosphoglucomutase and adenylate kinase or in the enzymatic assay of glucose, glucose 1-phosphate, ATP, ADP and other biologically important compounds.

In the next step (after hydrolysis of the lactone by a lactonase), phosphogluconate is oxidized and decarboxylated by another dehydrogenase (E.C. 1.1.1.44.):

$$\text{6-phospho-D-gluconate} + \text{NADP} = \text{D-ribulose 5-phosphate} + CO_2 + \text{NADPH}_2$$

This dehydrogenase is again found in animal tissues, plants, yeast and bacteria and, in general it accompanies the preceding enzyme.

Determination of activity is again based on spectrophotometric assay of the formation of reduced NADP, but in this case with 6-phosphogluconate as substrate.

Kirk *et al.* (1959) investigated the activity of these enzymes in homogenates of human aortas, and pulmonary and coronary arteries. No significant change in glucose 6-phosphate dehydrogenase activity was found in the ageing aorta, when calculated either on a wet weight or tissue nitrogen basis. Enzyme activities in aortas and pulmonary arteries from children (0–5 years) were significantly lower than those of adults. In atherosclerotic tissue activity decreased by about 28–35 per cent. The activity of phosphogluconate dehydrogenase does not significantly vary either with age or the development of atherosclerosis.

Kittinger *et al.* (1960, 1962) studied the same problem in aortic extracts from repeatedly-bred rats that develop spontaneous atherosclerosis. They found that significantly less $NADPH_2$ is produced by the combined action of glucose 6-phosphate and phosphogluconate dehydrogenases in extracts from aortas with moderate and severe atherosclerosis. The activity of 6-phosphogluconate dehydrogenase was only 15–20 per cent of such total activity in both normal and atherosclerotic aortas. Therefore, the reduced NADP production was interpreted as decreased glucose-6-phosphate dehydrogenase activity only. The results were referred to both wet weight and the protein content of the extracts.

Alekseeva (1964) obtained similar results with extracts of aortic homogenates from cholesterol-fed rabbits. Such activity, calculated on the protein content of the extracts, decreased as early as the 5th day from the start of the feeding experiment.

It is interesting that in early rat DOCA hypertension, Gardner and Laing (1965) observed in very small visceral arteries an increased glucose 6-phosphate dehydrogenase activity and an association between blood pressure levels and enzyme acitivity.

Ribosephosphate Isomerase, E.C. 5.3.1.6.
(D-ribose-5-phosphate ketol-isomerase)

This widely distributed enzyme catalyses the interconversion of ribose 5-phosphate and ribulose 5-phosphate. The reaction is very similar to that catalysed by glucosephosphate isomerase.

Kirk (1959c) investigated the activity of this enzyme in homogenates of human arterial tissue using a colorimetric carbazol-cysteine method. The activity in the coronary arteries expressed on both wet weight and tissue nitrogen bases was higher than the activity in either aorta or pulmonary artery. The activity of the aortic specimens showed a tendency to increase with age but significantly higher activity was found in atherosclerotic regions of the vessel than in normal parts. This increased activity could, however, be due to the presence of capillary red cells in the samples.

Chapter V

ENZYMES OF THE RESPIRATORY CHAIN

BIOLOGICAL oxidation is essentially a stepwise transfer of hydrogen (electrons) from the substrate to intermediate carriers and finally to molecular oxygen. It is carried out by way of the intracellular oxidation system. The components of this system are enzymes catalysing the transfer of electrons and highly efficient carriers with the properties of specific coenzymes.

In previous sections we have seen that, through the action of various dehydrogenases, a pair of hydrogen atoms is transferred from substrate to specific coenzyme (NAD or NADP)—either directly or by intermediary reactions.(The latter category is illustrated by the conversion of pyruvate to acetyl CoA). In other hydrogen transfer reactions the prosthetic group of flavoprotein enzymes acts as the immediate acceptor for hydrogen, such as in the reaction catalysed by succinate dehydrogenase (see also Chapter IX).

The hydrogen of the reduced forms of specific coenzymes is then shuttled along the various components of the respiratory chain, as shown schematically in Fig. 5. It should be mentioned that the direction of hydrogen (electron) transfer is regulated by the gradient of redox potentials between the components of the respiratory chain, and that it is directed from the lower (more negative) to the higher (more positive) potential.

The normal redox potential (E'_0) of the $NADH_2/NAD$ system is -0.32 volts, while the redox potential of O_2 (the oxygen electrode) is $+0.81$ volts.

A mathematical relationship exists between the redox potential and the free energy change*; the above difference of potential corresponds to a free energy change of about 57 kcal. This is the amount of theoretically available energy per mole of water produced, when starting from $NADH_2$. In the succinate/fumarate system with an E'_0 of 0.02 volts we get a free energy change of only 36.3 kcal, which suggests that less energy is produced when succinate is the starting substance.

The great biological importance of the intracellular oxidation system is its ability to convert a comparatively high fraction of the "chemical energy" in the reaction chain into useful energy. This high efficiency results from the coupling of intracellular oxidation to the system for oxidative phosphorylation that generates high-energy phosphate bonds. The detailed mechanism of oxidative phosphorylation is not at present understood. It is known, however, that the complete oxidation of one mole of succinate or $NADH_2$ yields two or three moles of ATP respectively. This represents an efficiency of about 42–44 per cent.

The energy balance of the TCA cycle, as discussed in Chapter III, is based on such calculations. In investigating the relationship between the respiratory chain and oxidative phosphorylation, much information has been gained by the use of certain inhibitors that inhibit intermediate steps in the system. The specific

* See footnote p. 47.

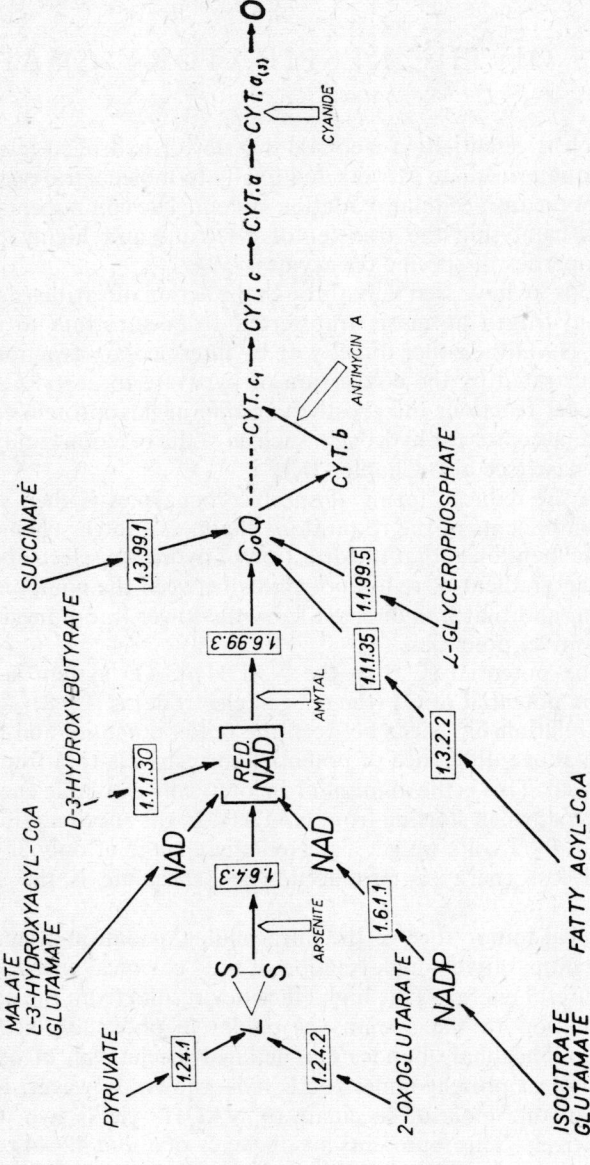

Fig. 5.—Schematic diagram of the respiratory chain. (Code numbers of enzymes according to Enzyme Nomenclature of the IUB—see footnote p. 15).

property of all these inhibitors is that they act only on mitochondrial preparations. A widely used inhibitor is amytal, which inhibits the oxidation of $NADH_2$ by ubiqinone. Antimycin A, BAL (in the presence of oxygen) and urethane all inhibit the transfer of electrons between ubiquinone or cytochrome b and cytochrome c_1. Cytochrome oxidase is inhibited by cyanide and carbon monoxide.

A final comment has to be made about the localization of the intracellular oxidation system. It is important to realize that flavoprotein enzymes and the cytochrome oxidase system are confined to the inside of mitochondria. $NADH_2$ can be produced inside mitochondria by the action of some dehydrogenases. This internally-produced $NADH_2$ can readily be oxidized, because mitochondria contain a complete system for the transfer of both electrons and energy. However, certain cytoplasmic enzymes also produce $NADH_2$ and it is not quite clear how this compound is oxidized. In some tissues there seem to exist separate pathways also for the oxidation of cytoplasmic $NADH_2$. However, it must be anticipated that most tissues are endowed with a mechanism for transporting hydrogen into mitochondria because the intact mitochondrial membrane is not permeable to $NADH_2$. A plausible mechanism has been suggested by Bücher and Klinkenberg (1958); it consists of the coupling of an external glycerol-3-phosphate dehydrogenase ("Baranowski enzyme", E.C. 1.1.1.8.) and internal glycerol-3-phosphate dehydrogenase ("Meyerhof-Green enzyme", E.C. 1.1.99.5.). The first enzyme transfers the hydrogen pair from $NADH_2$ to glycerol 3-phosphate. The second enzyme is confined to the mitochondria (mitochondrial membrane?) and transfers the hydrogen to an internal carrier of the respiratory chain, while the dihydroxyacetone phosphate produced by the reaction remains in the cytoplasm.

Both reactions can be written as follows:

$$\text{Dihydroxyacetone phosphate} + NADH_2 = \text{L-glycerol 3-phosphate} + NAD$$

$$\text{L-glycerol 3-phosphate} + \text{acceptor} = \text{dihydroxyacetone phosphate} + \text{reduced acceptor}$$

The mitochondrial enzyme is a flavoprotein whose redox potential is higher than that of $NADH_2/NAD$ (see p. 39). This circumstance provides a physicochemical basis for electron transport by this "glycerol-3-phosphate cycle".

It has been suggested that other similar "shuttle" mechanisms exist (see Chapter I).

In connection with cytoplasmic glycerol-3-phosphate dehydrogenase, conditions of inadequate oxygen supply would be expected to result in only partial consumption of $NADH_2$ and production of glycerol 3-phosphate. According to Klinkenberg and Bücher (1960) this may prevent lowering of the redox potential of extramitochondrial NAD-dependent systems. Because glycerol 3-phosphate ("active glycerol") is a key compound in the synthesis of triglycerides and phospholipids, its greater activity under hypoxic conditions seems to be of considerable interest. In this connection it is interesting that for example in late or severe stages of experimental *canine atherosclerosis* the aortas histochemically show conspicuous increase of glycerol-3-phosphate dehydrogenase in addition to loss of oxidative enzyme activity (Rubinstein *et al.*, 1966). However, in early stages of rabbit atherosclerosis and in human atherosclerotic aortas the activity of AGPDH is either decreased or unchanged (see pp. 26 and 128).

In the arterial wall three enzymes of the respiratory chain have so far been investigated.

$NADH_2$ Cytochrome c Reductase, E.C. 1.6.2.1.*

and

Lipoamide Dehydrogenase, "Diaphorase," E.C. 1.6.4.3.
($NADH_2$: cytochrome c oxidoreductase and $NADH_2$: lipoamide oxidoreductase)

The history of both enzymes is closely related to Warburg's old yellow enzyme, (E.C. 1.6.99.1.) which has a high "diaphorase" activity (with methylene blue as acceptor) but reacts only slowly with cytochrome c. Both enzyme activities can be clearly distinguished. The first enzyme is a metalo-flavoprotein with flavine-adenine dinucleotide as a probable prosthetic group and with four atoms of Fe per mole of flavine. It catalyses the reaction:

$NADH_2$ + oxidized cytochrome c = NAD + reduced cytochrome c

It is found in animal tissues and some bacteria.

Recent evidence indicates that cytochrome c acts as an acceptor only after enzyme preparations have been subjected to rather drastic treatments. However, digestion of beef heart "electron transport particles" with phospholipase A (*Naja naja* venom) and subsequent purification by methods that avoid organic solvents yield a highly purified, soluble and native form of the enzyme. It is maintained that this preparation represents the flavoprotein component of the respiratory chain that links $NADH_2$ oxidation to the cytochrome system (Ringler *et al.*, 1960). This purified enzyme readily reduces ferricyanide and some dyes, such as phenazine methosulphate, but it reduces only very slowly, if at all, oxygen, lipoate or cytochrome c and other intermediate carriers of the electron transport system. It was suggested that the enzyme should be classified as $NADH_2$ dehydrogenase ($NADH_2$: acceptor oxidoreductase, E.C. 1.6.99.3.) and the term "cytochrome c reductase" is no longer recommended. (Concerning other problems of $NADH_2$ dehydrogenases see Hatefi, 1963 and Lehninger, 1965.)

"Diaphorase", is the second of the enzymes investigated in the arterial wall. It was isolated by Straub in 1939 from the heart; it was found only to react with some dye acceptors and not with cytochrome c. However, the natural substrate of the enzyme was recently identified as protein-bound lipoic acid or lipoamide (Massey, 1960); it was shown that it catalyses the reaction:

$NADH_2$ + oxidized lipoamide = NAD + dihydro-lipoamide

This reaction is of great biological importance in the oxidative decarboxylation of pyruvate and α-ketoglutarate. (For details see reviews by Massey, 1963 and Sanadi, 1963.) The enzyme is a flavo-protein found in animal tissues, plants, yeast and bacteria.

"Diaphorase" may be readily distinguished from $NADH_2$ dehydrogenase by many of its properties. Thus, under the conditions recommended by Massey

* For other recommended name and code number see text.

(1960) for the assay of lipoamide dehydrogenase at pH 6·5, the ratio V_{max} ferricyanide/V_{max} lipoamide is 56 for purified $NADH_2$ dehydrogenase but only 0·1 for lipoamide dehydrogenase (Ringer *et al.*, 1960). On the other hand, the controversies concerning the identity of diaphorase and lipoamide dehydrogenase appear to be explained by the sensitivity of the physiological (i.e. lipoamide) dehydrogenase activity to copper ions. Veeger and Massey (1960) reported that the pig heart enzyme, when incubated at 0° with one atom of Cu^{++} per mole of flavin, loses all lipoamide dehydrogenase activity in the course of a day. At the same time the "diaphorase" activity (with 2, 6-dichlorophenolindophenol as acceptor) increases to a value approximately thirty times higher than that of the native enzyme.

Kirk (1962*a*) investigated the activity of both $NADH_2$ cytochrome *c* reductase and "diaphorase" in homogenates of human arteries. The estimation of the former enzyme was based on the colorimetric measurement of the reduction of oxidized cytochrome *c* (ferricytochrome *c*) at 550 mμ. "Diaphorase" activity was determined by measuring the decolourization time of 2, 6-dichlorophenolindophenol in the presence of $NADH_2$ and NaCN. The activities were assayed in aortas, pulmonary arteries and coronary arteries. With advancing age, the activity of aortic cytochrome *c* reductase decreased but in the other vessels no such change could be observed. Likewise, "diaphorase" activity did not decline in senescent vessels. The activity of both enzymes was reduced in atherosclerotic segments of the aorta and somewhat lower in the coronary arteries. Activity was calculated on both wet weight and tissue nitrogen bases.

Cytochrome Oxidase, E.C. 1.9.3.1.
(Ferrocytochrome *c*: oxygen oxidoreductase)

This is a cytochrome which is classified as of *a*-type because it absorbs light of wave length greater than 585 mμ.

Keilin and Hartree (1939) identify cytochrome oxidase with cytochrome a_3, Warburg's "Atmungsferment" and with indophenol oxidase.

However, this topic has provoked considerable controversy. In a recent review, Yonetani (1963) emphasized that the name cytochrome oxidase should be used for purified preparations containing "Atmungsferment". The latter term refers to the enzyme (or enzymes) that directly transfers electrons derived from the cytochrome chain to molecular oxygen; it exists in an insoluble state in cells. On the other hand, there are also soluble oxidases in the cell which catalyse aerobic oxidation of various intracellular substances, including reduced cytochrome *c*. Both types of enzymes catalyse the "NADI reaction" (hence the term indophenol oxidase—see below) and both types are sensitive to respiratory inhibitors. However, only the inhibition of the "Atmungsferment" by carbon monoxide is photoreversible and this is a distinctive feature of the latter enzyme.

Cytochrome oxidase catalyses the following reaction:

$$4 \text{ ferrocytochrome } c + O_2 = 4 \text{ ferricytochrome } c + 2H_2O$$

The enzyme is a haemoprotein with a haem group that—unlike the porphyrin of blood haemoglobin—contains one formyl, one ethylenic and two alkyl groups.

The enzyme isolated from heart muscle contains one atom of firmly bound copper per haem group. It is interesting that the activity in insoluble heart preparations is enhanced by non-ionic synthetic detergents, certain phospholipids and mitochondrial lipoproteins.

Almost all tissues contain cytochrome oxidase which is able to bring about the final rapid oxidation of ferrocytochrome c by molecular oxygen.

Cytochrome oxidase activity can be measured by the rate of oxygen consumption in the presence of reduced cytochrome c, using a manometric or polarographic method. Ascorbic acid, cysteine or hydroquinone can also serve as substrates. The oxidation rate can be followed by the decrease in optical density of reduced cytochrome c at 550 mμ. It is also known from Keilin's studies that the oxidation of dimethyl-p-phenylene diamine by oxygen is catalysed by cytochrome oxidase. This forms the basis of assays where, for example, activity is measured by the formation of indophenol blue; such oxidation is conducted in the presence of α-naphthol ("NADI reaction").

Maier and Haimovici (1957) investigated the activity of the enzyme in tissue slices of intact vascular tissue and liver from man, rabbit and dog. They measured oxygen consumption in the presence of p-phenylene diamine. The results were somewhat similar but not identical with those obtained in succinoxidase estimations (see p. 32). The activity of aortic tissue was on an average about 1/15th of that of the liver. In man activity was much higher in the vena cava than in the aorta, but in the dog the converse applied. The activity of the aorta was approximately identical in the rabbit and dog, but about 3–4 fold higher than in the human aorta. Three segments were compared in each aorta, viz. the arch, including the ascending aorta, thoracic aorta and abdominal aorta. In the dog the activity of the abdominal segment was lower than that of the thoracic segments, but in the rabbit and human aortas no such "gradient" could be observed. In human aortas the activity in the age group 1 day–17 years was about twice as high as in the age group 21–73 years. The activity in the vena cava and liver did not change with advancing age. All activities were expressed as oxygen consumption per mg. *initial dry weight* per hour.

In a subsequent paper Maier and Haimovici (1958) investigated the same problem in homogenates prepared from aortas and liver. In contrast to the findings with slices, the activity in homogenates of rabbit aortas was about twice as high as in dog aortas and about 8 to 10 times as high as in human aortas. A "gradient" of activity between the thoracic and abdominal segments could be found not only in the dog, but also in rabbit aortas. No such gradient was, however, observed in human aortas.

In a more recent paper (Maier and Haimovici, 1965b), the authors investigated the same problems in slices of aortas from rabbits and dogs subjected to an atherogenic regimen. They observed an increased oxidative response to succinate and p-phenylene diamine at an early stage in the atherosclerotic process, whereas at a later stage the oxidative capacity declined. It is interesting that uninvolved portions of the aorta displayed decreased activity of the cytochrome oxidase system. This was interpreted as a pre-atherosclerotic manifestation.

Chapter VI

ENZYMES LINKING ENERGY-RICH BONDS WITH MUSCULAR CONTRACTION

In muscle chemical energy is transformed to mechanical work. It is not possible to go into details of the complicated topic of muscular contraction: it will suffice to mention that striated muscle is particularly rich in glycolytic enzymes and the anaerobic pathway is the main mechanism which produces energy for muscular contraction in this tissue.

In other muscles whose function requires constant and prolonged work, for example the heart or the flying muscles of insects, the aerobic pathway is the main producer of energy-rich bonds.

Most energy-utilizing processes depend on the hydrolytic breakdown of ATP and this substance is also the immediate donor of energy-rich bonds for muscular contraction. However, in muscular tissue, particularly in striated muscle, most of the energy is stored in the form of a "phosphagen", which is creatine phosphate in vertebrates and arginine phosphate in invertebrates. The energy stored in the form of phosphagen can be very quickly utilized to regenerate ATP (see below).

It is interesting that striated muscle contains about 5 mM ATP and 20 mM creatine phosphate per kilogram of tissue, while in the heart the corresponding values are 1·5 and 2, and in smooth muscle 2 and 0·7 respectively (Rapaport, 1964).

Although the role of actin and myosin in contraction is widely accepted, there are still large gaps in our understanding of the biophysical nature of the process. Many of the older views have been challenged by the results obtained from new techniques, in particular electron microscopy.

Three enzymes play an important role in the transfer of energy to muscular contraction and all of them have been studied in vascular tissue.

Creatine Kinase, E.C. 2.7.3.2.
(ATP: creatine phosphotransferase)

In 1934 Lohmann made the important discovery that muscle extracts are able to convert creatine phosphate (phosphagen) to creatine and inorganic phosphate.

It was demonstrated that phosphagen is not hydrolyzed, as had been formerly supposed, but that two separate reactions take place, the first being the hydrolysis of ATP by adenosine triphosphatase and the second a transphosphorylation reaction between creatine phosphate and ADP:

$$ATP + H_2O = ADP + orthophosphate \qquad (I)$$
$$ADP + creatine\ phosphate = ATP + creatine \qquad (II)$$

Overall:
$$Creatine\ phosphate + H_2O = creatine + orthophosphate$$

Only reaction II ("the Lohmann reaction") is catalysed by creatine kinase and is freely reversible with a very slight change in energy. Whenever ATP is rapidly broken down, the reaction swings towards the right and ATP will be regenerated from the phosphagen-reserve of energy-rich bonds. On the other hand, when new energy-rich bonds are produced during the period of anaerobic recovery in muscle tissue, the direction of the reaction will be reversed to the left so that new stores of phosphagen are formed.

As pointed out in a recent review (Kuby and Noltmann, 1962) "... in spite of all the efforts expended over the last three decades, no reaction of major importance has been discovered and substantiated which could substitute for ATP—creatine phosphotransferase in the synthesis of creatine phosphate".

Creatine kinase is present in animal tissues, particularly muscle; it has been crystallized from rabbit tissue. Magnesium, calcium or manganese ions are required for its full activity and they also influence the position of the equilibrium. The equilibrium is also to a large extent pH dependent and a higher pH favours the phosphorylation of creatine. Sulphydryl reagents (e.g. p-mercuribenzoate) and chelating agents such as EDTA or citrate inhibit the enzyme.

In animal as well as in human tissues maximum activity is confined to skeletal muscle, brain, heart and diaphragm, whereas the activity is lower in the lung, kidney and liver. There is no activity in normal human serum, but in myocardial infarction and myopathies serum enzyme activity is raised.

Activity can be measured either by the formation of creatine phosphate or creatine. Activity can also be determined spectrophotometrically (Tanzer and Gilvarg, 1959) by coupling the creatine kinase reaction to an auxiliary reaction catalysed by pyruvate kinase. The product of the latter reaction is then assayed by means of lactate dehydrogenase as the indicator enzyme (see "Optical Tests", Chapter XI).

Kirk (1962c) investigated activity of creatine kinase in human vessels using creatine phosphate as substrate and measuring the amount of the free creatine colorimetrically. He observed that the activity of the enzyme is unequivocally higher in human arteries of the muscular type (brachial and femoral artery) than in the aorta. Enzyme activity decreases with increasing age, while in atherosclerotic segments of the aorta the activity is about 50 per cent of that in intact parts. The decrease is significant when calculated both on a wet weight or tissue nitrogen basis. According to the author these changes may be related to atrophy of smooth muscle tissue in the diseased artery.

Adenosinetriphosphatase, ATPase, E.C. 3.6.1.3. and E.C. 3.6.1.8.
(ATP phosphohydrolase and ATP pyrophosphohydrolase)

The chemical energy produced by metabolic processes is stored mainly in the pyrophosphate bonds of nucleoside triphosphates, especially adenosine 5'-triphosphate (ATP). The other related energy-rich nucleoside triphosphates are CTP, ITP, UTP and GTP, which respectively contain the nucleosides cytidine, inosine, uridine and guanosine. The energy of exergonic reactions trapped in this way is used to drive endergonic biological processes. The high energy content of these compounds is due to their free energy of hydrolysis* and amounts in the case of ATP and creatine phosphate to -8 kcal/mole at pH 7·0. For comparison the energy released by hydrolysis of an ordinary phosphate ester, such as glucose 6-phosphate, is only -3 kcal/mole.

It is believed that the energy-rich bonds are the result of several distinct phenomena. One factor which seems to be clearly established is the difference in stability between the energy-rich compound and its hydrolytic products. If the latter are much more stable than the energy-rich compound, then the change of free energy will attain a large negative value. (In the case of pyrophosphates their instability may be caused by electrostatic repulsion from the negatively charged acid groups).

From what has been said it is clear that hydrolysis of ATP is of extreme biological importance.

The enzyme ATP phosphohydrolase (E.C. 3.6.1.3.) splits off the terminal phosphate group from ATP to give ADP (adenosine 5'-diphosphate) and inorganic phosphate:

$$ATP + H_2O = ADP + \text{orthophosphate}$$
(see also reaction I on p. 45)

This important ATPase is found in animal tissues, mainly in mitochondria but also in nuclei, microsomes and cell membranes. In muscle one form of ATPase is associated with myosin (Engelhardt and Ljubimova, 1939) and actomyosin. This ATPase also acts on ITP and other nucleoside 5'-triphosphates; it has an optimum pH of 9·0 to 10, is activated by calcium ions and inhibited by magnesium ions. (Myosin ATPase has an additional optimum pH at pH 6·4 and it is now generally agreed that the optimum at alkaline pH is due to selective inactivation—see Mommaerts and Green, 1954). The enzyme is very unstable but ATP seems to protect it.

In animal tissues there is another ATPase activated by magnesium ions and inactivated by calcium ions. This form of ATPase in muscle is associated with sarcosomes (i.e. not bound to myosin) and appears in aqueous extracts of this tissue. Formerly it was separately coded as ATPase E.C. 3.6.1.4. and it was believed possibly to act also on ADP. There is still another form of ATP phosphohydrolase, the so-called transport ATPase located in cell membranes (see reviews by Hoffman, 1962, and Skou, 1964). The enzyme is stimulated by sodium and potassium ions and is specifically inhibited by ouabain. Reports that this ATPase and active cation transport systems have common characteristics led to the hypothesis (Skou, 1964) that they are identical or that transport ATPase is the main component of a larger so far unresolved transport system of the cell membrane.

In addition to the above forms of ATP phosphohydrolase, animal tissues and snake venom contain a further ATPase (E.C. 3.6.1.8.). The systemic name of this enzyme, "ATP pyrophosphohydrolase", indicates that it catalyses the

* Of all the energy contained in a system, that part which can be converted into other forms of energy is termed "free energy". The maximum useful work that can be obtained from a chemical reaction is termed its change in free energy (ΔF). Negative ΔF denotes that energy is released by the system to the surroundings.

hydrolysis of ATP to AMP (adenosine 5'-monophosphate) and pyrophosphate:

$$ATP + H_2O = AMP + pyrophosphate$$

The pH optimum of the enzyme is between 8·4 and 8·8; it also acts on other nucleoside 5'-triphosphates (UTP, ITP, CTP and GTP). It has to be distinguished from ATP diphosphohydrolase or apyrase (E.C. 3.6.1.5.), a calcium-stimulated enzyme that is present in plants. This plant enzyme hydrolyses ATP to ADP and orthophosphate or AMP and two equivalents of orthophosphate.

The most convenient assay of ATPase activity is by the determination of free inorganic phosphate after incubation with adenosine 5'-triphosphate. Details of the method will be given in Chapter XI.

Baló et al. (1948-1949) investigated the ATP-splitting ability of human and bovine aortas. In agreement with the failure to detect any myosin or actomyosin in the aortas by viscosity studies, the authors were not able to find appreciable myosin-bound ATPase activity in the insoluble protein fractions of the vessel wall. On the other hand, high ATPase activity was present in the soluble protein fraction. As the enzyme simultaneously split two inorganic phosphate groups off ATP, it was regarded by the authors as adenylpyrophosphatase. Its pH optimum was shown to be pH 9·0; it was not activated by calcium and magnesium ions or by extracts of muscle and aorta. These characteristics indicate that the authors were probably dealing with that form of ATPase which is at present coded as the above-mentioned ATP pyrophosphohydrolase, E.C. 3.6.1.8.

The activity of the enzyme was higher in bovine than human aortas. However, only 2 hours elapsed between death and assessment of enzyme activity in bovine aortas, compared with 24 hours in human vessels.

In contrast to these findings, further work from the same laboratory (Banga and Nowotny, 1951a) revealed 8–10 per cent of myosin in the human aorta and 20–22 per cent in the femoral artery. Measurement of ATPase activity at pH 7·0 and pH 9·0 indicated that both the aorta and the femoral artery contain two ATP-splitting enzymes: the insoluble myosin- and actomyosin-bound ATPase (assayed at pH 7 with magnesium as activator) and ATP pyrophosphatase (assayed at pH 9). The total activity of the femoral artery was found to be much higher than that of the aorta; the actomyosin-bound ATPase content of the femoral artery was also considerably higher. Calculations based on the data presented by the authors indicate that the actomyosin-bound ATPase activity in the femoral artery and aorta is about 41 per cent and 20 per cent, respectively. In contrast to their previous paper, "adenylpyrophosphatase activity" proved to be clearly magnesium-dependent; it was lower in human than rabbit aortas.

Atherosclerotic human aortas and femoral arteries displayed a marked decrease in total enzyme activity both at pH 7 and at pH 9, calculated on the basis of protein content of the aortic suspensions (Banga and Nowotny, 1951a). In the aorta the decrease, as compared with control vessels, averaged 45 per cent at pH 7 and 26 per cent at pH 9.

Carr et al. (1952) measured ATPase activity in aortas of several species at pH 7·4 and in the presence of calcium ions as activators. The activity in the canine aorta was 2–3 fold higher than the activity of the carotid, coronary, renal or femoral arteries. These observations do not accord with the results of Banga's group, but this discord could be explained by the different species examined.

Carr *et al.* found the highest activity in the rat's aorta, much lower levels in chicken, rabbit, guinea-pig and frog aortas. In the turtle's aorta no activity could be observed. The ATPase activity of rat aortas was not affected by age or captivity. It is interesting that the ATPase activity of the rabbit's aorta is inhibited by rapidly-acting vasodilators, such as amyl nitrite (Krantz *et al.*, 1951).

Kirk (1959*b*) also investigated the activity of ATPase ("adenylpyrophosphatase") in human arteries. In contrast to the findings of Baló *et al.* (1948–1949), enzyme activity was stimulated by magnesium ions and was highest at pH 8·1 instead of 9·0. The values observed in these studies were about 100 per cent higher than those reported by Banga's group. These discrepancies seem to reflect probable differences in the time that elapsed between death and enzyme assessment as performed by the two groups of authors.

In Kirk's laboratory enzyme activity did not change with age, whether referred to wet weight or tissue nitrogen content. However, activity decreased by about 20 per cent in atherosclerotic aortas and coronary arteries.

It is of interest that inorganic pyrophosphatase activity (E.C. 3.6.1.1.), determined with sodium pyrophosphate as substrate, revealed similar changes to those of ATPase activity. Nevertheless, Kirk (1963*a*) stressed the fact that the pH optimum and optimum magnesium concentration were rather dissimilar in these systems, indicating that two different enzyme activities are present with ATP or sodium pyrophosphate as substrates.

Sandler (1960) and Sandler and Bourne (1960*a*, *b*; 1963) observed a histochemical "gradient" in ATPase activity in the aorta of the cat, rat and dog; activity was highest in the arch and thoracic aorta and decreased to a very low level in abdominal segments. In the human aorta no such gradient was found, but areas in the intima and media showed a considerable decrease of enzymatic activity. These workers also observed a similar gradient and distribution of 5'-nucleotidase activity.

Sandler and Bourne (1960*a*) were not able to find any histochemical differences between aortas from young and old individuals. They found, however, decreased ATPase and 5'-nucleotidase activity in atherosclerotic regions and in the above-mentioned histologically normal areas of aortas. Similar findings were also reported in aortas of cholesterol-fed dogs (Sandler and Bourne, 1963). Higginbotham and Higginbotham (1967) observed a "longitudinal gradient" of ATPase activity in the aorta of purebread beagles and particularly low activity in areas susceptible to arteriosclerotic changes.

In other experiments Sandler and Bourne (1962*a*, *b*) investigated aortas of female breeder rats fed on a semisynthetic "essential fatty-acid-free" diet containing 5 per cent of cholesterol. Such a diet had previously been used by Thomas and Hartroft (1959*a*, *b*) to produce experimental atherosclerosis in rats (see Chapter XIV). After feeding this diet for 4 weeks, areas of decreased enzymatic activity were observed in the animals' aortas, but typical atheromatous changes were only seen after 12 weeks on the diet. The authors speculated that the decreased ATPase activity in these areas in rat and human aortas ("pre-atherosclerotic changes") may result in an excess of ATP that leads to increased production of long-chain fatty acids.

It must be remembered however, that the aortas of breeder rats may suffer from spontaneous arteriosclerotic changes (see p. 147) which could interfere with the interpretation of the results. In addition, the authors (Sandler and Bourne, 1963) expressed reservations about the specificity of the histochemical technique used.

Results obtained in our laboratory will be described in subsequent parts of this book.

Adenylate Kinase, E.C. 2.7.4.3
(ATP: AMP phosphotransferase)

This enzyme (formerly usually known as "myokinase") catalyses the transfer of the terminal phosphate group from one mole of ADP to another mole of the same substance, so that ATP and AMP are produced.

$$ADP + ADP = ATP + AMP$$

It is assumed that this way of producing ATP is probably used only as a "last resort" when the tissue is *in extremis*.

The enzyme is widely distributed in animal tissues, plants, yeast and bacteria. It has been described, for example, in rat kidney, liver, brain, muscle, nerves, bone marrow and in erythrocytes. In some tissues (e.g. the liver) it is located mainly in the mitochondria. However, in heart muscle it cannot be recovered from these subcellular particles and it appears on fractional centrifugation in the supernatant fraction. Crude preparations withstand prolonged heating and even boiling in 0·1 N HCl, but the purified enzyme is extremely heat-labile.

The enzyme requires magnesium for its action and cysteine (or other SH compounds) for maximum activity. Its optimum pH is about 7·5. Fluoride and citrate as well as contact with glass or cellophane inactivate the enzyme. It can be isolated and crystallized from rabbit skeletal muscle.

Carr *et al.* (1954) demonstrated adenylate kinase activity in bovine coronary arteries; they based their observations on the ratio of AMP to ADP following incubation with ADP. They also demonstrated inhibition of activity by posterior pituitary extract (Carr *et al.*, 1955).

The demonstration of this enzyme activity in aortic homogenates is of considerable interest. It is quite possible that in crude extracts, adenylate kinase activity interferes with ATP phosphohydrolase activity, thus simulating the presence of ATP pyrophosphohydrolase activity in aortic tissue.

Chapter VII

THE GLYOXALASE SYSTEM

It has been known since 1913 that lactic acid can be formed from methylglyoxal in yeast, muscle, other animal tissues and in plants. It was suggested that the reaction is catalysed by the enzyme glyoxalase, glutathione acting as a specific coenzyme. The actual mechanism for converting methylglyoxal to lactic acid was elucidated by the work of Racker (1952) and Crook and Law (1952). The enzymatic system responsible for the reaction consists of two separate enzymes. The first enzyme is lactoyl-glutathione lyase (E.C. 4.4.1.5), formerly known as glyoxalase I. It catalyses the reaction:

$$\text{methylglyoxal} + \text{reduced glutathione} = \text{S-lactoyl-glutathione}$$

This is a typical lyase reaction and, hence, the systemic name of the enzyme is S-lactoyl-glutathione methylglyoxal-lyase (isomerizing).

The second enzyme is hydroxyacylglutathione hydrolase (E.C. 3.1.2.6), formerly known as glyoxalase II. It catalyses the hydrolytic cleavage of the thiolester S-lactoyl-glutathione, whereby glutathione is regenerated.

$$\text{S-lactoyl-glutathione} + H_2O = \text{lactate} + \text{glutathione}$$

(The systemic name of the enzyme is S-2-hydroxyacylglutathione hydrolase, indicating that other 2-hydroxyacid anions can also take part in the reaction).

"Glyoxalase I" also catalyses the production of thiolesters from other ketoaldehydes and glutathione; such esters are hydrolysed by "glyoxalase II". However, the second enzyme does not catalyse the hydrolysis of the thiolester formed from phenylglyoxal. This compound seems to be hydrolysed spontaneously and, therefore, "phenylglyoxalase" is a single enzyme (see Dixon and Webb, 1964).

Although extensively studied, the physiological significance of this pathway is so far not known. It is possible that the pathway is important as a detoxicating system.

The measurement of the activity of both enzymes is based on spectrophotometric assay at 240 mμ. of the amount of thiolester produced or degraded.

Kirk (1960a) observed that the activity of glyoxalase I decreases with age in supernatants of homogenates from human arteries. Activity was found to be lower in atherosclerotic segments of aortas and coronary arteries than in intact parts. (Activity was calculated on a wet weight basis.) It is of interest that the activity of the enzyme in arterial tissue is rather high, particularly in the pulmonary artery; the significance of these findings is not clear.

Chapter VIII

OTHER ENZYMES: LISTED ACCORDING TO THE NATURE OF THE CATALYSED REACTION

In the preceding sections enzymes of well established metabolic pathways were considered. The enzymes which follow are either difficult to fit into generally accepted pathways, or have so far been only studied in the vessel wall independently from other related enzymatic reactions. Therefore, they will be presented according to the type of reaction that they catalyse.

OXIDOREDUCTASES

Many members of this large group of enzymes have been mentioned on the preceding pages; they belong to definite metabolic pathways. Other oxidoreductases that have been studied in the vascular wall act on the CH-OH groups of donors (enzyme E.C. 1.1.1.35), on the CH-NH_2 group of donors (enzymes E.C. 1.4.1.2.–4. and E.C. 1.4.3.4.) or on reduced NAD or NADP (enzyme E.C. 1.6.4.2.).

3-Hydroxyacyl-CoA Dehydrogenase, E.C. 1.1.1.35.
(L-3-Hydroxyacyl-CoA: NAD oxidoreductase)

This enzyme catalyses the reversible oxidation of L-3-hydroxyacyl-CoA by NAD to the corresponding 3-oxo-acyl-CoA:

$$\text{L-3-hydroxyacyl-CoA} + \text{NAD} \rightleftarrows \text{3-oxo-acyl-CoA} + \text{NADH}_2$$

As can be seen from Fig. 6, this reaction is an important step in the breakdown of fatty acids. The enzyme has been identified, for example, in bovine and sheep liver and in hog heart; it is in fact widely distributed in animal tissues. It is relatively specific for NAD, but the cardiac enzyme also slowly acts with NADP.

The enzyme assay is based on the disappearance rate of reduced NAD, measured spectrophotometrically at 340 mμ. or 366 mμ. The best substrate is acetoacetyl CoA but, according to Lynen and Wieland (1955), S-acetoacetyl-N-acetylcysteamine—which is similar in structure and can readily be synthesized—may also be used for the enzyme assay.

Sanwald and Kirk (1965b) determined the activity of the enzyme in supernatants of homogenates from various types of human arteries. There was no difference in activity between samples of intact thoracic aortas and pulmonary arteries, when expressed on a wet weight basis (about 0·215 mM. substrate metabolized per g. tissue per minute). However, on a tissue nitrogen basis the activity was higher in the pulmonary artery. Activity in the coronary arteries

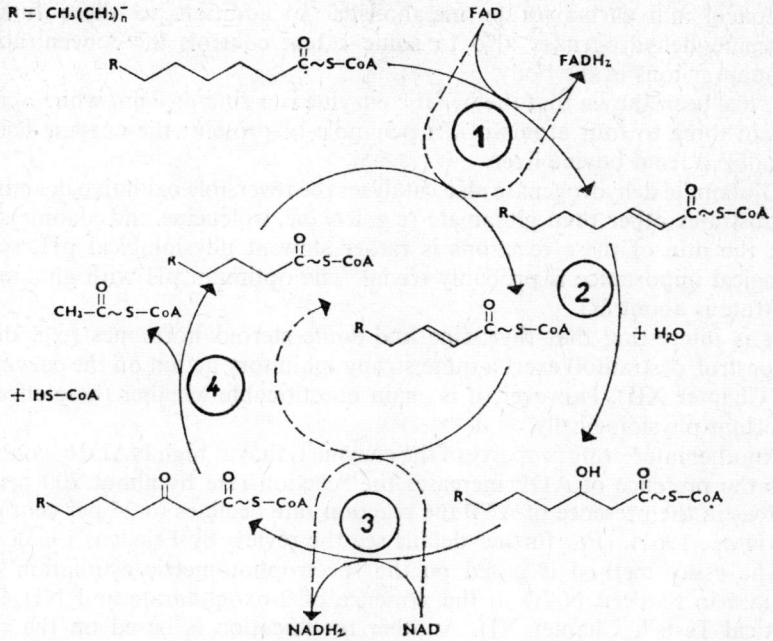

Fig. 6.—The "fatty acid oxidation spiral". Enzymes: (*1*) Acyl-CoA dehydrogenase (E.C. 1.3.2.2.). (*2*) Enoyl-CoA hydratase (E.C. 4.2.1.17). (*3*) 3-hydroxyacyl-CoA dehydrogenase (E.C. 1.1.1.35). (*4*) Acetyl-CoA acyltransferase (E.C. 2.3.1.16).

was even higher, but the cerebral arteries showed less activity than the intact aorta or pulmonary artery. Atherosclerotic segments of the thoracic aorta displayed somewhat lower activity than intact parts of the same vessels (87·7 per cent on a tissue nitrogen basis).

GLUTAMATE DEHYDROGENASE, E.C. 1.4.1.2.–4.
(L-glutamate: NAD(P) oxidoreductase [deaminating])

This enzyme catalyses the reversible oxidative deamination of L-glutamate to 2-oxoglutarate and ammonium ions:

L-glutamate + H_2O + NAD(P) = 2-oxoglutarate + NH_3 + $NAD(P)H_2$

The enzyme is widely distributed and occurs in animal tissues, plants and bacteria. The form of the enzyme that is found in the liver (E.C. 1.4.1.3.) is active with either NAD or NADP, whereas the plant enzyme (E.C. 1.4.1.2.) requires NAD and the yeast enzyme (E.C. 1.4.1.4.) depends on NADP.

The enzyme plays an extremely important role in the nitrogen metabolism of all living systems. Many aminotransferase reactions (see p. 56) involve glutamate and, if they are coupled with glutamate dehydrogenase, a variety of aminoacids can be formed from the 2-oxoacids (ketoacids) and ammonia. In addition, the reversible conversion of L-glutamate to 2-oxoglutarate—an intermediary product of the citric acid cycle—serves also as a link between aminoacid and carbohydrate metabolism. In addition to these functions, glutamate dehydrogenase also to some extent controls the concentration of ammonium ions in the body.

It has been shown that the hepatic enzyme is a zinc-protein, which seems to contain three to four atoms of Zn per mole of protein: the enzyme has been crystallized from bovine liver.

Glutamate dehydrogenase also catalyses the reversible oxidative deamination of substrates other than glutamate (e.g. leucine, isoleucine and alanine). However, the rate of these reactions is rather slow at physiological pH, so their biological importance is probably trivial. The optimum pH with glutamate as substrate is about 8·3.

It is interesting that thyroxine and some steroid hormones (e.g. diethylstilboestrol, oestradiol) exert a quite strong inhibitory action on the enzyme (see also Chapter XII). However, it is again questionable whether these effects are important physiologically.

Another interesting property of the enzyme is that at high $NADH_2$ concentrations the presence of ADP increases the reaction rate by about 200 per cent, whereas in the presence of ATP the reaction rate declines to 25 per cent (Tomkins *et al.*, 1961). (For further details see the review by Frieden, 1963).

The assay method is based on the spectrophotometric estimation of the decrease in reduced NAD in the presence of 2-oxoglutarate and NH_4Cl (see "Optical Tests", Chapter XI). Another modification is based on the reverse reaction, i.e. the estimation of the increase in NAD in the presence of glutamate (Olson and Anfinsen, 1952).

Kirk (1965a) investigated the activity of glutamate dehydrogenase in supernatants of homogenates from various human arteries. Rather contradictory results were obtained. Activity was lower in the aorta than in the pulmonary and coronary arteries. Lipid-rich atherosclerotic lesions in the aorta showed increased activity, whereas no such change was detected in fibrous lesions. In contrast to the aorta, lipid-rich lesions of the coronary arteries exhibited lower activity than normal coronary arterial tissue. In normal aortic and pulmonary arterial tissue activity decreased with age.

Monoamine Oxidase, E.C. 1.4.3.4.
(Monoamine: oxygen oxidoreductase [deaminating])

The enzyme catalyses the reaction:

$$Monoamine + H_2O + O_2 = Aldehyde + NH_3 + H_2O_2$$

It is a widely distributed enzyme of animal and plant tissues and acts on primary, secondary and tertiary amines. There are considerable interspecies differences in the distribution of the enzyme. For example, the enzyme is active in the heart of man but not in the pig and sheep. In most mammals the highest activity is, however, found in the liver and kidney. The enzyme has been recovered from human submaxillary and parotid glands; it is also present in the brain where its activity is about a third of that in the liver. The enzyme is confined mainly to the mitochondria, but the microsomes also have some monoamine oxidase activity.

Several important functions have been ascribed to this enzyme, in particular the breakdown of some pharmacologically-active amines. The most important groups of naturally occurring substrates are catecholamines (epinephrine, norepinephrine, metanephrine, dopamine) and especially tryptamine derivatives, such as serotonin. In some species the enzyme also acts on histamine, but it does not appear to act on the toxic amines putrescine and cadaverine. (For a detailed review see Blaschko, 1963).

The most interesting and much discussed function of monoamine oxidase is the deamination of catecholamines. Epinephrine and norepinephrine are metabolized by two pathways, and in both of them amine oxidase is involved. The main pathway leads to the formation of methylated amines (metanephrine and normetanephrine, respectively), which are then further metabolized by amine oxidase (Axelrod, 1959).

The discovery of potent *in vivo* inhibitors of the enzyme (e.g. hydrazine derivatives such as iproniazid) provided useful tools for the investigation of the role of monoamine oxidase in the inactivation and metabolism of catecholamines. Monoamine oxidase inactivates norepinephrine within cells and thereby controls the concentration of stored norepinephrine. However, there is strong evidence that the enzyme does not play a major role in the response to sympathetic nerve stimulation (Burn *et al.*, 1954). The apparent value of certain modern monoamine oxidase inhibitors in some cardiovascular diseases (e.g. angina pectoris, hypertension) seems to be brought about by other mechanisms, perhaps through their effect on serotonin metabolism.

A satisfactory method of determining enzyme activity is by measuring oxygen consumption in the presence of substrates such as tyramine or tryptamine. Methods based on the reduction of tetrazolium salts are also useful, particularly in their histochemical modifications.

Thompson and Tickner (1951) demonstrated monoamine oxidase activity by an oxygen consumption method in the aorta, pulmonary artery and muscular arteries of the rabbit, rat and guinea-pig. They also detected the enzyme in small muscular branches of the human popliteal artery. The larger arteries generally revealed higher activity than the muscular arteries or the great veins. Adrenaline or noradrenaline were oxidized at about one third the rate of tyramine, the oxidation of adrenaline being slightly higher than that of noradrenaline.

Spinks (1952) observed that monoamine oxidase activity against tyramine in aortic homogenates of thyroid-fed rabbits was 288 μl. O_2/g. wet wt./hr. in comparison with 370 μl. in aortas of control rabbits. Spinks and Burn (1952) claimed that the greater response to catecholamines in hyperthyroidism is caused by diminished amine oxidase activity.

GLUTATHIONE REDUCTASE, E.C. 1.6.4.2.
(Reduced-NAD(P): oxidized-glutathione oxidoreductase)

It is known (see p. 77) that glutathione undergoes enzymatic oxidation and reduction and can act as a biological hydrogen carrier. Glutathione reductase catalyses the following irreversible reaction:

$NAD(P)H_2$ + oxidized glutathione = $NAD(P)$ + 2 glutathione

The reaction rate with reduced NADP is about hundred-fold higher than with reduced NAD. The enzyme is highly specific for glutathione and this property has been used for assaying it.

Glutathione reductase is widely distributed in animal tissues, plants, yeast and bacteria; it is mainly recovered from the soluble supernatant fraction after centrifugation. The enzyme from bacteria has flavin adenine dinucleotide as its prosthetic group. The enzyme from the liver exhibits a distinct absorption band corresponding to porphyrin (Racker, 1955).

Increased enzyme activity is found in the sera of patients with lymphoma or leukaemia and in the plasma of rodents with transplanted tumours.

The activity of the enzyme can be assayed spectrophotometrically by an optical test, where the decrease in optical density of reduced NADP at pH 7·4 is measured in the presence of oxidized glutathione (see p. 96).

Kirk (1965a) used a modification of the above principle to study the activity of the enzyme in supernatants of homogenates from human arteries. Activity, expressed in micromoles of glutathione reduced per g. of tissue nitrogen per hour, was lower in the aorta (143·5 units) than in the coronary or pulmonary arteries. Atherosclerotic aortic segments (with either fatty or fibrous lesions) exhibited lower mean activity than intact parts of the same vessel.

TRANSAMINASES OR AMINOTRANSFERASES

In 1937 Braunstein and Kritsman discovered the important reaction of amino-group transfer from aminoacid to an α-ketoacid (2-oxoglutarate). The reaction was catalysed by enzymes for which the name aminopherases was suggested.

It was shown that the majority of known transaminases react with 2-oxoglutarate and its amino-derivative L-glutamate. Thirteen of the twenty transaminases coded in the 1964 report of the Commission on Enzymes of the IUB have this function. Therefore, 2-oxoglutarate can be regarded as a common carrier of amino-groups (see Dixon and Webb, 1964). There are also transaminases specific for glutamine (see enzyme E.C. 2.6.1.16.below). The important role of glutamate in the reaction is its link with basic metabolic routes through the TCA cycle. Glutamate can therefore be regarded as a distributor for amino-groups in the formation of aminoacids. This "ammonia-carrier function" (see Dixon and Webb, 1964) of glutamate by some transaminases also plays an important role in the pathway by which urea is formed in the liver.

Most transaminases require pyridoxal phosphate for their action, this compound forms their prosthetic group ("pyridoxal-phosphate proteins"; see also glycogen phosphorylase). It seems that in the course of the transamination reaction the aldehyde group of pyridoxal phosphate serves as acceptor for the amino-group of the aminoacid, so that a Schiff base linkage is formed. The latter is then hydrolysed and the amino-group is transferred from pyridoxamine phosphate to the keto acid.

From this evidence it is clear that transaminases are important enzymes of aminoacid and protein metabolism. They are widely distributed in animal and plant tissue and in bacteria.

OTHER ENZYMES

In the vessel wall only three enzymes of this group have so far been investigated. The first two are:

(a) **Aspartate aminotransferase,** formerly known as glutamic-oxaloacetic transaminase (GOT), E.C. 2.6.1.1.

(b) **Alanine aminotransferase,** formerly known as glutamic-pyruvic transaminase (GPT), E.C. 2.6.1.2.

(L-Aspartate: 2-oxoglutarate aminotransferase and L-alanine: 2-oxoglutarate aminotransferase).

In human tissues the richest source of the first enzyme is the heart and brain and of the second enzyme the liver and kidney. However, they can be detected in practically all human and animal tissues. As is well known the serum levels of both enzymes have been widely used in clinical diagnosis. SGOT activity is particularly used for the early differential diagnosis of myocardial infarction and SGPT activity is of value in the diagnosis of liver diseases. The activity of these transaminases in the serum is normally low, but when the tissues are damaged their transaminases leak into the blood. There are several methods for measuring such activity and two of them will be described in connection with our own results (see Chapter XI).

Mandel and Kempf (1960) determined aspartate aminotransferase in saline extracts from homogenates of bovine aortas. In contrast to lactate dehydrogenase (see p. 24) and malate dehydrogenase (see p. 33) activities, they could not find any difference between aortas of young and old animals. The authors used a method based on the reduction of oxaloacetate by malate dehydrogenase, coupled with spectrophotometric determination of the decline in $NADH_2$ concentration during incubation (see Chapter XI). The results were calculated on a wet weight basis.

Chattopadhyay (1961) investigated the activity and characteristics of transaminases in the rabbit's aorta by a chromatographic method. He found comparatively high activity of aspartate transaminase in this vessel. The interaction of glutamate and pyruvate (i.e. alanine aminotransferase activity) could not be detected by this technique.

Aleksejeva and Nekrasova (1963) studied transaminase activity in extracts of aortas from cholesterol-fed rabbits. In the early stages, after five days of cholesterol feeding, the activity of alanine aminotransferase (and perhaps also that of aspartate aminotransferase) significantly rose. However, in the later stages after about 100 days on diet, activity decreased to subnormal values. (Calculations were based on the protein content of the extracts). These findings were essentially in agreement with Ryu's (1959) previous results.

Sanwald and Kirk (1965c) used the "optical tests", as described on page 96, to measure the activity of both aminotransferases in supernatants of homogenates from human aortas, coronary and pulmonary arteries. Grossly intact segments of the coronary artery exhibited about 90 per cent higher activity than that in normal aortic tissue. In the latter the average activity was 0·565 and 0·237 units/g. tissue N_2 for GOT and GPT respectively. Atherosclerotic segments revealed distinctly higher activity than intact segments of the same vessel.

GLUTAMINE-FRUCTOSE-6-PHOSPHATE AMINOTRANSFERASE or
HEXOSEPHOSPHATE AMINOTRANSFERASE, E.C. 2.6.1.16.
(L-glutamine: D-fructose-6-phosphate aminotransferase)

This enzyme catalyses the formation of glucosamine 6-phosphate, in which reaction an amino group from glutamine is transferred to D-fructose 6-phosphate:

L-glutamine + D-fructose 6-phosphate = glucosamine
6-phosphate + L-glutamate

The reaction seems to be important for synthesizing 2-amino-2-deoxy hexoses (hexosamines) for various mucopolysaccharides. Leloir and Cardini (1953) partially purified the enzyme from extracts of the mould *Neurospora crassa*. (Glucosamine is an essential constituent of fungal cell walls). Later, the enzyme was detected in the epiphyseal cartilage of growing rabbits, in rat liver extracts and other tissues. Whereas fructose 6-phosphate is the immediate acceptor of amino-groups with the mould enzyme (Blumenthal *et al.*, 1955), it is replaced by glucose 6-phosphate in the reaction catalysed by the liver enzyme (Pogell and Gryder, 1957). However, glucose 6-phosphate possibly exerts a protective effect on the liver enzyme and thus, in this case, fructose 6-phosphate is the immediate hexose precursor (Gryder and Pogell, 1960). The activity of the enzyme can be assayed by measuring the amount of hexosamine before and after incubation of the reactants at an optimum pH of 7·4.

Haruki and Kirk (1965) studied the activity of the enzyme in human arteries using a modification of the above principle. The mean activity of the pulmonary artery (12·78 mg. hexosamine/g. tissue nitrogen/20 min.) was about the same as in the coronary arteries, but much lower than in the aorta. No difference could be found in activity between the ascending and abdominal aorta, while atherosclerotic parts exhibited the same activity as intact segments of the same vessels.

PENTOSYLTRANSFERASES

Only one enzyme of this group has so far been studied biochemically in vascular tissue.

PURINE NUCLEOSIDE PHOSPHORYLASE, E.C. 2.4.2.1.
(Purine-nucleoside: orthophosphate ribosyltransferase)

It has been known for some time that some nucleosides like guanosine, inosine or xanthosine are readily cleaved by extracts of animal tissues to yield D-ribose 1-phosphate and the relevant purine (i.e. guanine, hypoxanthine and xanthine, respectively). The enzyme has been purified from bovine liver and it catalyses the reaction:

purine nucleoside + orthophosphate = α-D-ribose 1-phosphate + purine

The equilibrium of the reactions catalysed by purine nucleoside phosphorylase is such as to favour the synthesizing reaction. The enzyme does not act on *pyrimidine* nucleosides which are metabolized by similar reactions catalysed by other pentosyl-transferases (e.g. uridine phosphorylase, E.C. 2.4.2.3.; thymidine phosphorylase, E.C. 2.4.2.4.; guanosine phosphorylase, E.C. 2.4.2.15.).

The enzyme does not convert adenine to adenosine but it seems to catalyse the synthesis of nicotinamide mononucleoside from nicotinamide and ribose 1-phosphate (Rowen and Kornberg, 1951). This may be an important step in the biosynthesis of the coenzymes NAD and NADP. The enzyme is found in animal tissues, yeast and bacteria.

Kirk (1961b) used a spectrophotometric method for determining purine nucleoside phosphorylase, where hypoxanthine liberated from inosine is oxidized by excess xanthine oxidase to uric acid. The uric acid is oxidized by urate oxidase (formerly known as uricase) to allantoin. The latter reaction is accompanied by a fall in optical density at 292·5 mμ. and this is proportional to the amount of hypoxanthine in the incubating medium.

Kirk could not find any relationship between age and enzyme activity in homogenates of human arteries. However, atherosclerotic segments, especially in the aorta, revealed significantly higher activity (on a wet weight and tissue nitrogen basis) than intact parts of the same vessel.

The significance of this interesting finding remains obscure.

HYDROLASES

The "hydrolysing" enzymes catalyse the transfer of a part of the substrate molecule to a hydroxyl group from water (Hoffmann and Ostenhof, 1954). In addition, many of the hydrolysing enzymes can also transfer the group that is split off to other molecules. For these and other reasons the dividing line between transferring and hydrolysing enzymes is not sharp (Dixon and Webb, 1964).

According to the type of bond hydrolysed, several sub-classes of hydrolases can be distinguished such as those that act on ester bonds, peptide bonds, acid anhydride etc. Those hydrolases that do not participate in metabolic cycles but, nevertheless, have been investigated in vascular tissue will be reviewed in the following pages. However, the topics of carboxylesterase, lipase (including lipoprotein lipase), phospholipase A, cholesterol esterase and elastase will be dealt with in part three of this volume.

ACETYLCHOLINESTERASE AND CHOLINESTERASE, E.C. 3.1.1.7. AND E.C. 3.1.1.8.
(Acetylcholine hydrolase and acylcholine acyl-hydrolase)

It is well known that neural tissue contains the enzyme acetylcholinesterase (formerly known as "true" cholinesterase), which catalyses the hydrolysis of acetylcholine and some other acetyl esters:

$$\text{Acetylcholine} + H_2O = \text{choline} + \text{acetate}$$

Interest in this enzyme primarily arises from the role of acetylcholine as a mediator of nerve stimuli. Many animal tissues and serum contain another enzyme (Mendel and Rudney, 1943) that catalyses the hydrolysis of a variety of choline esters and some other esters. The enzyme responsible for this action is cholinesterase, formerly known as "pseudocholinesterase". There are important specificity differences between these enzymes. With acetylcholinesterase, there is a sharp drop in activity with increasing chain length of the acyl group attached to choline. For example the rate of hydrolysis of acetylcholine is 50 to 100-fold higher than that of butyrylcholine. On the other hand, the opposite relationship holds for cholinesterase; butyrylcholine is hydrolysed by this enzyme at least twice as fast as acetylcholine. Enzymes from different sources, however, display somewhat different properties (for references see Augustinsson, 1960).

Both enzymes are inhibited by eserine (physostigmine), certain organophosphorus compounds and other natural or synthetic substances, especially those having a quaternary or basic nitrogen atom. However, the two enzymes show differential sensitivity to certain inhibitors. Thus, compounds Nu-1250, 3116 CT and BW 62C47 seem to be specific competitive inhibitors for acetylcholinesterase, whereas "Mipafox" and "Astra 1397" inhibit cholinesterase activity (for references see Augustinsson, 1960; Dixon and Webb, 1964).

The distribution of the enzymes is also different, although many tissues contain both. In addition to neural and conducting tissue, acetylcholinesterase activity is high in erythrocytes, thymus, suprarenals and striated muscle. Cholinesterase activity is particularly found in serum, pancreas, heart, intestine, skin and placenta. Both enzymes are mainly present in microsomes. Whereas the physiological function of acetylcholin esterase is well known, the significance of the other enzyme remains so far unresolved.

There are numerous methods for determining these enzyme activities. They are based on the estimation of the liberated acetic acid by either the Warburg technique or potentiometric and indicator titration. A reliable colorimetric method involves conversion of unhydrolysed acetylcholine to acetylhydroxamic acid with hydroxylamine and the subsequent formation of an orange-red complex with ferric chloride.

Thompson and Tickner (1953) manometrically determined cholinesterase activity in human and animal vessels, using as substrates acetylcholine, acetylmethylcholine, propionylcholine and tributyrin. Their findings indicate that such enzyme activity in these blood vessels is largely, but not entirely, of the "pseudocholinesterase" type, i.e. cholinesterase, E.C. 3.1.1.8. The ratio of butyrylcholine to acetylcholine hydrolysis varied from 1·4/1 to 3·3/1; it was highest in the guinea pig's aorta and in other species' aortas it decreased in the order rabbit—human—rat. It is interesting that the rabbit carotid artery split tributyrin about 4 times as fast as the aorta. With tributyrin as substrate, inhibition by 10^{-7}M eserine was only about 25 per cent in the rabbit aorta and 5 per cent in the human carotid artery. These findings suggest that carboxylesterase (E.C. 3.1.1.1.) activity was present in the investigated vessels (see Chapter XXIII).

Rosenberg and Dettbarn (1965) also presented evidence that cholinesterase activity in the rabbit aorta is of acylcholine acyl-hydrolase type (E.C. 3.1.1.8.).

The relationship between lipolytic activity and cholinesterase activity will be discussed later (see p. 225).

ALKALINE PHOSPHATASE AND ACID PHOSPHATASE, E.C. 3.1.3.1. AND E.C. 3.1.3.2.
(Ortho-phosphoric monoester phosphohydrolases)

These enzymes catalyse the hydrolytic cleavage of orthophosphoric monoesters:

$$\text{Ortho-phosphoric monoester} + H_2O = \text{alcohol} + H_3PO_4$$

They also catalyse transphosphorylations, as for example, transfer of a phosphate group from one alcohol to another, or from a phosphoamide to an alcohol.

Formerly these enzymes were termed phosphomonoesterase I and II, but it is now preferable to refer to the whole group 3.1.3. as phosphoric monoester hydrolases or simply phosphomonoesterases. They are classified according to their optimum pH as "alkaline" or "acid" phosphatases. The specificity of both types of enzyme is low; they act on a wide range of monoesters of orthophosphoric acid.

Alkaline phosphatases have an optimal pH of 8–9. They are present in practically all animal tissues. The highest activity is found in the intestinal mucosa, but bone, kidney, milk and, under certain circumstances, bile and blood serum also contain this enzyme. There is strong evidence that alkaline phosphatase participates in bone formation. The enzyme also seems to play some role in absorption of maternal blood glucose by the placenta. Of special interest is the increase of alkaline phosphatase activity in the mammary gland during pregnancy and lactation, as well as its increase in rapidly growing tissues such as neoplasms. Its role in tissue calcification and the formation of fibrous proteins is a much disputed function; these functions will be discussed in more detail in Chapter XIII.

Alkaline phosphatases require the presence of divalent cations such as Mg^{++}, Mn^{++}, Ca^{++} or Zn^{++} for enzymatic activity and are inhibited by orthophosphate, arsenate, borate and other anions. Raised serum values are due either to osteoblastic bone disorders, or to diseases of the liver or bile duct.

Acid phosphatases have an optimum pH near 5. Particularly rich sources of acid phosphatases are the prostate, kidney, spleen and erythrocytes. It is also abundant in human seminal fluid. Only very small amounts are found in skeletal and heart muscle and in the mucosa of the small intestine. The enzyme in serum is inhibited by fluoride ions and that in erythrocytes by formaldehyde. The enzyme from the prostate is activated by citrate and selectively inhibited by α-hydroxycarboxylic acids, particularly tartrate (0·02 M). This is important in the diagnosis of prostatic carcinoma which is accompanied by raised acid phosphatase activity in the serum.

The physiological significance of these types of phosphatase is not known. The prostatic phosphatase seems to be important in spermatozoal metabolism; in plants it plays a role in the formation of sucrose during photosynthesis. The hypothesis has been advanced that 70–80 per cent of acid phosphatase, together with some 17 other acid hydrolases, is concentrated in the subcellular particles known as lysosomes. In these particles these enzymes are thought to be in a relatively inactive latent state and are "activated" by membrane-rupturing procedures, such as freezing and thawing, exposure to hypotonic media, immunogenic damage etc. (De Duve, 1959; Novikoff, 1961*a* and *b*). Although the real existence of lysosomes has been challenged by some authors (Conchie and Levvy, 1963; Levvy and Conchie, 1964), biological evidence indicates that lysosomes are specifically involved in intracellular digestion, autolysis and cell injury.

De Duve, who introduced the lysosomal concept, enumerated in 1964 the functions and disorders of these subcellular particles in cell injury and cell death:

1. **Heterolysis.**—This refers to cellular "digestion" of exogenous material and includes the well-known phagocytic activity of cells, as well as extracellular secretion of lysosomal enzymes.

2. **"Digestive dysfunctions"** of cells caused by overloading of the lysosomal system by undigestible material as in silicosis and lipid storage diseases, or caused by enzymatic deficiency of the lysosomes.

3. **Cellular autophagy** which is localized autolysis or "cell suicide".

4. **Generalized autolysis,** which is usually a secondary event in necrotic cells and acts as a kind of tissue-clearing mechanism. However, the rupture of lysosomes may be a primary phenomenon in living cells and play a causal role in cell injury and cell death. The participation of similar phenomena in some types of cell injury is of considerable interest in view of the protective action of hydrocortisone which appears to stabilize the lysosomal membrane (for details see Jacobson, 1964). It must be mentioned, however, that there are considerable differences, at least in the liver (Nelson, 1966), between lysosomal and cytoplasmic forms of acid phosphatases. In contrast to the lysosomal phosphatase, the cytoplasmic enzyme is only slightly inhibited by tartrate and fluoride, has a slightly different pH optimum, is more heat-labile and is strongly bound to DEAE-cellulose. Such differences suggest different physiological roles for the two forms of acid phosphatase in the liver and, perhaps, in other tissues.

In 1939 Gomori detected alkaline phosphatase in vascular tissue by histochemical means, but it was located only in the adventitia of medium-sized arteries. On the other hand, Kirk and Praetorius (1950) used disodium phenylphosphate as substrate to show that extracts of human intima-media homogenates contain definite acid phosphatase activity with a maximum at pH 5·7–5·8. A small peak was also observed at about pH 9·5.

Schlief et al. (1954) studied acid and alkaline phosphatase activity at pH 4·9 and 8·9 in homogenates of human arteries obtained shortly after amputations. They observed an increase of acid phosphatase activity in thromboangiitis and a decrease in atherosclerosis. In atherosclerotic vessels with fibrous lesions, alkaline phosphatase activity increased.

In a later paper Kirk (1959b) studied acid phosphatase activity in human arterial homogenates at pH 5·6, with p-nitro-phenyl phosphate as substrate. Enzyme activity was lower in the child's aorta than in the adult's aorta. In atherosclerotic aortas no change of activity was observed, but activity increased in atherosclerotic portions of coronary arteries. The mean activity of the pulmonary artery was lower than that of the aorta. (Results were calculated on a wet weight and tissue nitrogen basis).

Malinow et al. (1959) studied the distribution of alkaline phosphatase at pH 9·0 using disodium phenylphosphate as substrate. They found much higher activity in cerebral vessels than in the aortas of the animal species studied. The aorta and other arteries of the rat revealed higher activity than those of the chicken; the lowest activity was observed in rabbits. (These results are very similar to our own findings concerned with interspecies differences in vascular lipolytic activity, which will be presented in Chapter XXIII).

Mandel et al. (1959) observed high phosphatase activity at pH 8·9 and low activity at pH 6·2 in bovine aortic extracts. This observation is contrary to Kirk and Praetorius's findings in the human aorta. Mandel et al. used β-glycerophosphate as substrate in these experiments. In contrast to human vessels, the activity—calculated on a wet weight basis—was significantly lower in the aortas of old animals.

Newman et al. (1950) was able to show by histochemical means that alkaline phosphatase is primarily of cytoplasmic origin and is present in the endothelial cells of capillaries and other small vessels of most organs. The latter findings agree well with the results of Paterson et al. (1957) who used Gomori's method to demonstrate alkaline phosphatase in the earliest plaques of human atherosclerotic aortas. The presence of this enzyme in such lesions was interpreted by these authors as evidence of extremely early vascularization of the intima in the course of atherosclerosis. In the normal aorta the activity is mainly located in the *vasa vasorum* (Lojda and Zemplényi, 1961; Hess and Stäubli, 1963; Adams, 1964b; Stein and Harris, 1964; Hashimoto and Kobernick, 1964 and others).

Levonen et al. (1960) detected histochemically phosphatase (pH 5·8–6·2) only in atheromatous plaques of human aortas, while no activity could be demonstrated in normal aortas. This problem has also been studied by Gonzales (1963), Stein and Harris (1964) and others. Very recently Kahn and Slocum (1967a) reported on an increased alkaline and acid phosphatase activity (together with increased ATPase activity) in aortas of chickens fed "atherogenic" diets.

In our own work we paid considerable attention to acid and alkaline phosphatase activities and the results will be presented in detail in the second part of this book.

5'-Nucleotidase, E.C. 3.1.3.5.
(5'-Ribonucleotide phosphohydrolase)

This enzyme has a wide specificity for 5'-nucleotides and catalyses their hydrolytic cleavage:

5'-ribonucleotide + H_2O = ribonucleoside + orthophosphate

5'-Nucleotidase does not act on 3'-nucleotides. It is specific for the substrate as a whole (Dixon and Webb, 1964), in contrast to acid and alkaline phosphatases which are specific for the phosphate group only. The enzyme requires manganese ions for its full activity and is inhibited by zinc and nickel ions; its pH optimum is 7·5–7·8. The enzyme is found in many animal tissues. Activity is particularly high in the neurohypophysis and testes, whereas its activity in other organs is much lower. It is also found in snake venom and bacteria. According to Reis (1951) and Ahmed and Reis (1958) 5'-nucleotidase is an important factor in the regulation of tissue phosphate and adenylate concentrations, and also in the regulation of tissue calcification. There is indeed evidence that indicates a specific role for this enzyme in the synthesis and breakdown of NAD and NADP in some animal tissues. According to Dixon and Purdom (1954) and others, the serum activity of the enzyme is increased in liver disease and biliary obstruction.

Reis (1951) observed that human aortic extracts display considerable activity at pH 7·5 when incubated with AMP. 5'-Nucleotidase activity is concentrated in the media and is here about 40 times higher than the activity of alkaline phosphatase measured at pH 9·0. The results of Reis were confirmed by Antonini and Weber (quoted by Kirk, 1959d), who further found a 21 per cent reduction in activity in atherosclerotic samples.

Kirk (1959a) observed a marked age dependent increase in the activity of 5′-nucleotidase activity at pH 7·8 in homogenates of human aortas and pulmonary arteries. (Results were expressed on both wet weight and tissue nitrogen bases). In atherosclerotic portions of coronaries no significant change was observed, but in atherosclerotic aortas activity slightly decreased when calculated on a wet weight basis. It is interesting that the activity found in the aorta is of the same magnitude as that observed in ossifying cartilage. Hamoir et al. (1961) were also able to confirm the comparatively high arterial activity of this enzyme in bovine carotid vessels. Such estimations, however, were made at pH 9·0.

By a histochemical method based fundamentally upon Gomori's technique, Newman et al. (1950) observed significant 5′-nucleotidase activity in the media of human, rabbit and guinea-pig coronary arteries. Sandler and Bourne's histochemical results were mentioned in connexion with the ATPase activity of the vascular wall, while our own findings will be reported in part two of this book.

ARYLSULPHATASE, E.C. 3.1.6.1.
(Aryl-sulphate sulphohydrolase)

This enzyme belongs to the sulphuric ester hydrolases and catalyses the following reaction:

$$\text{Phenol sulphate} + H_2O = \text{Phenol} + H_2SO_4$$

Enzymatic activity of this type has been known since Derrien's work in 1911 and has been found in animal tissues, plants, moulds and bacteria. Snails are a rich source of the enzyme, which is also present in the enzyme mixture (takadiastase) prepared from *Aspergillus oryzae*. Some authors claim that the liver of some mammals contains three distinct sulphatases (A, B, C), which differ in specificity, optimum substrate concentration, pH optima, response to chloride ions and other features. Sulphatases A and B were said to be associated with mitochondria, while sulphatase C was recovered from microsomes. However, the existence of these three different types is not accepted by all workers in the field.

The biological significance of arylsulphatases is not known. Perhaps in some tissues they form an essential part of a sulphate donor system. (See also lysosomes, p. 61.)

Arylsulphatase activity can easily be determined using phenolphthalein disulphate as substrate and measuring the amount of liberated phenolphthalein at 545 mμ. Another method uses p-nitrophenyl sulphate as substrate; enzyme activity can be assayed from the increase of absorption at 330 mμ.

Kirk and Dyrbye (1956) measured the activity of the enzyme in homogenates of human aortas, pulmonary and coronary arteries. A modification of the p-nitrophenyl sulphate method was used. There was definite, but low, arylsulphatase activity in the vessels studied. The highest activity was observed in the pulmonary arteries, the lowest in the coronary arteries. The activity in the aorta declined with age when calculated on a wet weight basis. In aortic samples from individuals aged 40–79 years there was a negative correlation between the degree of atherosclerosis and arylsulphatase activity.

β-Glucuronidase, E.C. 3.2.1.31.
(β-D-Glucuronide glucuronohydrolase)

This enzyme belongs to the large group of glycoside hydrolases and catalyses the following reaction:

$$\beta\text{-D-glucuronide} + H_2O = \text{alcohol} + \text{D-glucuronic acid}$$

In addition, the enzyme catalyses the transfer of β-glucuronosyl residues to aliphatic alcohols and glycols.

The enzyme is widely distributed in vertebrate and invertebrate tissues; the highest activity is exhibited by the spleen, liver, kidney and epididymis. It has been purified from liver, kidney, brain, bacteria and molluscs and isolated commercially from snails (*Helix pomatia*). The enzyme is confined chiefly to the mitochondria and microsomes, but a small fraction is also in a free state in the cytoplasm. Some of its properties have led to an assumption that the enzyme is lysosomal in character. β-Glucuronidases of mammalian origin have multiple pH optima in the range between 3·4 and 6·3 suggesting the existence of different proteins with identical specificity (Smith and Mills, 1953; Bartalos and Gyorky, 1963 and others).

A well-established physiological action of β-glucuronidase is that it catalyses the hydrolysis of conjugated glucuronides, such as the steroid glucuronides or the glucuronide of bilirubin. There is also good evidence that some degradation products of hyaluronic acid, i.e. oligosaccharides of varying chain length, can be further split by this enzyme (Meyer, 1958). This suggests that the enzyme has a physiological role in mucopolysaccharide catabolism. The synthetic role of β-glucuronidase in the formation of glucuronides, e.g. those of oestrogens, seems to be questionable, because saccharo-1, 4-lactone (a competitive inhibitor of the enzyme) fails to influence glucuronide synthesis by tissue preparations (Lathe and Walker, 1958). Some authors relate increased β-glucuronidase activity to cellular proliferation (Levvy *et al.*, 1948) or to neoplastic growth (Fishman, 1947, 1951).

Findings indicating a rise in β-glucuronidase activity accompanying atrophic tissue changes or following tissue injury (e.g. after irradiation) would be in line with the alleged lysosomal character of this enzyme.

Organic acids such as citric and saccharic acids inhibit β-glucuronidase. Inhibition by the latter acid appears to be due to the formation of traces of saccharo-1, 4-lactone, which acts as a competitive inhibitor as also does D-glucuronic acid. Other important inhibitors are heparin and ascorbic acid.

The activity is enhanced by deoxyribonucleic acid and some diamines. At higher dilutions preparations require the addition of DNA or albumin for full activity (Levvy and Marsh, 1960).

The assay of activity is based on the use of synthetic phenolphthalein mono-β-glucuronide as substrate and the colorimetric determination of the liberated phenolphthalein. (For details see Chapter XI).

Dyrbye and Kirk (1956) determined β-glucuronidase activity in homogenates of human aortas and pulmonary arteries, with phenolphthalein glucuronide as substrate. Arterial enzyme activities were found to be very low in comparison with those reported for parenchymatous organs. The authors found a tendency for enzyme activity to increase up to the age of 50–69 years, after which a decrease occurred. Enzyme activities in arterial specimens from women were in general lower than the activities found in the samples from men of the same age group. There was a tendency for the enzyme activity to increase with the degree of atherosclerotic change. (Activities were calculated on wet weight and tissue nitrogen bases).

Branwood and Carr (1960) investigated the activity of β-glucuronidase in the tunica intima of 160 male and 160 female aortas. They were able to confirm most results of Dyrbye and Kirk and found, moreover, a relationship between enzyme activity and the quality of atherosclerotic changes. Lesions rich in mucopolysaccharides and undergoing fibroblastic proliferation exhibited highest enzyme activity. In lesions with much lipid accumulation activity was also high in comparison with the normal intima, but the difference was not so great as in the preceding types of lesions. The activity in normal as well as pathological female aortas was higher than in male aortas. This differs from the results of the preceding authors.

In contrast to the above authors Kayahan (1960) reported decreased β-glucuronidase activity in ten human atherosclerotic aortic intimas as compared with normal vessels. These results were, however, calculated on a wet weight basis and obviously no attention was paid to the increased amount of metabolically inert material in such vessels.

Miller *et al.* (1966) investigated the activity of β-glucuronidase in supernatants of homogenates from human atherosclerotic arteries. Activities were determined over the pH range 2·2 to 7·0, at intervals of 0·2 pH units. Maximum activity occurred at a more acid pH (3·6–3·8). Normal segments of atherosclerosis-resistant vessels, such as the internal mammary artery or thoracic aorta revealed higher β-glucuronidase activity than the more susceptible abdominal aorta and coronary arteries. This seemed to suggest that arteries with high initial β-glucuronidase activity can more readily mobilize the enzyme and thus prevent atherosclerotic lesions by possibly increased degradation of mucopolysaccharides. On the other hand, atherosclerotic segments from the aorta showed definitely increased activity and a similar trend was noted in the other three vessels. This was interpreted as "an unsuccessful attempt to remove the excess mucopolysaccharide".

Our own findings will be presented in part two of this book. It should be mentioned here that our results in experimental rabbit atherosclerosis show an increase of arterial β-glucuronidase activity and accord with the above findings in human atherosclerosis. Similar results have also been obtained by Wexler and Judd (1966) in rats. These authors studied such activity in supernatants from aortic homogenates of female breeder rats progressively developing severe spontaneous arteriosclerosis. The aortas exhibited a significant rise in β-glucuronidase activity as compared with intact aortas of virgin female rats of comparable age. Enzyme activity consistently ran parallel with the anatomical progression of the disease, i.e. aortas with moderate or severe arteriosclerosis displayed the highest enzyme activity. When plaques could only be detected in the abdominal aorta and arch, β-glucuronidase activity was significantly increased only in these two segments. The unaffected thoracic aorta in such rats showed no significant increase in enzyme activity. Likewise—as in our own experiments—histoenzymatic studies agreed well with the biochemical findings.

Leucine Aminopeptidase, E.C. 3.4.1.1.
(L-Leucyl-peptide hydrolase)

The group of peptide bond-cleaving enzymes comprises exopeptidases (or peptidases) and endopeptidases (or proteinases). Exopeptidases hydrolyse peptide bonds adjacent to terminal α-amino or terminal α-carboxyl groups, whereas endopeptidases can act on centrally located peptide bonds as well as terminal peptide bonds (Bergmann, 1942). It must be stressed that, contrary to previous ideas, the size of the substrate molecule is unimportant. Whether or not a particular enzyme of this group will attack a substrate depends on the nature of the other chemical groups in the neighbourhood of the peptide link. The main function of the *endo*peptidases may be to produce a large number of "free ends" where the *exo*peptidases can act (Dixon and Webb, 1964).

Leucine aminopeptidase is a typical exopeptidase that hydrolyses many L-peptides to split off an N-terminal residue with a free amino-group:

$$\text{L-leucyl-peptide} + H_2O = \text{L-leucine} + \text{a peptide}.$$

The preferred substrates are L-leucine-amide, L-leucyl-glycine and L-leucyl-glycyl-glycine. Hydrolysis is, however, not restricted to leucyl compounds, because a wide variety of synthetic amides and peptides can be attacked by the enzyme (for review see Smith and Hill, 1960). Leucine aminopeptidase requires a free α-amino group; blockade of this group prevents enzymatic action (Dixon and Webb, 1964). The enzyme is used as a valuable analytical tool in the structural analysis of proteins and peptides. It is widely distributed in animal tissues, intestinal secretions, erythrocytes and yeast.

The enzyme has been purified from pig intestine and especially kidney. It requires magnesium or manganese ions for full activity, whereas agents which bind these ions—such as citrate, pyrophosphate and EDTA—are strong inhibitors. The optimum pH of the enzyme from pig intestinal mucosa is between 7·8 and 9·3. Straight-chain aliphatic alcohols also inhibit the enzyme and their effectiveness increases with increasing chain length (Smith and Hill, 1960). Heavy metals completely inhibit the enzyme.

Electrophoretic analysis of normal human serum reveals only one component of the enzyme migrating with α_1-globulins. However, in some liver diseases four separate fractions can be detected (Behal et al., 1962). Increased activity is found in the early stages of biliary obstruction, in hepatitis, in pancreatitis, in pregnancy and in carcinoma. The activity of leucine aminopeptidase can be accurately assayed using L-leucinamide as substrate and measuring the hydrolytic products, e.g. titration of the free hydroxyl groups. Other widely used methods are based on the hydrolysis of L-leucyl-β-naphthylamine. The liberated β-naphthylamine is coupled with a diazonium salt (for example 3-chloro-4-nitroalanine) to give a dye that can be readily extracted and measured colorimetrically. This principle is also the basis of the histochemical method for the enzyme (see Glenner et al., 1965). However, some caution is necessary with the use of L-leucyl-β-naphthylamine as substrate, because other peptidases also catalyse the hydrolytic cleavage of this compound (Smith and Hill, 1960).

Green *et al.* (1955) studied the activity of leucine aminopeptidase in human organs, using L-leucyl-β-naphthylamine as substrate. The activity in homogenates of intestine, kidney and spleen was about 2½-fold higher than that in aortic homogenates.

Kirk (1960b) employed the same assay method in homogenates of human vessels. The mean activities found in samples of normal aorta, pulmonary artery and coronary artery were 1·34, 1·32 and 1·57 mg. β-naphthylamine/g. wet wt./hr. This activity was about 10 times higher than that reported by Green *et al.* (1955) for normal human serum. This enzymatic activity showed a moderate trend to decrease with age when calculated on a wet weight basis, but no difference was found between activities of atherosclerotic and normal segments of the same vessel.

Levonen *et al.* (1960) investigated the histochemical activity of this enzyme in atheromatous human aortas; moderate activity was detected in early plaques, mainly in the thickened intima. In severely affected aortas many foci of activity were present, even in the media. In our own studies on experimental rabbit atherosclerosis we observed a somewhat higher activity in early plaques (Lojda and Zemplényi, 1961).

CATHEPSINS
(No systemic names can yet be used in the sub-group E.C. 3.4.4. of peptidyl peptide hydrolases that comprise cathepsins).

It has been known for a long time that animal tissues contain intracellular protein-splitting enzymes (see Vernon, 1913, 1914). These proteinases are similar to, but not identical with, the well-known enzymes found in secretions of the mammalian gastro-intestinal tract (pepsin, trypsin, chymotrypsin). The term cathepsin covers a number of similar enzymes, but much uncertainty still exists concerning their specific properties.

As the properties of proteinases such as pepsin, trypsin and chymotrypsin have been defined by their action on synthetic substrates of known structure, similar compounds have been used to try to differentiate the cathepsins. On the basis of such studies three different types of cathepsins have been distinguished (see review by Fruton, 1960): cathepsin A which hydrolyses carbobenzoxy-L-glutamyl-L-tyrosine at about pH 5 and has similar substrate specificity to pepsin; cathepsin B which acts on benzoyl-L-argininamide at about pH 5 and is similar in specificity to trypsin; and cathepsin C which attacks glycyl-L-tyrosinamide at pH 4 to 8 and exhibits specificity similar to chymotrypsin.

However, there is uncertainty about the exact identity of these cathepsins. Hence the most recent recommendations on nomenclature and classification of enzymes from the IUB (see Florkin and Stotz, 1965) refer only to cathepsin C (E.C. 3.4.4.9.). The crystalline and highly purified enzyme hydrolyses peptides, especially at bonds where an aromatic aminoacid is adjacent to a free α-amino-group. The need for this free α-amino-group, which is provided by another aminoacid attached to the amino-group of the aromatic aminoacid, is a distinctive feature of cathepsin C substrates in contrast to the requirements of pepsin and chymotrypsin (Dixon and Webb, 1964). These views of the latter authors are based on an analysis of the available literature on the action of the enzyme on a large number of synthetic peptides and related compounds.

In addition to the above-mentioned cathepsins, Press *et al.* (1960) identified a further enzyme in ox spleen. This enzyme—known as cathepsin D (E.C. 3.4.4.23.)—has a specificity somewhat similar to, but more restricted than that of pepsin. The enzyme does not act on substrates for cathepsin A, B or C, but it hydrolyses the B chain of oxidized insulin.

The confusion about the identity of cathepsins is further confounded by the fact that the activity of crude preparations is most frequently determined by their action on denatured haemoglobin. Although activity measured in this way is similar to that ascribed to cathepsin A, such activities are not identical. For example the "Hb-splitting enzyme" usually has a pH optimum at 3·5 and also differs in other respects from cathepsin A (see Fruton, 1960). Denaturated haemoglobin is also used, however, in the assay of catheptic activity in ox spleen and, in this case, the activity is considered to reflect that of cathepsin D.

Despite this confusing situation about classification, it is important to realize that the class of cathepsins with their hydrolysing action on proteins is an integral part of the general and ubiquitous enzyme equipment of all animal tissues. It is unlikely that their function is restricted to autolytic and other processes ascribed to lysosomal enzymes (see p. 61). It has been suggested that, in addition to a general catabolic role in the living cell, they may also participate in protein biosynthesis (see Fruton, 1960).

As mentioned above, in addition to assay with specific substrates, cathepsin activity is usually measured at pH 3·5 with denatured haemoglobin as substrate (Anson, 1938). The tyrosine- and tryptophan-containing reaction-products are determined colorimetrically by means of the Folin-Ciocalteu reagent. In a convenient modification of the method heat-inactivated human serum is used as substrate and activity is polarographically measured by the Brdička reaction (Janoušek, 1959; Homolka and Angerová, 1963).

Kritsman and Bavina (1955) investigated "proteolytic" activity in aortas of rabbits suffering from experimental cholesterol atherosclerosis. Denaturated haemoglobin was used as substrate. *Post-mortem* autolysis (without substrate) was determined by measuring the amount of free tyrosine in the aortas. Both "proteolytic" activity and autolysis were found to be increased as compared with control animals. Proteosynthesis, as measured by the rate of incorporation of ^{14}C-labelled aminoacids into aortic proteins, was also decreased in the aortas of such atherosclerotic rabbits.

Kirk (1962*d*) obtained similar results in studies on the human aorta, pulmonary artery and coronary artery. He measured catheptic activity in essentially the same way as did Kristman and Bavina in their investigations on "proteolytic" activity (see above). The difference between activity in the presence and absence of substrate (i.e. "total proteolysis" minus autolysis) was ascribed to catheptic activity.

The activity in the aortic samples (expressed on a wet weight basis) was much higher than in the other arterial samples. In atherosclerotic parts of the aorta and coronary arteries activity referred to tissue nitrogen was on average 45 per cent and 78 per cent higher than that in normal parts of these vessels. On a wet weight basis the difference between atherosclerotic and normal tissue was lower, being 18 per cent for the aorta and 64 per cent for the coronary arteries.

It is interesting to note that in Kirk's studies increased catheptic activity in atherosclerotic parts of the coronaries ran parallel with the increase of another lysosomal enzyme, namely acid phosphatase. However, this relationship did not apply in the aorta.

LYASES

The main group of enzymes termed lyases comprises "enzymes that remove groups from substrates non-hydrolytically, leaving double bonds (or which add groups to double bonds)". (Recommendations of the IUB Commission on the nomenclature and classification of enzymes, see Florkin and Stotz, 1964.)

Lyases, therefore, usually catalyse a reaction of the type

$$A \rightarrow B + C$$

When it is desired to emphasize the synthetic aspect of the reaction, the term "synthase" is used (e.g. the enzyme previously known as "condensing enzyme" is designated citrate synthase).

Some enzymes of this group—such as fructose diphosphate aldolase, fumarate hydratase, aconitate hydratase and lactoyl-glutathione lyase—have been mentioned in the preceding sections dealing with specific metabolic pathways. Carbonate dehydratase is an important enzyme of this group, but it is not part of a well-defined enzyme system.

CARBONATE DEHYDRATASE, CARBONIC ANHYDRASE, E.C. 4.2.1.1.
(Carbonate hydro-lyase)

The enzyme catalyses the hydration of carbon dioxide to carbonic acid and bicarbonate.

$$CO_2 + H_2O \rightleftharpoons H_2CO_3 \qquad (1)$$

$$CO_2 + H_2O \rightleftharpoons H^+ + HCO_3^- \qquad (2)$$

In the absence of the enzyme the *uncatalysed reaction* is of type (1) below pH 8, whereas bicarbonate is produced above pH 10. However, careful kinetic studies indicate that the direct product of the *enzyme-catalysed reaction* even at pH 7·0 is the HCO_3^- ion (see Davis, 1961).

The rapidity of the enzyme-catalysed reaction, as well as the production of bicarbonate at a physiological pH, is of considerable importance in many fundamental functions of living cells (e.g. secretion of hydrogen ions in the kidney, elimination of carbon dioxide from the lungs, production of gastric acid, etc.).

Carbonate dehydratase is widely distributed in animal tissues; it is particularly concentrated in erythrocytes, gastric mucosa and renal cortex. It is also found in plant leaves, where it probably facilitates diffusion of carbon dioxide during photosynthesis. The role of the enzyme in calcification processes is at present uncertain.

Carbonate dehydratase requires zinc ions for full activity and is specifically inhibited by sulphonamides.

The activity of carbonate dehydratase is usually determined by measuring the time required for the conversion of CO_2 to HCO_3^- as indicated by a substantial fall in pH (Brinkman, 1934; and others). The end-point is determined with an indicator, such as bromthymol blue or phenol red. In contrast to the methods based on estimating carbon dioxide *hydration*, manometric techniques measure the *dehydration* of carbonic acid as revealed by the CO_2 tension over an appropriate bicarbonate-containing substrate (Roughton and Booth, 1938, 1946; and many others). A rapid and reliable electrometric method is provided by electrometric measurement of changes in hydrogen ion concentration in weakly buffered solutions (Davis, 1958).

Kirk and Hansen (1953) measured the activity of carbonate dehydratase in homogenates of human aortas using a modification of the above colorimetric method. A manometric method was used to demonstrate carbonic acid dehydration by the catalytic action of the arterial enzyme. Low but definite carbonate dehydratase activity was observed which amounted to approximately 1 per cent of that in whole blood. (For comparison, values for gastric mucosa and kidney are 500 and 20 per cent, respectively, of the activity of whole blood).

The physiological significance of carbonate dehydratase in arterial tissue is not known. Kirk and Hansen suggest that, as in the lens, the comparatively large amount of lactic acid produced in the aorta during aerobic glycolysis (see Chapters I and XVI) is neutralized by bicarbonate with evolution of carbonic acid. Hence, the function of the enzyme would be catalytically to convert carbonic acid to more readily diffusible carbon dioxide.

ISOMERASES

Apart from isomerases already discussed in the sections on specific metabolic pathways, only one other enzyme of this group has been studied in the vessel wall.

Mannosephosphate Isomerase, E.C. 5.3.1.8.
(D-Mannose-6-phosphate ketol-isomerase)

This enzyme catalyses the following interconversion:

$$\text{D-mannose 6-phosphate} = \text{D-fructose 6-phosphate}$$

It also acts on D-xylose and rhamnose. The enzyme has been found in animal tissues, including muscle and red blood cells, and in yeast. It has been purified from the last source.

While the muscle enzyme does not need to be activated by metals, the red cell enzyme is metal-dependent and requires divalent cations (Mg^{++}, Mn^{++}, Ca^{++}) to achieve full action.

Kirk (1966a) studied the activity of the enzyme in homogenates of human arteries using a modification of the assay described by Bruns *et al.* (1958). This assay is based on the determination of fructose 6-phosphate by Roe's (1934) colorimetric resorcinol method at 490 mμ.

Moderate activities of this enzyme were observed in aortic tissue (0·397 mM substrate isomerized/g. N_2/hr.). The activities in the pulmonary artery and coronary artery were 0·526 and 0·580, respectively. Atherosclerotic regions exhibited reduced enzyme activity.

Chapter IX

ENZYME COFACTORS

It is well known that many enzymes are unable to exert their catalytic function except in the presence of certain cofactors which may be either specific coenzymes or "activators" (e.g. certain inorganic ions). Like enzymes most cofactors can be conveniently classified according to the reaction in which they take part (see Dixon and Webb, 1964).

In *oxido-reduction reactions* hydrogen carriers play a most important role. Such carriers are exemplified by NAD, NADP, FMN, FAD, cytochromes, lipoate, quinones, ascorbate, glutathione etc.

The *transport* of phosphate is effected by nucleoside 5'-diphosphates and nucleoside 5'-triphosphates; the important carrier for acyl groups is coenzyme A; tetrahydrofolate is the carrier for methyl, hydroxymethyl, formimino and formyl groups; biotin is the cofactor of reactions involving incorporation or transport of carbon dioxide; pyridoxal phosphate is an important amino-carrier; thiamine pyrophosphate is involved in decarboxylation reactions producing active acetaldehyde. Many other examples could be cited.

Hydrolases work without specific coenzymes, but are often activated by metal ions.

In *decarboxylation* reactions that are catalysed by lyases, thiamine pyrophosphate, lipoate, coenzyme A, FAD and NAD are involved as cofactors.

In *isomerisation* reactions, coenzyme B_{12} is involved in the intramolecular transfer at C—C bonds, while the action of some isomerases requires UDP as cofactor.

Finally the *ligases* are by definition "enzymes which catalyse the joining together of two molecules coupled with the breakdown of a pyrophosphate bond in ATP or a similar compound".

In considering the enzymology of arterial tissue it is important to take account of the following points:

1. Many of the cofactors are closely related to vitamins, for example the flavin prosthetic groups contain riboflavin; NAD and NADP contain nicotinamide; pyridoxal phosphate and thiamin pyrophosphate are related to the vitamin B group, while biotin itself also belongs to this group.

2. Most coenzymes* are either nucleotides (i.e. compounds in which a

---

* It is difficult to draw a sharp line of distinction between two types of cofactors, namely coenzymes (carriers) and prosthetic groups. The most logical distinction seems to be that suggested by Dixon and Webb (1964): "A true prosthetic group undergoes its whole catalytic cycle while attached to the same enzyme protein molecule: a carrier like NAD must migrate from one enzyme protein to another to fulfil its catalytic function". Typical examples were encountered in the preceding chapters. The flavin-adenine dinucleotide (FAD) moiety of succinate dehydrogenase is a prosthetic group (see p. 31). The reduced NAD produced in the reaction with glyceraldehydephosphate dehydrogenase is "regenerated" under anaerobic conditions in the reaction with lactate dehydrogenase; NAD being thus the coenzyme of both enzyme proteins, but "working" in a reverse direction in the latter instance (see Fig. 2, p. 16).

nitrogenous base, D-ribose, and phosphoric acid are linked to one another) or have some structural analogy to nucleotides. Examples may be cited in NAD, NADP, the flavin nucleotides, ADP, ATP and the related nucleoside 5'-di- (or tri-) phosphates.

The structures of some cofactors that are relevant to vascular enzymology are shown in Figs. 7 and 8.

NICOTINAMIDE NUCLEOTIDE COENZYMES

The oxidized and reduced forms of nicotinamide-adenine dinucleotide (NAD, formerly known as DPN or coenzyme I) and of nicotinamide-adenine dinucleotide phosphate (NADP, formerly known as TPN or coenzyme II) are of fundamental importance in many common metabolic pathways and they have already been mentioned many times in connection with various enzyme-catalysed reactions.

The mechanism of action of these coenzymes depends on the reversible reduction of the nicotinamide ring (for a comprehensive review see Kaplan, 1960). Simultaneously the nitrogen of the base loses its positive charge and the ultraviolet absorption spectrum changes from 260 mμ. to 340 mμ. This last feature forms the basis of many enzymological "optical tests", in which the appearance or disappearance of reduced NAD can be followed spectrophotometrically (see Chapter XI).

The amount of NAD in rat tissues (Glock and McLean, 1955) varies between about 100 to 600 μg./g. wet wt., the highest levels were recorded in the liver and heart. There is less NADP in tissues, except in the liver and adrenals which contain about 200 and 150 μg./g. wet wt., respectively. NAD is mainly present in tissues in the oxidized form, whereas the converse applies to NADP.

Most of these coenzymes are localized (at least in the liver) in the soluble subcellular fraction. Approximately 10 per cent of the total amount is confined to mitochondria, where NAD is predominantly in the oxidized form and NADP in the reduced form.

These distribution characteristics seem to be related to the different general metabolic functions of NAD and NADP. The hydrogen of the reduced form of NAD is most frequently transferred to enzymes of the respiratory chain (see p. 39 and Fig. 5) to produce high-energy bonds. The hydrogen of reduced NADP, on the other hand, is predominantly used for reductive biosynthetic processes.

The methods for determining the oxidized and reduced forms of NAD and NADP have been reviewed in detail by Ciotti and Kaplan (1957) and more recently by Klingenberg (1963). Most of the methods use specific dehydrogenase reactions where the coenzymes are oxidized or reduced and such changes are followed spectrophotometrically. Other methods are based either on the strong autofluorescence of the reduced forms or on the induction of fluorescence after addition of compounds such as methyl ethyl ketone (Carpenter and Kodicek, 1950).

In all such estimations great care must be paid to the rapid preparation of the tissues as these coenzymes are very easily destroyed. Thus, in most tissues there are several enzymes that act on oxidized and/or reduced forms of NAD or

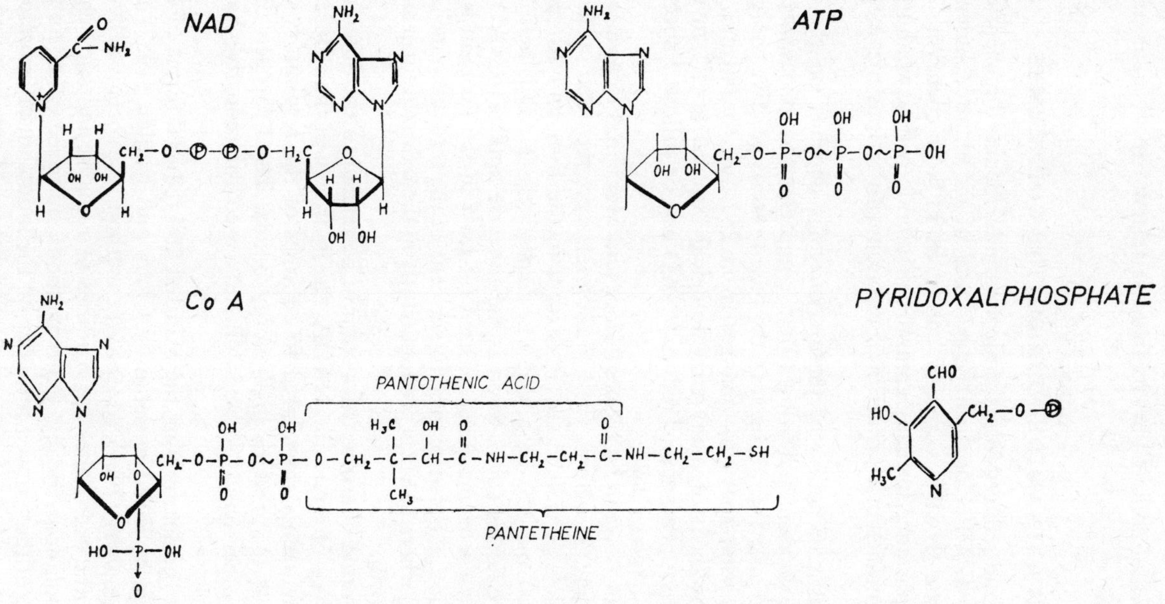

FIG. 7.—The structure of some enzyme cofactors. The symbol P denotes phosphoric acid residue and the symbol ~ stands for energy-rich bonds.

NADP—such as phosphatases, pyrophosphatases, NAD- and NADP-nucleosidases.

Chang *et al.* (1955) tried to measure the total NAD and NADP content of human thoracic aortas obtained fresh at autopsy They used a modification of the above methyl ethyl ketone method where the oxidized forms of NAD and NADP react with the reagent in alkaline solution and, after acidification, yield a fluorescent product. In the same samples the total nicotinic acid content was also estimated by a colorimetric method with cyanogen bromide. The average total nicotinic acid content was found to be about 19 μg. and the total NAD and NADP content about 26 μg./g. wet wt. Assuming that all the nicotinic acid found on acid hydrolysis represented material initially present as coenzyme, then the values for the total NAD and NADP content accounted for only about 25 per cent of what would be expected from the nicotinic acid values. This discrepancy was explained by the constant finding of less NAD and NADP during

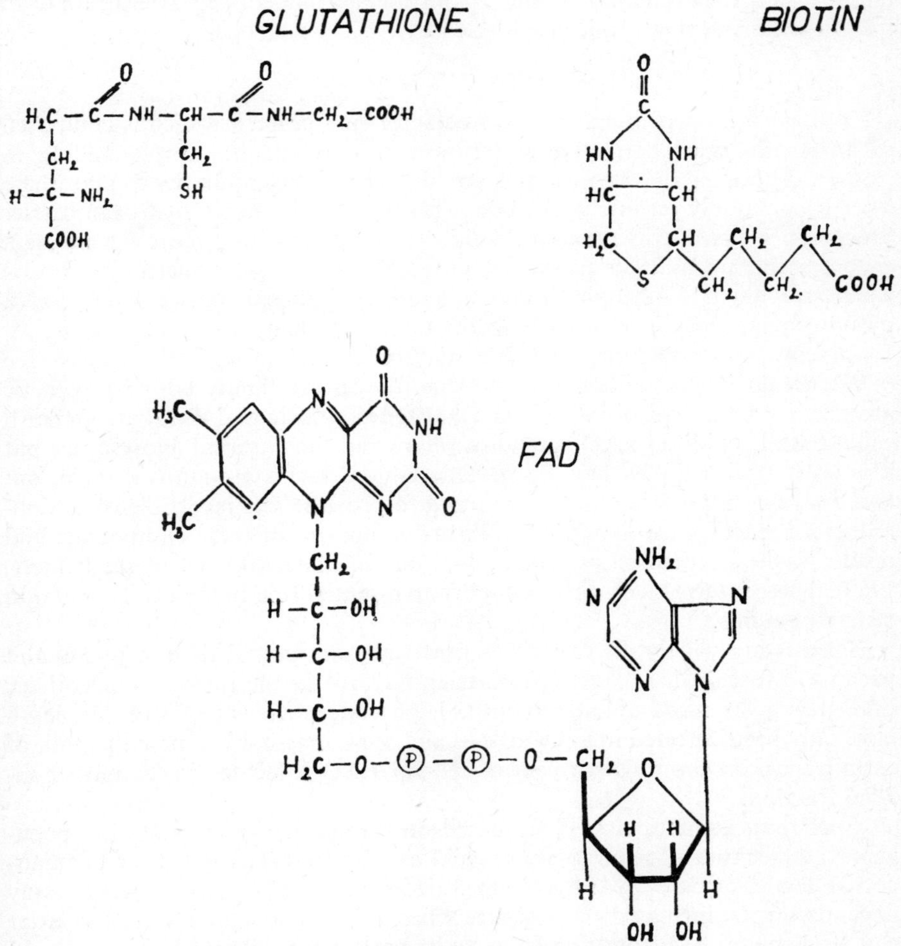

FIG. 8.—The structure of some enzyme cofactors.

storage of the samples; on an average as much as 27 per cent was lost after 30 min. incubation. This suggested—as was anticipated—that NAD and NADP destroying enzymes are present in the vessel wall (see above). Therefore, it can be assumed that the aortic concentrations of these coenzymes are higher *in vivo*, perhaps in the range suggested by the nicotinic acid values. The latter amounted to 30 per cent of the concentration of nicotinic acid in the human liver and were, thus, surprisingly high for a tissue with a comparatively low metabolic activity. There was a tendency for the nicotinic acid content of the aorta to decrease with advancing age.

Mandel and Kempf (1961) determined the free nucleotide content of bovine aortas (see p. 81) and also presented some data on their NAD content. In young bovine aortas they found 1·92 μM./coenzyme/100 g. wet wt. (about 12.7 μg./g. wet wt.).

Unfortunately, no data are yet available about the nicotinamide-nucleotide enzymes of arterial samples obtained from animals immediately after death or in human vessels during surgical operations.

FLAVIN NUCLEOTIDES

The essential constituent of these cofactors is riboflavin which is composed of an isoalloxazine derivative and ribitol (a sugar-alcohol corresponding to ribose). Although riboflavin is not strictly a true nucleoside, its 5'-phosphate ester is commonly termed nucleotide. This is the well-known hydrogen carrier prosthetic group flavin mononucleotide (FMN). However, most flavoprotein enzymes contain as their prosthetic group flavin-adenine dinucleotide (FAD) where riboflavin 5'-phosphate and adenosine 5'-phosphate are linked by a pyrophosphate bond—in analogy to the nicotinamide nucleotide coenzymes.

In contrast to the nicotinamide nucleotide "migrating" coenzymes (see p. 73), the flavin nucleotides of mammalian tissues are tightly bound prosthetic groups that act as hydrogen carriers. Oxidized riboflavin and FMN are greenish yellow, while oxidized FAD is reddish yellow: all show typical fluorescence but it is greater with FMN than FAD. The reduced forms are autoxidizable, but this does not apply when they are bound to protein. The reversible oxidation-reduction reaction involves the isoalloxazine nucleus of these compounds and results in the production of semiquinoid intermediates (for review see Beinert, 1960). These intermediary products play an essential role in the catalytic action of flavoproteins.

Flavins are widely distributed in practically all animal tissues, plants and bacteria. Most of this naturally-occurring flavin is in the form of nucleotides, FAD being the more abundant form. High concentrations of free riboflavin have only been detected in a few tissues and secretions, for example the milk of certain species. About 65 per cent of cellular FAD is located in the mitochondrial fraction.

Spectrophotometric assays are based on the specific reactivation of apoenzymes; either that of renal D-amino acid oxidase by FAD or that of pneumococcal lactic oxidase by FMN (see Friedmann, 1963). Fluorometric assay depends on the increased fluorescence when FAD is split to FMN; the latter can be distinguished from riboflavin by its partition coefficient between benzyl alcohol and aqueous solutions (see Burch, 1957).

Flavin nucleotides have been studied in the aortic wall by Schaus *et al.* (1955) who used essentially the same fluorometric method as outlined above. In dog aortas removed from anaesthetized animals and immediately chilled with carbon dioxide snow, the total flavin content averaged 2·57 μg./g. wet wt. Of this amount 89·9 per cent was found to be FAD, 4·7 per cent FMN and the rest was present as free riboflavin.

In human aortas obtained fresh at autopsy the total flavin content averaged only 1·16 μg./g. wet wt., i.e. about 1/6th of that reported for the heart. The FAD fraction was lower than in dog aortas being only 65·5 per cent of the total riboflavin; the residue represented the sum of FMN and free riboflavin. The results of a few separate estimations showed that the ratio of free riboflavin to FMN was about 1·8:1). The higher percentage of FAD in the dog aortas than in the human vessels could be ascribed to decomposition by FAD-splitting enzymes, because in other investigations on stored dog aortic tissue the FAD content declined by 41 per cent. It is interesting to note that the total aortic riboflavin content, which is not subject to *post-mortem* changes, tended to decline with advancing age.

GLUTATHIONE

This compound, universally present in living cells, is a tripeptide of glutamic acid, cysteine and glycine. Although it was discovered as early as 1888, its functions are still not completely understood. As it is readily oxidized and reduced by enzymatic systems (particularly in the presence of NADP), there is little doubt that it could serve as a biological hydrogen carrier. The oxidation and reduction of glutathione depends on its thiol group; reduced glutathione (GSH) is easily oxidized to its disulphide (GSSG) in a reaction much the same as the conversion of cysteine to cystine. For a long time GSH was assumed to be physiologically autoxidizable, and it seemed to be the only compound with such properties in animal tissues. More recently, however, both oxidizing and reducing enzymes acting on glutathione have been detected (see p. 55). Its oxidation is also conducted *via* dehydroascorbate in the presence of a specific dehydrogenase (glutathione: dehydroascorbate oxidoreductase, E.C. 1.8.5.1.).

In addition, glutathione acts as a specific coenzyme for both the glyoxalase system (see p. 51) and formaldehyde dehydrogenase. In both cases, it acts by adding GSH to a carbonyl group. Glutathione is also the coenzyme of some *cis-trans* isomerases.

Glutathione is widely distributed in living cells, where it exists primarily in the reduced state: liver and yeast are rich sources. The glutathione content of tissues is subject to variations with nutritional state and hormonal influences. (For review see Knox, 1960.)

Many methods are now available for determinating glutathione. Chemical methods, such as iodometric titration and amperometric titration with Ag^+, are essentially based on the reactions of the thiol group. Enzymatic methods take advantage of the glyoxalase reaction, in which the conversion of methylglyoxal to lactic acid requires the presence of GSH. The amount of lactic acid is then determined manometrically by measuring the CO_2 produced from buffered bicarbonate. Since S-lactoyl-glutathione is produced, a more exact modification depends on the direct spectrophotometric measurement of this compound at

240 mμ. In a further method, oxidized glutathione is quantitatively reduced by glutathione reductase (E.C. 1.6.4.2) in the presence of $NADPH_2$ (see p. 96): the oxidaton of reduced coenzyme can be determined by the fall in optical density at 340 mμ (see Klotzsch and Bergmeyer, 1963).

Wang and Kirk (1960) measured the total glutathione content in supernatants of homogenates from human aortas and pulmonary arteries. The assay depended on electrolytic reduction of the oxidized glutathione and subsequent iodometric titration.

The average glutathione values recorded in intact aortic and pulmonary artery samples were 5·6 and 8·3 mg./g. tissue N_2. The glutathione concentration of aortic tissue tended to increase with age. Atherosclerotic and normal aortic tissue had the same glutathione content.

COENZYME A

The structure of this extremely important coenzyme has some similarity to an adenine dinucleotide. It can logically be regarded as a compound of 3'-phospho-ADP and pantetheine: the latter consists of pantothenic acid, L-alanine and cysteamine.

The mechanism of coenzyme A's action depends on its terminal -SH group. It functions in many reactions as a carrier for acyl groups according to the following general formula (see Jaenicke and Lynen, 1960):

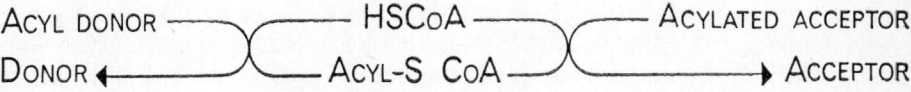

Energy is needed for forming thiolesters and this is derived either from ATP or from coupling the biosynthesis to an exergonic reaction such as oxidative decarboxylation. The free energy engendered by hydrolysis of the thiolester bond is about − 8·25 kcal. (see footnote, p. 47) and can be utilized for synthetic reactions.

Coenzyme A is implicated in many reactions. Thus Jaenicke and Lynen (1960) drew up a list of 63 enzymatic reactions that require this coenzyme or its derivatives.

Coenzyme A takes part in the final oxidation of pyruvic acid by the TCA cycle. CoA is an indispensable requirement for fatty acid metabolism: fatty acids are activated by formation of a thiolester with CoA. Many other metabolic pathways could be cited where CoA plays an important role.

Probably the best known CoA derivative is acetyl-CoA or "active acetyl", which primes the TCA cycle and is essential for a great variety of biosynthetic processes.

Coenzyme A is widely distributed in animal and plant tissues, and in bacteria. In most species the liver is the most abundant source of the enzyme. The content of CoA is usually also high in the adrenals, kidney, brain and heart. The usual biological source for preparing CoA is yeast.

Many methods exist for determining coenzyme A. Most of them are based on coenzyme A-dependent specific enzymatic reactions. For example, a useful method utilizes the ATP-dependent catalytic formation of acyl-CoA by the enzyme acyl-CoA synthetase (E.C. 6.2.1.3; formerly known as thiokinase). Sorbic acid is used as the acyl donor and, because sorbyl-CoA has a strong absorption maximum at 300 mμ., the reaction can be conveniently followed spectrophotometrically (Wakil and Hübscher, 1960). Another elegant method is based on the catalytic transfer of acyl groups from acetyl phosphate to CoA by the action of phosphate acetyltransferase (E.C. 2.3.1.8.). The acetyl-CoA produced can be determined spectrophotometrically at 233 mμ., or alternatively the disappearance of acetyl phosphate can be measured by hydroxamate formation. (For other methods see Novelli, 1957; Michal and Bergmeyer, 1963.)

In arterial tissue Paoletti *et al.* (1959) studied the coenzyme A content of aortas of different species. They used the well-known method of Kaplan and Lipmann (1948), which is based on acetylation of sulphanilamide by an aged bicarbonate extract of pigeon liver, in the presence of CoA and ATP. The rate of disappearance of sulphanilamide depends on the rate-limiting amount of CoA. They observed a higher coenzyme A content in the aortas of rabbits and chickens (on an average 17·6 and 18·2 units/g. wet wt.) than in the aortas of pigeons, hamsters and rats (9·3, 10·1 and 9·9 respectively). In young rabbits the aortic CoA content was much lower (7·6 units/g. wet wt.) than in adult animals. A similar trend was also observed in the aortas of young rats. In the livers of the animals studied the mean CoA content varied between 66 and 83 units. For reasons unknown these values are almost 50 per cent lower than those reported by Kaplan and Lipmann (1948) using the same method for determining CoA. In contrast to the CoA findings, parallel autoradiographic experiments with [1-^{14}C]-acetate showed that lipid biosynthesis is highest in the aortas of those species that are more susceptible to experimental atherosclerosis—such as the rabbit and chicken.

Sanwald and Kirk (1965*a*) estimated coenzyme A in intima-media preparations from human aortas, pulmonary and coronary arteries. As an appreciable fraction of tissue coenzyme A is rapidly split after death by autolysis, the authors preferred to use an elegant and sensitive, but time-consuming method. It depends on the quantitative release of pantothenic acid after treatment of the tissue samples with alkaline phosphatase and an acetone-extracted hog kidney preparation. Subsequently pantothenic acid is microbiologically assayed by the *Lactobacillus plantarum* procedure, which is similar to that described for biotin (see p. 83).

The average values of normal specimens were highest in the coronary arteries (229 units/g. tissue N_2). The coenzyme A content tended to increase from the ascending aorta (106 units) to the abdominal aorta (186 units). Lipid-rich atherosclerotic segments contained as much CoA as normal tissue, but markedly lower concentrations were found in fibrous aortic lesions (83 units).

In view of the great importance of coenzyme A in intermediary metabolism it is clear that these problems deserve much more attention in the future.

PURINE AND PYRIMIDINE NUCLEOTIDES AS PHOSPHATE CARRIERS

In contrast to the nucleotides derived from nucleic acids, where the phosphate groups are attached to either the 2' or 3'-position, purine and pyrimidine nucleotides are chemically characterized by the attachment of their phosphate groups to the 5'-position of the ribose component of the molecule.

Whereas the nicotinamide-containing coenzymes (p. 73) are intimately related to hydrogen transport, the nucleotides we are now concerned with have two main functions. Their energy-rich pyrophosphate bond (especially that of ATP) represents the most readily available form of energy in biological systems. In addition, these nucleotides act as coenzymes in trans-phosphorylation reactions. Such transfer reactions are closely linked to the first-mentioned function, for phosphate transfer is the mechanism for transporting energy from one metabolic process to another.

In the enzyme table of the IUB's Enzyme Commission about 75 enzyme-catalysed reactions (sub-groups E.C. 2.7.1.–4. and 6. 3) are listed where adenosine 5'-diphosphate can be considered as a coenzyme for phosphate transport. It is reasonable to regard this compound as the fundamental biological phosphate carrier (Dixon and Webb, 1964); it can take on a further phosphate group to give rise to adenosine 5'-triphosphate.

The phosphate carrier functions of the other nucleoside 5'-diphosphates (CDP, UDP, GDP and IDP, see p. 46) seem to be much more limited, although theoretically they may well form interlinked systems *in vivo* with ADP and ATP through the action of nucleoside-diphosphate kinase (Dixon and Webb, 1964). The latter enzyme (E.C. 2.7.4.6) catalyses the reaction

ATP + nucleoside diphosphate = ADP + nucleoside triphosphate.

Some nucleoside 5'-diphosphates take part in systems concerned with the exchange transfer of phosphate with other groups, such as glycosyl. The physiologically most important example is uridine 5'-diphosphate (UDP). The terminal phosphate group in uridine 5'-triphosphate can be replaced, for example, by glucose in a reaction with glucose 1-phosphate; the UDPglucose or "active glucose" (or other sugar compounds) produced in this way can then be transferred to appropriate acceptors. This type of reaction was discussed in connection with glycogen synthesis (see p. 26): analogous reactions play a similar role in the synthesis of sucrose, starch, cellulose, chitin, hyaluronic acid and, perhaps, other mucopolysaccharides.

Uridine nucleotides are widely distributed in animal tissues, plants and micro-organisms. Considerable amounts are found in the liver, brain, muscle, mammary gland and milk. (For a detailed review see Leloir and Cardini, 1960.).

Cytidine 5'-phosphates are coenzymes of enzymatic reactions leading to the synthesis of lecithin, phosphatidyl ethanolamine and sphingomyelin. For example in the synthesis of lecithin, phosphorylcholine is "activated" by CTP to form CDP-choline and, in a further step, phosphorylcholine is transferred to D-1,2-diglyceride to produce lecithin. The other phospholipids mentioned above are synthesized in an analogous way (for details see the review by Kennedy, 1960).

The cytidine nucleotides are probably ubiquitous constituents of animal and plant tissues; they have been isolated from yeast.

In contrast to the preceding classes of nucleotides whose role in various metabolic systems is fairly well-defined, the metabolic functions of guanosine and inosine nucleotides are somewhat less clear. It has been established that phosphorylation of ADP, coupled with the breakdown of succinyl CoA, proceeds *via* a two-step reaction where GDP acts as a specific coenzyme. In protein biosynthesis the reactions by which aminoacids are activated and allowed to combine successively with the ribosomes, are also catalysed by GTP-dependent enzymes (see Chapter I). According to Kleinzeller (1952) myosin acts on ITP at a rate exceeding that of ATP. Likewise, ITP can to a certain extent replace ATP in the hexokinase and creatine kinase reactions. Guanosine nucleotides appear to be effective in reaction systems at very low concentrations in comparison with other nucleotides that participate in reactions as coenzymes (Utter, 1960).

Whereas the natural occurrence of inosine nucleotides is still uncertain, guanosine nucleotides seem to be widely distributed in nature. Comparatively large amounts have been reported in embryonic tissue.

There are many methods for assaying nucleotide coenzymes, in particular adenosine and uridine nucleotides. Elegant enzymatic determinations are based on the use of coenzyme-dependent specific enzymes, as for example phosphoglycerate kinase or hexokinase in the assay of ATP, nucleoside diphosphokinase for UTP etc. The reactions are coupled with "indicator" systems which allow the reaction rate to be followed spectrophotometrically (see "Methods of Enzymatic Analysis", Bergmeyer, 1963). An interesting "biological" method is based on the linear luminescence response of aqueous firefly extracts to added ATP (Strehler and Totter, 1952). Such "Luciferase" preparations are commercially available.

Somewhat complex chromatographic methods are also available; they will be illustrated below by reference to investigations on arterial tissue.

Mandel and Kempf (1961) investigated the free nucleotide content of perchloric acid extracts from intima-media preparations of chilled bovine aortas. Fractionation was performed on anion-exchange columns (DOWEX-1) in the formate form; formic acid-ammonium formate was used as the displacing liquid. The fractions were subjected to paper chromatography, using specific solvent systems that allow the separate members of the purine and pyrimidine nucleotide classes to be identified. The assay of some components was also confirmed by special spectrophotometric techniques. (For further details the reader is referred to the original publication.).

In the aortas of young animals Mandel and Kempf found an average of 74 μM. of total free nucleotide/100 g. wet wt.; the adenosine nucleotides accounted for 52 per cent of this amount. The percentage distribution of uridine, cytidine and guanosine nucleotides was found to be 17·8, 13·7 and 9·2 per cent. An interesting feature was the relatively high content of CDP coenzymes, which perhaps reflects the ability of arterial tissue to synthesize phospholipids.

In a further paper Kempf and Mandel (1961) compared the free nucleotide content of adult (2 yr.) and old (11–12 yr.) bovine aortas. With advancing age the total amount of adenosine nucleotides fell from about 56 to 29 μg./g. wet wt. This decline was mainly due to the large fall in ATP—from 44 to 22 μg. There are doubtless major inaccuracies in the estimation of ATP when not performed immediately after death. Nevertheless, the same bias can be expected in both young and old animals, so it can be assumed that the figures reflect an actual decrease of both ATP and total adenosine nucleotide.

The aortas of old animals also contained less uridine and guanosine nucleotide, whereas the CDP coenzyme content did not significantly change. Kempf and Mandel believe that the lower nucleotide contents in ageing vessels initiate senescent metabolic changes in both protein synthesis (decreased adenosine nucleotides) and mucopolysaccharide synthesis (decreased uridine nucleotides).

BIOTIN

It is widely known that biotin is identical with "vitamin H"; it protects rats and probably all higher animals against avidin, the toxic protein of raw egg white. However, only comparatively recently has it been shown that biotin functions as a cofactor in certain enzymatic reactions. At present five "biotin enzymes" have been recognized; biotin forms their prosthetic groups. One of these enzymes is a carboxytransferase, while four are carboxylases that catalyse synthetase reactions characterized by incorporation ("fixation") of free CO_2 coupled with the breakdown of ATP. There is evidence that a carbonic acid derivative of biotin ("active CO_2") and carboxybiotin-enzymes are involved in these reactions. Avidin, which has a very high affinity for biotin, specifically inhibits such reactions.

Of extraordinary interest is the fact that the first step in fatty acid synthesis consists in the carboxylation of acetyl-CoA to form malonyl-CoA. This reaction is catalysed by one of the above-mentioned biotin-containing enzymes, namely acetyl-CoA carboxylase (E.C. 6.4.1.2.). This enzyme has been extensively studied by Wakil *et al.* (1958), Lynen *et al.* (1959) and other workers. Bicarbonate, ATP and manganese (or magnesium) are involved in this classical carboxylation reaction.

$$\text{Acetyl-CoA} + CO_2 + \text{ATP} \xrightleftharpoons{Mn + E.C.6.4.1.2.} \text{malonyl-CoA} + \text{ADP} + \text{orthophosphate}$$

The "malonyl-CoA pathway" initiated by this reaction seems to be the main metabolic pathway for fatty acid synthesis in animal tissues.

It is interesting that studies concerned with regulatory mechanisms of fatty acid synthesis indicate that the addition of citrate (or isocitrate) stimulates the formation of malonyl-CoA in the above reaction (Waite and Wakil, 1962; Martin and Vagelos, 1962). In this connection, studies by Numa *et al.* (1961) and Lynen *et al.* (1963) indicate that the purified biotin enzyme, acetyl-CoA carboxylase, requires citrate or isocitrate as cofactor. It was suggested that, in starvation or diabetes, the decreased citrate content of liver cells might be responsible for the reduced capacity of this organ for fatty acid synthesis (Lynen *et al.* 1963). In addition, long-chain acyl CoA derivatives, such as palmityl-CoA, have been shown strongly to inhibit this biotin enzyme-catalysed reaction. Thus, these acyl-CoA compounds participate in a self-regulatory mechanism by feedback inhibition. (See Chapter I.)

Further aspects of the biochemistry of biotin and of reactions catalysed by biotin-enzymes have been recently reviewed by Mistry and Dakshinamurti (1964). These authors also discussed the ill-understood indirect effects of biotin deficiency on protein synthesis, purine synthesis and carbohydrate metabolism.

Baker and Sobotka (1962) found that the biotin content of human liver is within the range of 230–660 mμg./g. lyophilized tissue, while that in the brain ranges between 210–240 mμg. The subcellular distribution studies of Mistry and Dakshinamurti (1963) show that biotin in normal rat liver cells is chiefly confined to the nuclear and supernatant (cytoplasmic) fractions, but in biotin deficiency it disappears from the cytoplasmic fraction.

The usual methods for assaying biotin are microbiological (see Baker and Sobotka, 1962), using biotin-requiring micro-organisms such as *Lactobacillus plantarum* (*arabinosus*). The growth of the microorganism, which is estimated turbidimetrically or by optical density readings, is linearly proportional to the concentration of biotin. The biotin content of the sample is computed from a curve obtained by assay of standard solutions of D-biotin.

Kirk and Sanwald (1966) estimated biotin in a large number of human aortas, pulmonary arteries, coronary arteries and venae cavae inferiores. Homogenates of intima-media preparations were treated with trypsin to remove biotin from protein-conjugation and were then assayed by a microbiological method, essentially similar to that outlined above. There was no difference in the biotin content of the aorta and pulmonary artery. The biotin content of the intact ascending aorta (4·41 mμg./g. wet wt.; 111·5 mμg./g. tissue N_2) was significantly lower than that of the intact descending thoracic aorta (5·02; 130·0) and abdominal aorta (5·96 mμg./g. wet wt.). The thoracic aorta and pulmonary arteries exhibited a notable increase in the biotin content from early childhood to adulthood. No unequivocal sex difference could be established in the biotin content of the vessels studied.

Lipid-rich atherosclerotic regions of aorta and coronary artery contained significantly more biotin but little change was observed in fibrous atherosclerotic plaques.

In view of the important part played by biotin in fatty acid synthesis, its investigation in arterial tissue seems to be of particular interest in connection with the role of vascular metabolism in atherogenesis.

As can be seen from this part of the present volume, a considerable body of evidence exists to show that the arterial wall contains enzymes catalysing the basic "mainstreams" of metabolism (Lehninger, 1959). Enzymes are also present that catalyse metabolic activities connected with specialized functions of the arterial wall, but other enzymes have been identified whose biological significance is as yet unresolved.

In the next chapter (the introduction to the second part of this volume) some of the vascular enzymatic findings will be briefly recapitulated, while other findings—such as sexual and regional differences in vascular enzyme activities—will be summarized in relation to topics to be discussed in subsequent chapters.

PART TWO

SPECIAL PROBLEMS OF VASCULAR ENZYMES AND ATHEROSCLEROSIS

Chapter X

INTRODUCTION

IN the second part of this volume it is intended mainly to concentrate on some special problems that have been studied in an attempt to elucidate the pathogenesis of atherosclerosis. It must be emphasized that many "special problems" closely related to this topic have been intensively investigated by many authors quoted in the preceding chapters (e.g. Banga, Kritsman, Maier, Malinow, Mandel, Miller, Sandler, Wexler and their co-workers) and especially by the outstanding work of Kirk and his co-workers. The contributions of all these investigators should logically be again included in this part of the present volume when referring to special problems related to atherosclerosis. They were, however, mentioned in the preceding chapters under the heading of "General Properties", and in the next chapters we shall try to discuss only those problems that have been studied in our own laboratory.

Nevertheless, in view of the comprehensive work performed by Kirk and his co-workers, it will be useful to present in this introduction a few synoptic figures dealing with their results on changes of enzyme activity with advancing age and development of atherosclerosis. They are based on the data discussed in the preceding chapters and on the material summarized in Kirk's more recent publications (Kirk, 1963a, 1964b, 1966b).

Figures 9a and b show the differences in enzyme activities and cofactor levels between aortic specimens of children (0–10 years) and those of adults (20–39 years or in some cases 18–35 years). The figures do not show aortic enzymatic changes at a more advanced age. They may, therefore, be supplemented by the data summarized by Kirk (1966b) indicating that in the age group 40–69 years the following further changes appear in aortic enzyme activities. The activities of lactate dehydrogenase, ribosephosphate isomerase and β-glucuronidase *rise*; whereas the activities of fumarate hydratase, lactoyl-glutathione lyase, glycogen phosphorylase, 3-hydroxyacyl-CoA dehydrogenase, "α-hydroxybutyric acid dehydrogenase", glutamate dehydrogenase, creatine kinase and arylsulphatase *decrease*. The level of total riboflavin and nicotinic acid also decreases with advancing age, whereas that of glutathione rises.

Figures 10a and b and 11 summarize the changes of enzyme activities and cofactor levels in atherosclerotic human aortas and coronary arteries as compared with normal tissue segments.

Further synoptic figures of the results obtained by Kirk and co-workers will be presented in some of the subsequent chapters. Some other topics, although of considerable interest, will not be discussed in detail. For example, one of such topics is the interesting interspecies difference of some enzyme activities as related to susceptibility to atherosclerosis. In the preceding chapters we mentioned the interspecies differences in alkaline phosphatase activity (Malinow *et al.* 1959) and ATPase activity (Carr *et al.* 1952). Our own results (Zemplényi

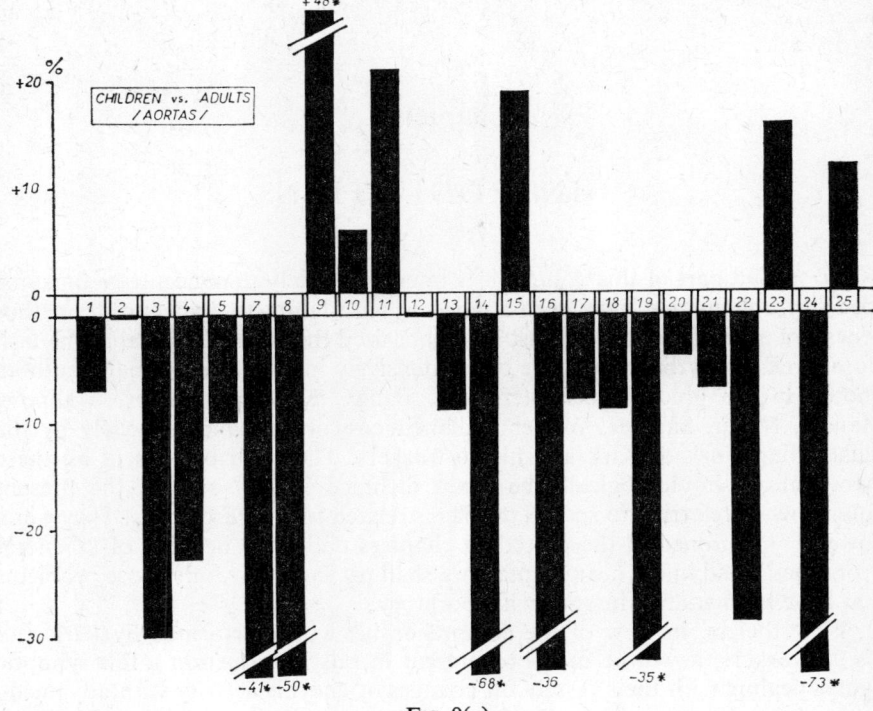

```
1 to 8        Enzymes of glycolysis and of the glycogen pathway
9 to 12       Enzymes of the TCA cycle
14 to 17      Enzymes of the pentose phosphate pathway
18, 19        Enzymes of the respiratory chain
20, 21        Enzymes linking energy rich bonds with muscular contraction
24, 31        Enzymes of fatty acid metabolism
13, 22, 23, 25  Other enzymes
```

FIG. 9(a) and (b). (*See opposite*).—Differences in some enzyme activities and cofactor levels between aortic specimens of children (0–10 years) and those of adults (20–39 years or 18–35 years). The enzyme activities or cofactor levels of adult aortas, calculated on a wet weight basis = 100 per cent. (Constructed according to data from Kirk, 1963a, 1965 and some more recent publications by Kirk and co-workers.) Symbols used in this and other synoptic Figures, constructed according to the data of Kirk's publications. (*1*) Glycogenphosphorylase. (*2*) Phosphoglucomutase. (*3*) Hexokinase. (*4*) Glucosephosphate isomerase. (*5*) Fructosediphosphate aldolase. (*6*) Phosphoglyceromutase. (*7*) Phosphopyruvate hydratase. (*8*) Lactate dehydrogenase. (*9*) Aconitate hydratase. (*10*) Isocitrate dehydrogenase. (*11*) Fumarate hydratase. (*12*) Malate dehydrogenase. (*13*) Malate dehydrogenase (decarboxylating). (*14*) Glucose-6-phosphate dehydrogenase. (*15*) Phosphogluconate dehydrogenase (decarboxylating). (*16*) Ribosephosphate isomerase. (*17*) Transketolase. (*18*) $NADH_2$ cytochrome c reductase. (*19*) Lipoamide dehydrogenase ("Diaphorase"). (*20*) Creatine kinase. (*21*) ATPase. (*22*) Inorganic pyrophosphatase. (*23*) Lactoyl-glutathione lyase (Glyoxalase I). (*24*) 3-Hydroxyacyl-CoA dehydrogenase. (*25*) Glutamate dehydrogenase. (*26*) Glutathione reductase. (*27*) Aspartate aminotransferase (GOT). (*28*) Alanine aminotransferase (GPT). (*29*) Glutamine-fructose-6-phosphate aminotransferase. (*30*) Purine nucleoside phosphorylase. (*31*) Carboxylesterase ("aliesterase"). (*32*) Acid phosphatase. (*33*) 5′-nucleotidase. (*34*) Arylsulphatase. (*35*) β-glucuronidase. (*36*) Leucine aminopeptidase. (*37*) Cathepsin. (*38*) Mannosephosphate isomerase. (*39*) Carbonic anhydrase. (*40*) "α-hydroxybutyrate dehydrogenase" (LDH isozyme). (*41*) Coenzyme A. (*42*) Total riboflavin. (*43*) Flavin-adenine dinucleotide. (*44*) Nicotinic acid. (*45*) Glutathione. (*46*) Pyridoxin. (*47*) Biotin.

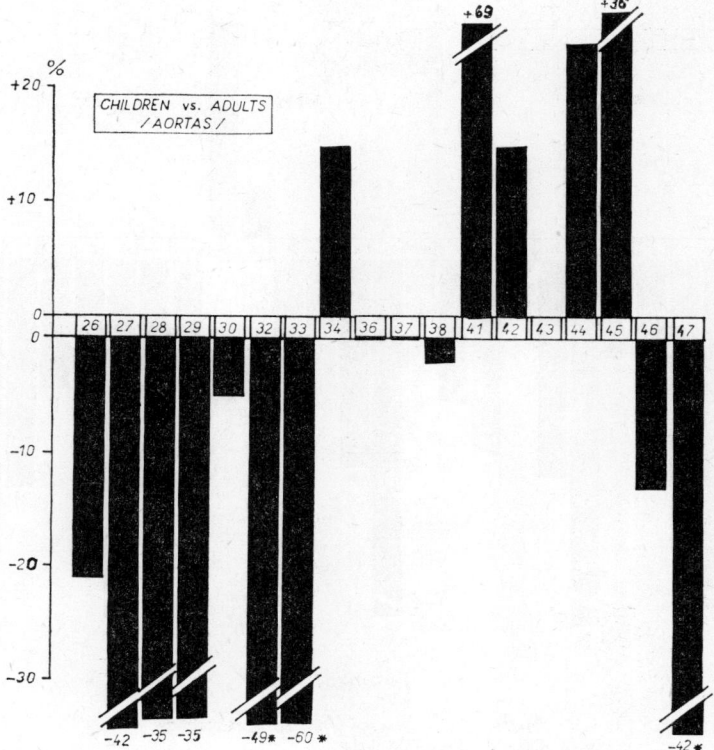

Fig. 9(b)

26 to 30, 32 to 39 Other enzymes
41 to 47 Enzyme cofactors

and Mrhová 1963a), revealed higher alkaline and acid phosphatase and ATPase activity in the rat aorta as compared to that of the rabbit or chicken. On the other hand, the activity of 5′-nucleotidase was found to be highest in the rabbit aorta. These problems will be discussed in connection with the lipolytic activity of the vessel wall in part three of the present volume.

Another interesting topic is the enzymatic activity of venous tissue in comparison with arterial tissue. Figure 13 summarizes pertinent information about this problem that has been provided by Kirk and co-workers.

As the structure of both types of vessels is somewhat different, the evaluation of these results is difficult. The intravascular pressure is also different. Nevertheless, it is tempting to speculate that to some extent they reflect the well-known difference in susceptibility of arteries and veins to sclerotic changes. It should be noted, however, that Stein et al. (1966) found by histochemical means much the same enzyme activities in human veins as in arteries but acid phosphatase activity was absent from venous tissue. Enzyme distribution in phlebosclerotic areas was similar to that in normal veins.

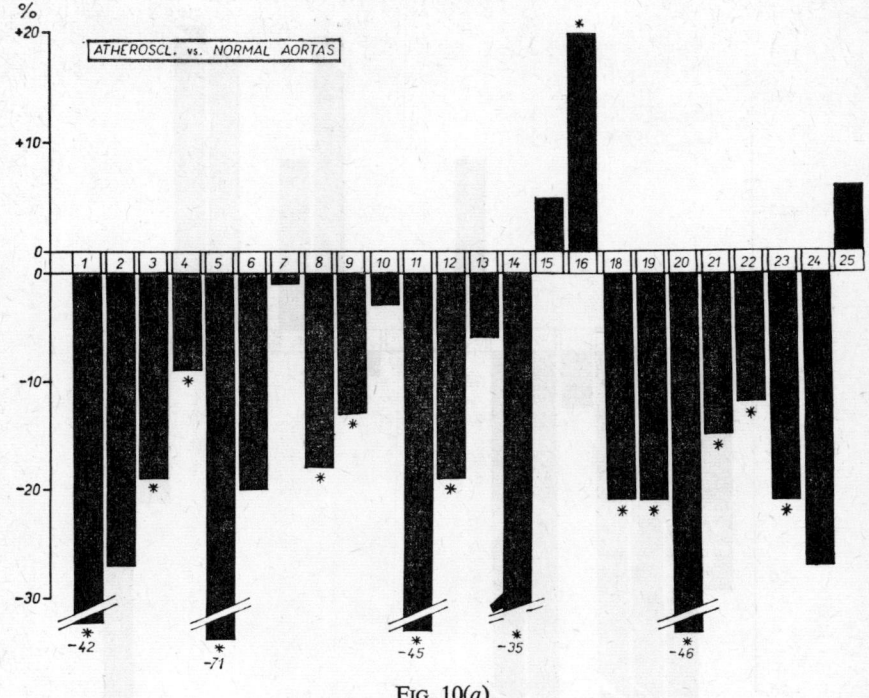

FIG. 10(a)

    1 to 8, 40   Enzymes of glycolysis and of the glycogen pathway
    9 to 12   Enzymes of the TCA cycle
    14 to 17   Enzymes of the pentose phosphate pathway
    18, 19   Enzymes of the respiratory chain
    20, 21   Enzymes linking energy rich bonds with muscular contraction
    24, 31   Enzymes of fatty acid metabolism
    13, 22, 23 25   Other enzymes

FIG. 10(a) and (b).—Differences in enzyme activities and cofactor levels between atherosclerotic and normal human aortas. The enzyme activities or cofactor levels of normal aortas, calculated on a wet weight basis = 100 per cent. (Constructed according to data from Kirk, 1963a and some more recent publications by Kirk and co-workers.) For symbols see Fig. 9(a) and (b).

INTRODUCTION

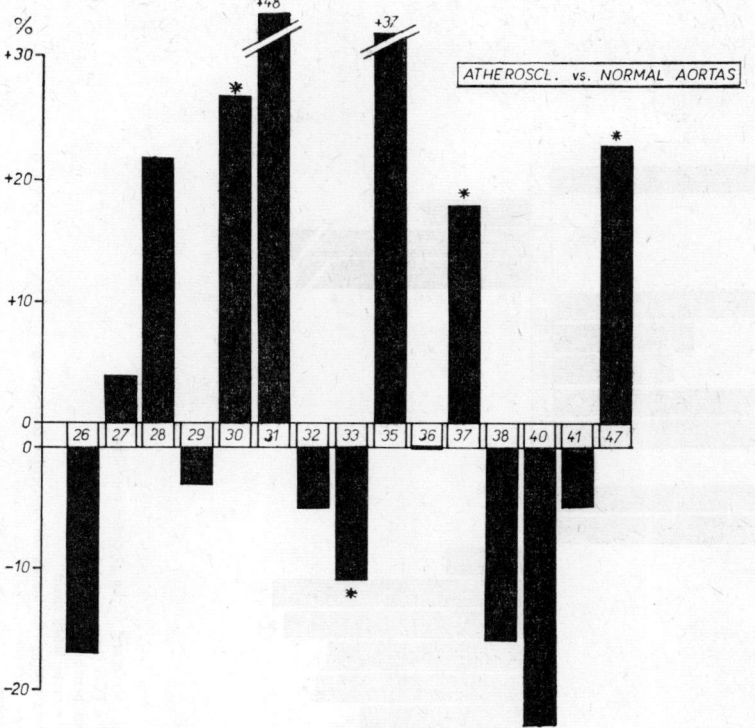

Fig. 10(b)

26 to 30, 32 to 38 Other enzymes
31 Enzyme of fatty acid metabolism
41 to 47 Enzyme cofactors

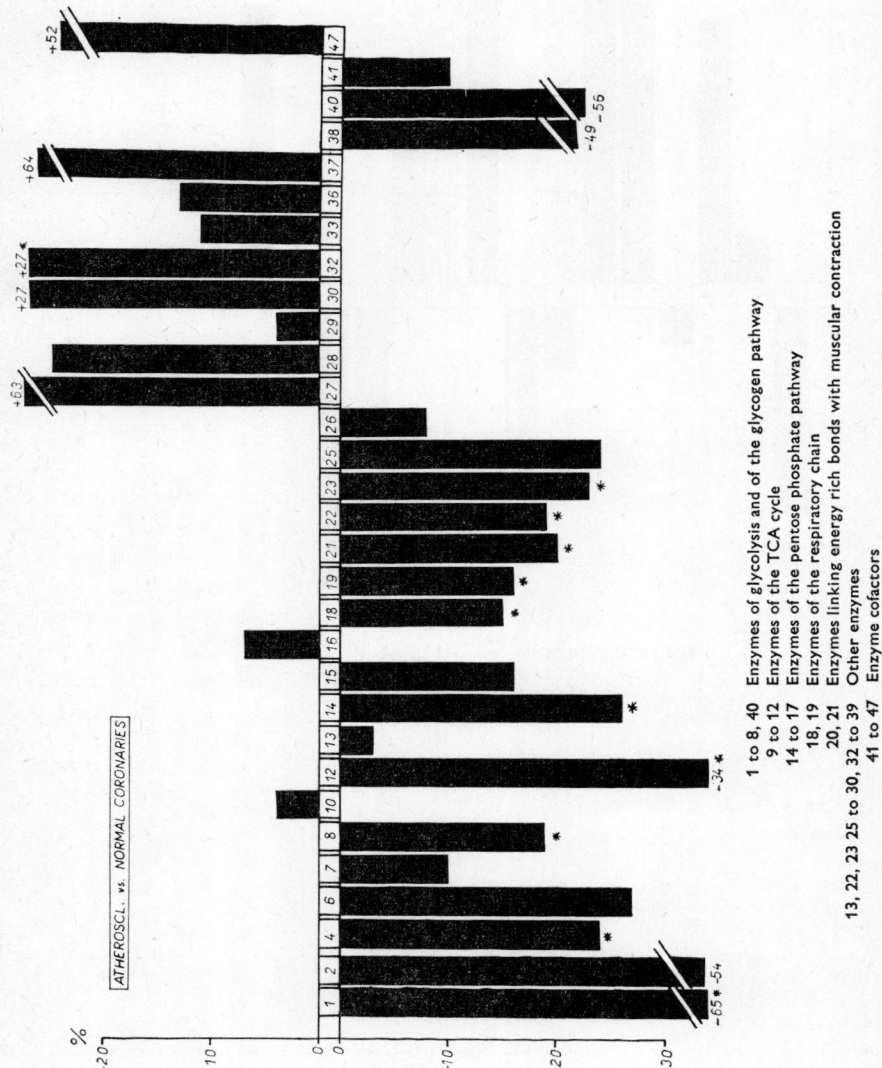

Fig. 11.—Differences in enzyme activities and cofactor levels between atherosclerotic and normal human coronary arteries. The enzyme activities or cofactor levels of normal coronary arteries, calculated on a wet weight basis = 100 per cent. (Constructed according to data from Kirk, 1963.) For symbols see Fig. 9(a) and (b).

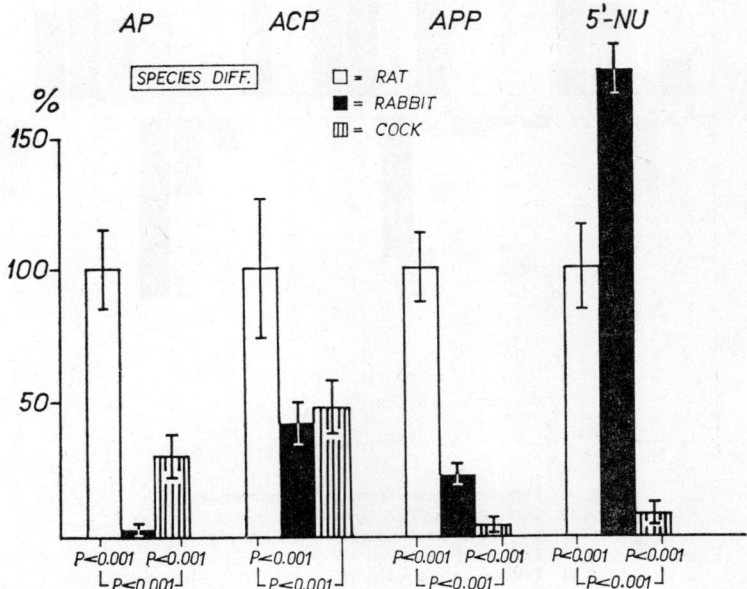

Fig. 12.—Species differences in the activities of aortic alkaline phosphatase (AP), acid phosphatase (ACP), ATPase (APP) and 5'-nucleotidase (5'-Nu). Results expressed as percentage differences. Upright line with bars ± S. D. (Data from Zemplényi and Mrhová, 1963).

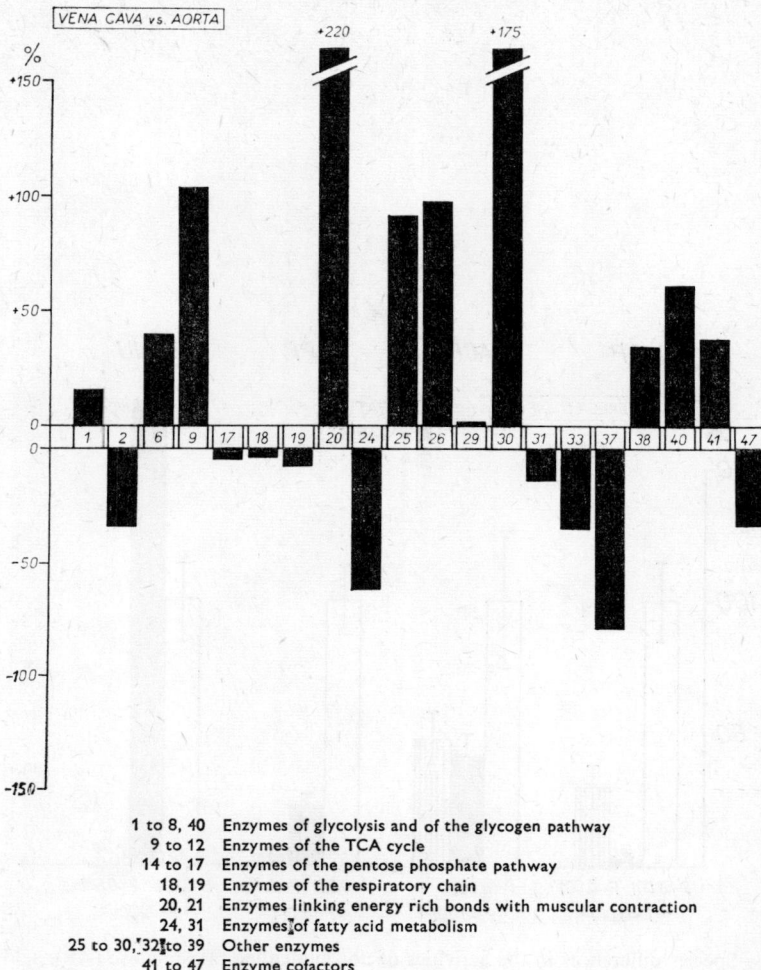

FIG. 13.—Differences in enzyme activities and cofactor levels between the human aorta and inferior vena cava. The enzyme activities or cofactor levels of aortas, calculated on a wet weight basis = 100 per cent. (Constructed according to data from Kirk, 1964b and some more recent publications by Kirk and co-workers.) For symbols see Fig. 9(a) and (b).

Chapter XI

NOTES ON METHODS USED IN THE AUTHOR'S LABORATORY FOR STUDIES ON VASCULAR WALL ENZYMES*

THE enzyme activity of the vascular wall can be assessed by both biochemical and histochemical methods. The next chapters will be concerned mainly with results obtained by biochemical techniques. In addition, however, some histochemical findings will be mentioned, because it is our experience that they are often of great importance in the proper evaluation of many biochemical data.

We must realize that all organs, including the arterial wall, are heterogeneous, being composed of different tissues. The cells of these tissues possess different metabolic activities. In these tissues there is also a variable amount of extracellular material, such as ground substance and connective tissue fibres, where enzyme activity is lacking. Estimation of enzyme activities by conventional biochemical techniques (in homogenates, extracts, etc.) yields valuable quantitative data about overall activity in tissues. However, due attention must be given to the above-mentioned problems of extracellular and metabolically inert substances and calculations should be based, for example, on the deoxyribonucleic acid (DNA) or soluble protein content of the tissue. The histochemical approach, although yielding less reliable quantitative data, can provide valuable information about enzyme activity at the cellular level. Careful comparative analysis of findings given by both approaches eliminates many of the inherent limitations in each of them and provides the most reliable results.

In our studies concerned either with human vessels (obtained fresh at autopsy) or animal vessels (removed immediately after death), we used the following methods: the vessels were carefully cleaned of perivascular and adventitial tissue with the exception of those parts that were destined for histological and histochemical study. After rinsing the vessel several times in ice-cold saline to remove as thoroughly as possible all adherent blood, samples of 50–200 mg. wet weight (depending on the species and vessel studied) were sectioned on a freezing microtome at 20–30 μ and were then prepared as a 1–2 per cent homogenate in ice-cold saline (adjusted to pH 7·0) using a Potter-Elvehjem glass homogenizer. The homogenate was stored—if needed—at 4°C and, after centrifugation at 3,000 rpm for 5 minutes, the supernatant crude extract was used for determining certain enzyme activities.

In some studies on dehydrogenases the preparation of vascular tissue was carried out in a different way. Samples of about 20–30 mg. weight were frozen in liquid nitrogen, immediately pulverized and quantitatively transferred to test tubes. However, in most studies only succinate dehydrogenase activity was

---

* The biochemical aspects of this chapter and of Chapters XII to XVIII were carried out in co-operation with Olga Mrhová, Chem. Eng., C.Sc.

assessed in such preparations. For other dehydrogenase assays the above-mentioned crude extract was used.

These procedures proved particularly satisfactory for comparative purposes, and allowed the simultaneous determination of a fairly large number of enzyme activities. This is an important advantage especially in experiments with smaller laboratory animals where only a very small sample of vascular tissue is usually available.

All enzyme activities were calculated either on the basis of the saline extractable protein content of the homogenate (in some cases nitrogen content) or on the deoxyribonucleic acid content of the vessel wall.

Before presenting details of the actual enzymatic assays, some general remarks are necessary about "optical tests" and tetrazolium methods.

GENERAL PRINCIPLES OF "OPTICAL TESTS"

The "optical tests" to be described and most of those mentioned in part one, are based on the fact that the coenzymes NAD and NADP exhibit an absorption band at 340 mμ. or 366 mμ. in the reduced but not in the oxidized state (Warburg and Christian, 1936, see p. 73). Consequently the enzymatic interconversion of reduced and oxidized nicotinamide-containing coenzymes can be very conveniently followed by means of the spectrophotometer. This enables the rate of the reaction catalysed by an NAD or NADP-dependent enzyme to be measured. An excess of both coenzyme and substrate must be present and the reaction must be followed under optimal pH and temperature conditions. However, some enzymes are inhibited by a large excess of substrate, so the optimal concentration has to be used to obtain estimations of the initial reaction rate that is a reliable measure of enzyme activity.

In some cases, as for example in the lactate dehydrogenase assay, the situation is very simple, because the oxidation of NADH is a direct measure of the reduction of pyruvate to lactate. (Such activity is usually expressed in units related to changes in NADH extinctions at 340 mμ.).

However, the enzyme-catalysed reaction to be measured (the "primary" reaction) has often to be coupled with another "indicator" reaction catalysed by an NAD or NADP-dependent enzyme:

$$A \to B$$
(Primary reaction catalysed by the enzyme to be determined)

$$B + \text{reduced NAD(P)} \to C + \text{NAD(P)}$$
(Reaction catalysed by the NAD- or NADP-dependent "indicator" enzyme)

The indicator reaction *indicates* the amount of the reaction product B of the primary reaction.

Bergmeyer (1963) emphasized that more than a 100-fold excess of the "indicator" enzyme is required to achieve a sufficiently rapid reduction of product B by the reduced coenzyme.

In some even more elegant assays a third "auxiliary" reaction between the primary and indicator reactions must be inserted (see for example creatine kinase, p. 46). Such assays are, of course, rare because it is difficult to achieve the appropriate ratios of enzyme activities (specific activities). (For further details of the general principles of "optical tests", see Bergmeyer, 1963).

NOTES ON METHODS USING TETRAZOLIUM SALTS FOR DETERMINATION OF DEHYDROGENASE ACTIVITIES

The common catalytic action of dehydrogenases is the transfer of hydrogen atoms (electrons) from a substrate (hydrogen-donor) to a hydrogen-acceptor (usually a cofactor of the nicotinamide or flavin type, see Chapter IX.).

In oxidation systems the transfer of hydrogen atoms (electrons) from the reduced cofactors to oxygen takes place through intermediate substances that act as carriers. (Most authors use the term "electron transport chain" for the whole mechanism. According to Dixon and Webb the theory on which the "electron" terminology is based lacks concrete evidence and should be considered as misleading. In this book we use both terminologies.)

In place of the biological carriers, hydrogen atoms can be transferred to artificial acceptors possessing a suitable redox potential (see Chapter V). The reduction of such artificial hydrogen-acceptors (usually dyes) is the basis of certain methods for determining dehydrogenase activity. Compounds such as phenazine alkylsulphates or 2,6 dichlorophenol indophenol are directly reduced by the cofactors, but others (for example methylene blue or tetrazolium salts) are reduced through intermediate carriers ("diaphorase", see p. 42) or carrier systems of the respiratory chain (see p. 39).

As histochemists employ various tetrazolium salts to display dehydrogenase activity, we tried in our studies to use biochemical methods based on a similar principle in order to make the two approaches as comparable as possible.

For our purposes 2, 2'-5, 5'-tetraphenyl-3, 3'-(p-biphenylene) ditetrazolium chloride (known as neotetrazolium, NT) proved to be very satisfactory. In older histochemical techniques tetrazolium salts were used without the addition of artificial acceptors such as phenazine methosulphate; for this reason in some of our early experiments on experimental rabbit atherosclerosis we also omitted phenazine methosulphate in dehydrogenase assays (see Chapter XIII).

For the proper interpretation of either biochemical or histochemical results obtained by such techniques the following reservations must be made. The precise sites of tetrazolium reduction in the hydrogen carrier system of biological oxidation is at present uncertain. For example Lester and Smith (1961) showed that in beef heart mitochondrial preparations the reduction of the tetrazolium salt is carried out directly by $NADH_2$ cytochrome c reductase (E.C. 1.6.2.1., see p. 42) or by another carrier of the hydrogen transport chain near to the dehydrogenase. However, in crude preparations, such as ours, the situation is more complex.

Nachlas et al. (1960a) investigated this problem in connection with the succinoxidase system (see p. 31). They studied the sites of hydrogen transfer to six commonly used tetrazolium salts, which mainly differ in redox potentials. These sites were tested by the effects of special inhibitors or incubating conditions, such as inhibition by antimycin A (see Fig. 5, p. 40), removal of cytochrome c by 0·85 per cent saline solution, inihibition by cyanide and anaerobiosis. Purified soluble succinate dehydrogenase was also used. These studies revealed that in homogenates neotetrazolium probably receives the majority of available electrons from reduced cytochrome oxidase (cytochrome a.$_{(3)}$.), with a small contribution from cytochrome c.

This finding indicates that for neotetrazolium the site of reduction is comparatively far along the hydrogen carrier system from the dehydrogenase (see Fig. 5, p. 40).

The situation is somewhat more favourable with the nitro-substituted tetrazolium salts (e.g. INT and nitro-BT) which are reduced in homogenates even in the presence of antimycin A, but are not reduced in these circumstances by soluble succinate dehydrogenase. The explanation appears to be that in the succinoxidase system these tetrazolium salts are reduced by cytochrome b or the succinate dehydrogenase-cytochrome b complex.

There is little doubt that, in cruder preparations as well as in work with histological sections, more intermediate respiratory carriers are involved in electron transfer from succinate dehydrogenase (reduced NAD[P] in the case of NAD- or NADP-linked dehydrogenases) to the tetrazolium salt. These intermediate carriers might be rate-limiting in dehydrogenase assays using tetrazolium salts and, therefore, in our studies we referred only to "dehydrogenase systems" whenever only neotetrazolium was used as the hydrogen acceptor (see also Zemplényi et al., 1963c; Mrhová et al., 1963a).

As mentioned before, phenazine alkylsulphates are directly reduced by reduced flavin- or nicotinamide-containing cofactors. The hydrogen atoms of the reduced forms of the phenazines can immediately be transferred to the tetrazolium salts. This was confirmed in Nachlas et al.'s (1960a) experiments: in the presence of phenazine methosulphate (PMS) antimycin A does not interfere with reduction of tetrazolium salts. In addition, with purified soluble succinate dehydrogenase all tetrazolium salts are reduced through the interaction of PMS, whereas no reduction occurs when it is absent from the reaction mixture

In view of these observations, in the great majority of dehydrogenase assays to be reported we used phenazine methosulphate in addition to neotetrazolium salt. In these circumstances it is reasonable to regard the results obtained as reflecting true dehydrogenase activities. This conclusion agrees with experience obtained in the biochemical assay of lactate dehydrogenase (Nachlas et al. 1960b), glycerolphosphate dehydrogenase, (Fried et al. 1961) and other dehydrogenases, where tetrazolium salts (usually INT) were successfully used with phenazine methosulphate as an intermediate hydrogen carrier.

For individual enzymes assay methods were used as follows.

Alkaline Phosphatases, Alkaline Phosphomonoesterases

There are many methods for determining alkaline phosphatase activity. The substrates that are most often used are p-nitrophenylphosphate, β-glycerophosphate, disodium phenylphosphate, phenolphthalein diphosphate and β-naphthyl phosphate. The assays are based on measurement of either the phosphate liberated or the dephosphorylated residue.

We used the slightly modified method of Kaplan and Narahara (1953a) in which disodium phenylphosphate serves as substrate and the phenol residue is measured by coupling with a diazo reagent.

The incubation mixture consists of 1·9 ml. of the buffered substrate solution (i.e. 0·005 M disodium phenylphosphate in a borate buffer at pH 9·8), 0·1 ml. of 20 per cent $MgCl_2$ and 1 ml. of the tissue extract. Incubation was carried out in a 37° C water bath for 30 or 60 min.—depending on the blood vessel studied—and the reaction was stopped by adding either 2 ml. of 1·5 M formaldehyde solution or 2 ml. of 0·25 M NaOH solution. Blank estimations were run and handled precisely in the same way, with the exception that either the tissue extract was added after the incubation had been stopped or the extract was substituted by saline during the whole incubation period. Afterwards, 4 ml. of a saturated alcoholic borax solution (35 g. borax/150 ml. 95 per cent ethyl alcohol/850 ml. water) was added to each sample or blank and the colour reaction developed by adding 1 ml. of freshly prepared diazo reagent (0·25 per cent *para*nitroaniline, water and 0·2 per cent $NaNO_2$: v/v/v/ = 1:2:1). After 10 minutes the amount of phenol liberated was determined spectrophotometrically at 445 mμ., using a standard curve obtained from increasing concentrations of phenol solution.

Activity was expressed in μM of phenol liberated per hour per 100 μg. of protein (or nitrogen in a few experiments).

Acid Phosphatase, Acid Phosphomonoesterase

The methods for determining activity of this enzyme are based on the same principles as those for the preceding enzyme.

We used Kaplan and Narahara's (1953b) procedure with slight modifications. This assay differs from the determination of alkaline phosphatase in that a different buffer solution is used, namely 0·05 M acetate buffer of pH 5·0, and the incubation is stopped by the addition of 2 ml. of a saturated sodium fluoride solution. However, in place of fluoride, we used 2 ml. of a 0·25 M solution of NaOH. The further steps were identical with those in the preceding method and activity was expressed in the same way.

5'-Nucleotidase

The assay methods for this enzyme are mainly based on determining inorganic phosphate released from adenosine 5'-monophosphate (AMP).

In our studies we used the slightly modified procedure of Ahmed and Reis (1958). For each estimation 1·2 ml. of 0·05 M veronal-HCl buffer at pH 7·5, 0·2 ml. of 0·003 M $MgCl_2$ or $MnSO_4$, and 0·2 ml. of 0·01 M adenosine 5'-monophosphate were mixed in a test-tube. After equilibration in a water bath at 38° C for 5 minutes, 0·4 ml. of the tissue extract was added and the sample incubated for 30 minutes. Enzyme activity was stopped by adding 4 ml. of 5 per cent trichloracetic acid. After 5 minutes the sample was filtered through a Schleicher and Schnell filter No. 589 and a 5 ml. aliquot of the filtrate was used for estimation of free phosphate by the method of Fiske and Subba Row (King and Allott's modification, 1947). An unincubated blank, containing the same constituents, was run with each sample. The results were expressed as μg. of phosphate liberated per hour per 100 μg. of protein or nitrogen in the extract.*

ADENOSINETRIPHOSPHATASE, ATPASE

The assay methods for this enzyme depend on colorimetric estimation of inorganic phosphate liberated by the enzyme from ATP under specified conditions.

We estimated the activity of this enzyme in essentially the same way as 5'-nucleotidase; the difference being that 1·4 ml. of a veronal-acetate buffer at pH 9·0 and 0·2 ml. of 0·0075 M ATP (adenosine 5'-triphosphate) were used. Only 0·2 ml. of tissue extract was added in each test or blank estimation. The results were expressed as for the previous enzyme.

"NON-SPECIFIC" CARBOXYLESTERASE

The physiological substrates for most of these ubiquitous "simple" (non-specific) esterases are as yet unknown.

In histochemical work, it is customary to use substrates such as α-naphthol acetate, naphthol AS acetate (i.e. 2-hydroxy-3-naphthoic acid anilide acetate) and indoxyl acetate for detecting these enzymes. For purposes of easy comparability with these histochemical techniques we used the slightly modified method described by Seligman et al. (1949) as follows: 10 mg. of β-naphthol acetate, synthesized by the method of Nachlas and Seligman (1949a), were dissolved in 2 ml. of acetone and tipped into a swirling mixture of 20 ml. veronal buffer (at pH 7·4) and water to a final volume of 100 ml. This solution served as substrate; the incubation vessels contained 2·5 ml. of the freshly prepared substrate solution, 0·5 ml. of tissue extract and 1·5 ml. of water or 10^{-5} M aqueous physostigmine. Incubation was carried out for 30 minutes in a water bath at 27° C. At the end of this period, 1 ml. (4·5 mg.) of a freshly prepared cold solution of Fast Blue B or benzidine tetrazonium fluoroborate (Lojda, 1958b) was added with shaking to each incubation vessel. An azo dye formed immediately; after adding 1 ml. of 40 per cent trichloracetic acid the dye was extracted by vigorous shaking with 10 ml. ethyl acetate. The mixture was centrifuged for 15 min. at 1500 rpm and the supernatant was then measured spectrometrically at 490 mμ. against a blank treated in the same manner but containing saline instead of the aortic extract. A calibration curve was prepared from dilutions of pure β-naphthol and used for estimating the liberated naphthol.

Activity was expressed in μg. of β-naphthol liberated per 1 mg. of protein or nitrogen in the extract. When working with acetone powder, activity was expressed per 100 mg. of powder (see Chapter XXIII).

---

* The Committee on Enzymes of the IUB recommends the use of "standard units". This is the amount of any enzyme that catalyses the transformation of 1 μM of the substrate per min. under standard conditions (at 30° C., at optimal pH and optimal substrate concentration). In more recent work—wherever possible—we also express our results in such units.

Lactate Dehydrogenase (Optical Test)

We used Kubowitz and Ott's (1943) method as modified by Bergmeyer et al. (1963). It is based on the estimation of changes in the concentration of $NADH_2$ estimated photometrically at 340 or 366 mμ. according to Warburg and Christian's principle (see above).

The following solutions were prepared according to Bergmeyer et al.'s (1963) specifications:

i. Either 2·75 or 2·85 ml. of $3·1 \times 10^{-4} M$ pyruvate in 0·05 M phosphate buffer (pH 7·5) was pipetted into the 1 cm. cuvette of a Zeiss universal spectrophotometer VSU_1.

ii. Then 0·05 ml. of 8×10^{-3} M $NADH_2$ was likewise added.

iii. Next either 0·1 or 0·2 ml. of tissue extract was added to make a final volume of 3 ml.

Finally optical density was read at 1 minute intervals for 5 minutes. Activity was calculated from the extinction coefficient of $NADH_2$ ($6·3 \times 10^3$ M^{-1} at 340 mμ. and $3·4 \times 10^3$ M^{-1} at 366 mμ.). From the optical density difference activity was expressed in Wróblewski and LaDue's (1955) units, i.e. the amount of lactate dehydrogenase in 1 ml. that changes the optical density of $NADH_2$ at 340 mμ. by 0·001 in 1 min.). The results were referred to the protein content of the extract. (For calculation of units see p. 105).

Malate Dehydrogenase (Optical Test)

We used the assay method described by Ochoa (1955) and modified by Bergmeyer and Bernt (1963a). It is based on the estimation of changes in $NADH_2$ concentration (read at 340 or 366 mμ.) during the conversion of oxaloacetate to L-malate (see p. 33). As oxaloacetate is unstable in aqueous solution, oxaloacetate is enzymatically produced—just before activity determination—by the action of aspartate aminotransferase on 2-oxoglutarate and L-aspartate (see p. 57).

The following solutions, prepared according to Bergmeyer and Bernt (1963a),* were successively pipetted into the 1 cm. cuvette of the spectrophotometer:

0·05 ml. 2-oxoglutarate (0·06 M)
0·05 ml. $NADH_2$ (0·012 M)
0·05 ml. of an aspartate aminotransferase suspension (Boehringer, 1 mg. protein per ml.)
2·55–2·65 ml. of 0·042 M aspartate in 0·1 M phosphate buffer, pH 7·4

After 8–10 min. incubation at 25° C, 0·2–0·3 ml. of tissue extract was added to make a final volume of 3 ml. Optical density was then read at 1 minute intervals for 5 minutes. Activity was calculated in Wróblewski units per 100μg. protein in the extract—as for lactate dehydrogenase.

* Complete reagents as premixed solutions or pre-weighed dry substances are available commercially, e.g. from Boehringer.

TETRAZOLIUM METHODS FOR DETERMINATION OF DEHYDROGENASE ACTIVITIES

Neotetrazolium hydrochloride (NT) was used as electron acceptor and the amount of diformazan produced after incubation, was estimated photometrically.

(a) "NT" method for dehydrogenase systems (see p. 97).

In all determinations the following materials were added to the incubation vessels:

1·5 ml. of Krebs-Ringer phosphate buffer (pH 7·4); 1·5 ml. of the appropriate substrate (0·2 M sodium lactate, succinate, malate or glycerol 3-phosphate), 2 ml. of 0·5 per cent NT and approximately 10 mg. of aortic powder prepared by means of liquid nitrogen. In assays of the lactate dehydrogenase, malate dehydrogenase and glycerol-3-phosphate dehydrogenase systems 0·5 ml. of 0·002 M freshly prepared NAD, and 0·5 ml. of 0·1 M NaCN were also added. After incubation under nitrogen for 45 minutes at 38° C and centrifugation at 3000 rpm for 5 minutes, the diformazan was extracted for 20 hours in two changes of 5 ml. redistilled acetone. The amount of diformazan was spectrophotometrically estimated at 470 mμ. from a standard calibration curve. Activity was expressed in μg. of diformazan per 100 μg. of aortic deoxyribonucleic acid (DNA).

(b) "NT-PMS" method for dehydrogenases

The same reagents as described above were added to the incubation vessels, into which was pipetted 0·2–0·4 ml. of the aortic extract as well as 0·2 ml. of an 0·001 M solution of phenazine methosulphate (see p. 98). For determining succinate dehydrogenase approximately 10 mg. of aortic powder, prepared by means of liquid nitrogen, was used instead of the aortic extract. Activities were expressed in μg. of diformazan per 100 μg. protein in the aortic extract or, in the case of succinate dehydrogenase, in μg. of diformazan per 100 μg. of aortic DNA.

LACTATE DEHYDROGENASE ISOZYMES

(a) Separation by paper electrophoresis

We originally used the slightly modified method described by Raabo (1963) and we adapted it for tissue extracts. The principle of the method consists in the staining of electrophoretically separated LDH fractions by means of a tetrazolium salt. Recently, Homolka et al. (1966) introduced a modification where "nitroblue monotetrazolium" (NBMT, i.e. 2-p-nitrophenyl-5-phenyl-3-3, 3'-dimethoxy-4-diphenylene tetrazolium chloride) was used for staining. (For preparation of NBMT see Večerek and Křištanová, 1967; Večerek et al., 1967). We found that this modification is useful, because it enabled us to use only small amounts of phenazine methosulphate (PMS) and because the reaction product could be readily eluted by acetone and estimated photometrically.

A 20 per cent aortic homogenate was prepared for electrophoretic separation. After centrifugation at 3000 rpm for 10 minutes, 80–100 μl. of the supernatant were applied to a 3 cm. wide Whatman 3 paper strip and electrophoretic separation was then carried out at 6 volts per cm. in a barbital buffer (pH 8·6) for 4–5 hours.

After separation (and without previous drying) the paper strip is placed in the solution that is used for the histochemical detection of dehydrogenases (Pearse, 1960; Seligman, 1963; Lojda, 1965a). The stock solution has the following composition (Lojda, 1965a):

Tris — HCl (0·2 M, pH 7·2)	25 ml.
$MgSO_4$ (0·05M)	10 ml.
KCN (adjusted with N HCl to pH 7·2) 0·1 M	10 ml.
Distilled water	10 ml.

(This solution can be stored in the refrigerator for about two months.)

On the day of the experiment 25 ml. of an 0·1 per cent NBMT solution were prepared by dissolving 25 mg. of NBMT in 2·5 ml. of dimethylformamide and tipping this into 25 ml. of water. After thorough shaking the NBMT solution was added to 55 ml. of stock solution.

Immediately before the enzymatic reaction was started, 0·1 ml. of 0·2 M Na lactate, 0·1 ml. or 0·06 per cent phenazine methosulphate and 1 mg. of NAD were added per ml. of incubating medium. After 30 min. incubation at 37° C, the reaction was stopped by transferring the paper strip to a bath containing 5 per cent acetic acid. After rinsing in distilled water the strip was blotted with filter paper, the fractions were cut out and eluted for 60 min. in 3 ml. of acetone. Optical density was measured on a Unicam SP. 500 spectrophotometer at 490 mμ. against a blank paper eluate, and from the values obtained the percentage distribution of the individual isozymic fractions was calculated.*

(b) Differentiation by means of DEAE-Sephadex

We used the chromatographic "batch technique" described by Hess and Walter (1960, 1961) and Bergmeyer *et al.* (1963). It is based on the fact that, under defined conditions, DEAE-cellulose adsorbs the "heart-type" LDH isozyme protein (H subunit, see Chapter XVI), whereas the muscle-type LDH isozyme protein (M subunit) remains in solution. The LDH isozymes reveal analogous properties when DEAE-Sephadex is used as adsorbens (Wachsmuth and Pfleiderer, 1963).

In the investigations to be described, 1 ml. of the aortic extract—the supernatant of a 2 per cent homogenate—was mixed with 1 ml. of a 5 per cent suspension of DEAE-Sephadex in phosphate buffer (0·002 M, pH 7·5); Sephadex $_{A-50}$ Fine (Pharmacia, Uppsala, Sweden) was used for this purpose. The mixture was allowed to stand for 10 min. with occasional gentle stirring, was then centrifuged at 3000 rpm for 10 min. and, finally, the LDH activity in the supernatant was measured by the NT-PMS method (see p. 102). Aliquot control samples of the aortic extracts were usually diluted with 0·7 ml. of phosphate buffer (0·02 M, pH 7·5) and the LDH activity was measured as above. The dilution of the control samples was determined by the water-binding

---

*The NBMT used in our studies was kindly synthesized by Drs. Večerek and Štěpán from the First Institute of Medical Chemistry, Charles University, Prague.

capacity of the batch of DEAE-Sephadex, 1 ml. of which usually bound 0·3 ml. of water. (See also Ševela and Továrek, 1964.)

The difference between the activities of the Sephadex-treated and control samples indicates the amount of LDH activity adsorbed, i.e. the fast moving electrophoretic fractions that predominantly contain H subunits (see Chapter XVI).

To check the method, extracts of either 20 per cent aortic homogenates or 10 per cent heart homogenates were adsorbed on DEAE-Sephadex in the same way as above; the supernatants were then subjected to paper electrophoresis and LDH isozyme activity determinations as already described. Control aliquots of extracts were run simultaneously and the results compared both qualitatively and quantitatively.

Results of a typical experiment are presented in Figs. 50 and 51.

β-GLUCURONIDASE

The activity of this enzyme was assessed by the method of Talalay et al. (1946) with the slight modifications introduced by Dyrbye and Kirk (1956). The incubation tubes contained 0·1 ml. of phenolphthalein β-glucuronide (0·01 M) prepared from its cinchonidine salt, 0·8 ml. acetate buffer (0·1 M, pH 4·5), 0·5 ml. of the aortic extract and a small thymol crystal. Incubation was carried out overnight (16 hours) in a water bath at 38° C. The reaction was stopped with 1 ml. of 5 per cent trichloroaecetic acid. After centrifugation at 3000 rpm for 10 min., the supernatant was transferred into another test tube containing 0·5 ml. of 0·5 M sodium hydroxide and 2 ml. of a 0·2 M glycine buffer (pH 10·45). After washing the precipitate 3 times, the volume of the sample was adjusted to 6 ml. with redistilled water and then the amount of phenolphthalein liberated was estimated photometrically at 550 mμ. by the red colour produced at alkaline pH. An 0·01 per cent solution of phenolphthalein was used as a standard and enzyme activities were expressed in μg. of phenolphthalein liberated per 100 μg. protein or nitrogen in the extract per 16 hours.

AMINOTRANSFERASES

(a) Aspartate aminotransferase

This enzyme was estimated by either Reitman and Frankel's (1957) colorimetric method or Karmen's (1955) optical test as modified by Bergmeyer and Bernt (1963b).

The colorimetric method is based on the interaction of oxaloacetate (or pyruvate—see under b) and 2, 4-dinitrophenylhydrazine, whereby 2, 4-dinitrophenylhydrazone is produced and the amount of the latter is measured photometrically.

Incubation vessels containing 1 ml. of substrate (i.e. 0·002 M of 2-oxoglutarate and 0·2 M of L-asparate in a 0·1 M phosphate buffer, pH 7·4 were preheated for 10 min. in a water bath at 37° C. After adding 0·2 ml. of the aortic extract, incubation was carried out for 60 min. The enzymatic process was stopped by adding the ketone reagent—an 0·001 M solution of 2, 4-dinitrophenylhydrazine in N HCl. Blank samples were prepared in the same way but were not incubated. After stopping the reaction, all the samples were allowed to stand for 20 minutes at room temperature and then 10 ml. of 0·4 N NaOH are added to the experimental and control tubes. After 30 minutes the optical densities of the experimental tubes were read against the blanks at 510 mμ.

A standard curve was prepared using sodium pyruvate. Activity was expressed in Karmen units (see below) according to the data provided by Reitman and Frankel (1957) or by Bergmeyer and Bernt (1963b). Results were calculated in units per 100 μg. of protein or nitrogen in the extract.

The optical test is based on the estimation of oxaloacetate—the product of the enzyme catalysed reaction (see p. 57)—by means of an indicator reaction catalysed by malate dehydrogenase:

$$\text{Oxaloacetate} + \text{NADH}_2 \rightleftharpoons \text{malate} + \text{NAD}$$

The decrease in the amount of NADH_2, which is proportional to the oxaloacetate produced, is measured spectrophotometrically at 340 or 366 mμ.

The following solutions, which are prepared according to Bergmeyer and Bernt (1963b) from Boehringer reagents (see footnote p. 101), were successively pipetted into 1 cm. spectrophotometer cuvettes to make a final volume of 3 ml.:

2.10–2.30 ml. of 0·25 M aspartate in 0·1 phosphate buffer (pH 7·6)
0·05 ml. of NADH_2 (0·012 M)
0·05 ml. of a malate dehydrogenase suspension (Boehringer, 0·5 mg. protein per ml.)
0·50–0·70 ml. of aortic extract.

After equilibration at 25° C for 5 minutes, 0·10 of 2-oxoglutarate (0·02 M, adjusted to pH 7·6) was added to the cuvette and optical densities were read at 2 minute intervals over 10 minutes.

By analogy with serum determinations, activity can be expressed in Karmen units. This unit is the amount of aminotransferase in 1 ml. of the aortic extract that decreases the optical density of NADH_2 at 340 mμ. by 0·001 in 1 minute, when measured in a 3 ml. assay mixture at 25° C. With 0·5 ml. of extract the activity can be calculated as follows:

Units/ml. extract = (ΔE 340/min.) × 1000 × 2

However, the protein content varies in these extracts and for purposes of comparison the activity must be related to this factor. Therefore a modified unit (i.e. the amount of enzyme per 100 μg. of protein that decreases the optical density of NADH_2 at 340 mμ. by 0·001 in 1 min.) can be obtained by multiplying the above value by $\dfrac{100}{P}$, where P denotes the protein content (in μg.) of 1 ml. of the extract:

$$\text{Units/100 } \mu\text{g. protein} = (\Delta\text{E 340/min.}) \times \frac{2 \times 10^5}{P}$$

For measurements at 366 mμ. (see Bergmeyer and Bernt) it is necessary to multiply by 1·89 to take account of the different extinction coefficients of NADH$_2$ at 340 and 366 mμ.

(b) Alanine aminotransferase

The colorimetric method and the optical test for the assay of this enzyme are based on the same principles used in the estimation of the preceding enzyme.

In the colorimetric method a solution of 2-oxoglutarate (0·002 M) and DL-alanine (0·2 M) in phosphate buffer (0·1 M, pH 7·4) was used as substrate. The incubation period was only 30 min. The procedure is essentially the same as with aspartate aminotransferase.

In the optical test pyruvate, the product of the reaction, is estimated by means of an indicator reaction catalysed by lactate dehydrogenase.

The following solutions, prepared according to Bergmeyer and Bernt (1963c) from Boehringer reagents (see footnote p. 101), were successively pipetted into 1 cm. spectrophotometer cuvettes to make a final volume of 3ml.:

1·62–1·82 ml. of 0·11 M DL-alanine in 0·1 M phosphate buffer
0·04 ml. of NADH$_2$ (0·012 M)
0·04 ml. of a suspension of lactate dehydrogenase
            (Boehringer, 0·5 mg. protein per ml.)
1·0–1·2 ml. of aortic extract.

After equilibration at 25° C for 5 min., 0·1 ml. of 2-oxoglutarate (0·2 M) was added to the cuvette and optical densities were read at 2 min. intervals for a period of 10 min. Calculations were performed in the same way as for aspartate aminotransferase.

In our experience, however, the activity of this enzyme in arteries is very low, so that differences between vessels could not be reliably evaluated.

OTHER METHODS USED

The activity of glucosephosphate isomerase was determined in a few experimental series by Roe's method as modified by Slein (1955).

The method used for determining lipolytic activity is described in Chapter XXIII.

The protein content of the extracts was assessed by the method described by Lowry *et al.* (1951) The nitrogen content of the extracts was determined by a conventional Kjehldal procedure (Hořejší, 1948).

Deoxyribonucleic acid determinations were carried out according to the method of Schneider (1945) and Ceriotti (1952).

It is important in all the above enzymatic methods to check such aspects as optimum pH, optimum substrate concentration, proportionality of activity to tissue concentration and especially reproducibility.

It is beyond the scope of this volume to describe our *histochemical* methods in detail. In co-operation with Lojda and Urbanová the following methods were used:

Succinate dehydrogenase was demonstrated by Nachlas *et al.*'s (1957b) method, whereas the methods of Hess *et al.* (1958) and Lojda (1962, 1965a) were used for other dehydrogenases.

Non-specific esterases were detected by Gomori's (1952) and Gerebtzoff's (1953) methods, and by azo-coupling and indoxyl methods modified by Lojda (1958b).

Alkaline and acid phosphatases were detected by Gomori's method (1952) and by the azo-coupling method with α-naphthol phosphate, naphthol-AS-phosphate and naphthol-AS-MX-phosphate as substrates.

ATPase and 5'-nucleotidase were detected by Wachstein and Meisel's (1957) methods; ATPase was also demonstrated by Padykula and Herman's (1955) method.

β-glucuronidase was studied by Fishman and Baker's (1956) method using 8-hydroxiquinoline glucuronide as substrate, and aminopeptidase was detected by the method of Nachlas et al. (1957a).

Finally a few remarks must be made concerning the grading of experimental rabbit atherosclerosis. We could not stain our material with fat-soluble dyes, because the material for enzymatic analysis must be worked up as soon as possible; and such staining may interfere with enzymatic activities. Therefore, we were only able to determine the extent of the lesions by rough estimation.

The grading used was as follows (see Fodor et al., 1958b; Zemplényi et al., 1963c):

1. Changes seen only microscopically. These features included slight oedema of the intima, some extracellular intimal lipids and solitary lipophages.
2. Changes apparent macroscopically as slight fatty streaks or spots.
3. Small single plaques in the aortic ring and/or arch, and/or single plaques in the thoracic aorta.
4. Larger plaques in the arch and ascending part of the aorta with a tendency to confluence; single plaques in the descending part.
5. As for (4), but large confluent plaques in the descending part of the aorta.

Chapter XII

SEX DIFFERENCES

A TOPIC of great interest is the relationship between sex and atherosclerosis (and its complications).

It seems to be a well-established clinical experience that ischaemic heart disease and peripheral vascular disease are far more common among men than women, at least during middle age and in economically more developed countries. For example, in the Framingham study, Kannel et al. (1962) found that men, aged 30-39 years at the beginning of observation, developed new coronary disease at a rate of 21 per 1,000 over 8 years, whereas no new coronary disease was seen during this period in women of the same age. In the fifth decade the ratio of the incidence of ischaemic heart disease in men and women was 48:7, and in the sixth decade the corresponding figures were 94:20.

On the basis of data collected by Lukomsky and Tarajev, Myasnikov (1960) reported that in persons without hypertension the incidence of ischaemic heart disease is over three-fold higher in men than in women. Other authors report higher ratios, particularly in individuals below 40 years of age. However, in South African Bantu, Italians and Japanese, the sex differential ratio seems to be much lower and is sometimes only slightly above one. In the last two countries mentioned, the incidence as well as the prevalence of high grade ischaemic heart disease is said to be very low in both sexes. Stamler (1963) believes that such data are consistent with the "nutritional-metabolic" theory of atherogenesis, according to which a certain nutritional pattern is essential to bring about the metabolic and humoral prerequisites for atherogenesis.

It must be pointed out that morphological differences do not appear to be as striking as might be expected from the clinical data, and they are also somewhat contradictory. The detailed studies by Ackerman et al. (1950), White et al. (1950), Giertsen (1965), Falconer and Adams (1965) and many others indicate that atherosclerosis develops in men more quickly than in women and is more severe in men than women of the same age. The coronary arteries of women in the sixth and seventh decade exhibit a degree of atherosclerosis corresponding to that of men in the fourth or fifth decade. On the other hand, Roberts et al. (1959) claim that there is no difference in the degree and distribution of atherosclerosis between men and women after 40 years of age, but most authors who have autopsy experience would deny this view. Nevertheless, on the basis of autopsy studies, Roberts et al. also drew attention to the higher incidence of vascular catastrophes in men.

Maybe some of the discrepancies are due to the different criteria used for grading atherosclerosis. Mitchell and Schwartz (1965) found substantial sex differences in the distribution and perhaps also in the pathogenetic significance of the flat sudanophilic plaques (fatty streaks) as compared with the raised (sudanophilic and/or fibrous) plaques (see Chapter XVII). Mitchell et al. (1964)

reported that the area of fatty streaking in human aortas was unrelated to sex, whereas the area occupied by raised plaques exhibited a clear cut difference: women of a particular age showed arterial disease of the same severity as men who were 10–15 years older.

A relationship between sex (sexual hormones) and *coronary* atherosclerosis is also emerging from other morphological and clinical observations. Rivin and Dimitroff (1954) and Wuest *et. al.* (1953) found that, in women ovariectomized 1 to 42 years before death, the coronary arteries evinced a higher degree of atherosclerosis than in healthy women of the same age. On the other hand, the coronary arteries of elderly men, receiving prolonged high dosage stilboestrol treatment for prostatic carcinoma, revealed significantly less coronary atherosclerosis than a matched control group. In addition, the clinical studies of Oliver and Boyd (1959), Robinson *et al.* (1959), Ask-Upmark (1962), Sznajderman and Oliver (1963) and others indicate that bilateral ovariectomy in young women is followed by the premature development of ischaemic heart disease and, according to Robinson *et al.*'s (1959) study, by a significant increase of peripheral vascular disease.

The above sex differences and the effects of castration and of sexual hormones have been mostly explained in terms of differences or changes in circulating blood lipid levels, particularly of cholesterol, cholesterol/phospholipid ratios and lipoproteins (Katz and Stamler, 1953; Barr, 1953; Russ *et al.*, 1951; Jones *et al.*, 1951; Gertler *et al.*, 1953; Furman *et al.*, 1958; Marmorston *et al.*, 1958; Eder, 1958; Stamler, 1963 and many others).

There seems to be little doubt that sexual hormones can influence the spectrum of plasma lipids and that there are sex differences in circulating blood lipids. Nevertheless, other factors also play an important part in these differences between men and women. For example the predilection for the male coronary arteries is difficult to explain only by differences in plasma lipids. Dock (1946) attaches importance to his observation that the tunica intima of the coronary arteries is thickened (see p. 161) from early childhood in boys but not in girls. Although this finding has not been confirmed by all investigators in this field, it calls for a search for local vascular factors that may determine the higher susceptibility of males to atherosclerosis and its complications. One of these factors might be the difference in vascular metabolism between the two sexes and many aspects of these problems have been studied by Malinow's group in Buenos Aires (see Malinow, 1962). According to their findings (Malinow *et al.* 1961), gonadectomy increases aortic oxygen consumption in rats of both sexes, but this change is reversed after oestradiol or testosterone administration. In intact cockerels, Malinow and Moguilevsky (1961) observed a significant increase of arterial Qo_2 following the administration of oestradiol or stilboestrol, whereas testosterone was without effect. However, Rifkind and Munro (1963) were unable to confirm the influence of either sex hormones or gonadectomy on aortic oxygen consumption and could not detect any change in lactate production under these conditions (see Chapter I).

In our own studies on rats we paid attention to the relationship of sex and gonadectomy on the activity of some aortic enzymes (Mrhová and Zemplényi 1965). In a series of experiments concerned with gonadectomy, half of the animals were gonadectomized while the other half were sham-operated and

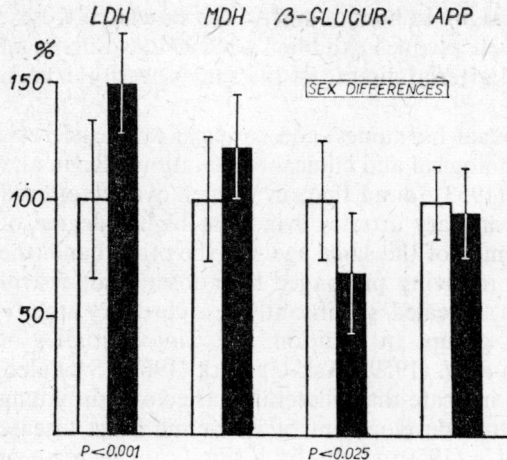

FIG. 14.—The activities of lactate dehydrogenase (LDH), malate dehydrogenase (MDH), ATPase (APP) and β-glucuronidase in 20 male (empty columns) and female (black columns) rat aortas. Upright line with bars ± S.D. Results expressed as percentage differences. (Data from Mrhová and Zemplényi, 1965.)

served as controls. Figure 14 shows higher activity of lactate dehydrogenase and a trend towards higher activity of malate dehydrogenase in the aortas of female rats. The activity of β-glucuronidase is higher in male rat aortas, but the activity of ATPase does not reveal any significant sex difference.

In other experiments (Fig. 15) significantly higher activity of alkaline phosphatase was found in the aortas of male rats and a significantly higher activity of 5'-nucleotidase was detected in the aortas of female rats. No difference in the activities of acid phosphatase and non-specific carboxylesterase was found between the two sexes.

In the experiments summarized in the next figures, animals 11–12 weeks old at the start of the study were used. Figures 16–18 demonstrate that 10 weeks after orchidectomy no clear-cut difference in the activities of the enzymes investigated could be detected as compared with sham-operated control animals. There was only a slightly significant increase of non-specific carboxylesterase

FIG. 15.—The activities of alkaline phosphatase (AP), acid phosphatase (ACP), 5'-nucleotidase (5'-Nu) and non-specific carboxylesterase in 20 male (empty columns) and female (black columns) rat aortas. Upright line with bars ± S.D. Results expressed as percentage differences. (Data from Mrhová and Zemplényi, 1965.)

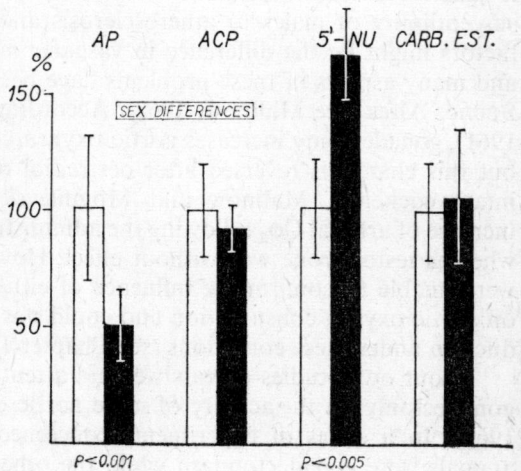

Fig. 16.—The activity of aortic ATPase and 5′-nucleotidase in rats 10 or 17 weeks after orchidectomy. Empty columns = sham operated animals (8). Black columns = orchidectomized animals (8). Upright line with bars ± S.D. Results expressed as percentage differences. (Data from Mrhová and Zemplényi, 1965.)

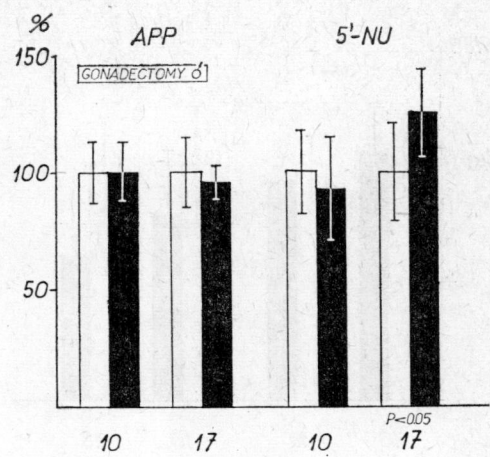

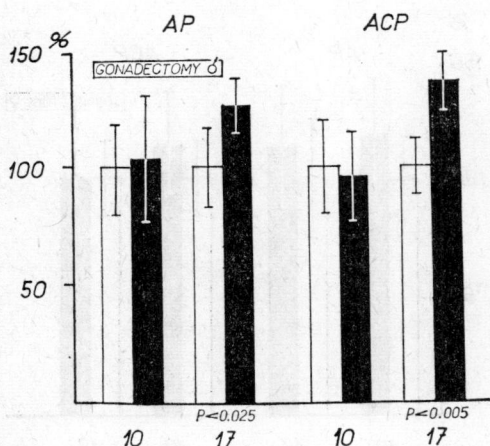

Fig. 17.—The activity of aortic alkaline and acid phosphatase in the same animals as in Fig. 16.

Fig. 18.—The activity of aortic malate dehydrogenase and non-specific carboxylesterase in the same animals as in Fig. 16.

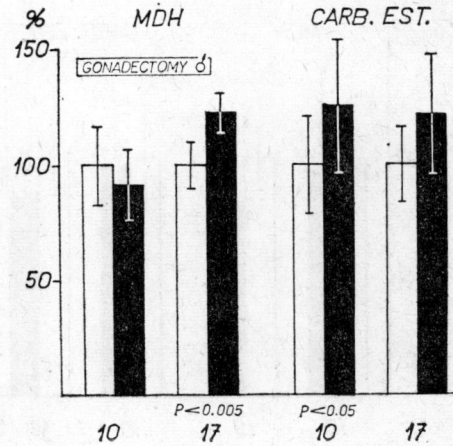

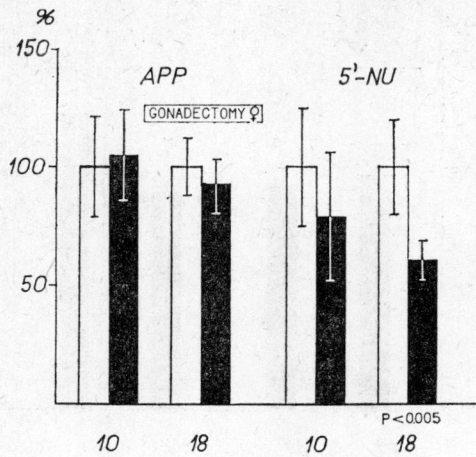

Fig. 19.—The activity of aortic ATPase and 5′-nucleotidase in rats 10 or 18 weeks after ovariectomy. Empty columns = sham operated animals (8). Black columns = ovariectomized animals (8). Upright line with bars ± S.D. Results expressed as percentage differences. (Data from Mrhová and Zemplényi, 1965.)

Fig. 20.—The activity of aortic alkaline and acid phosphatase in the same animals as in Fig. 19.

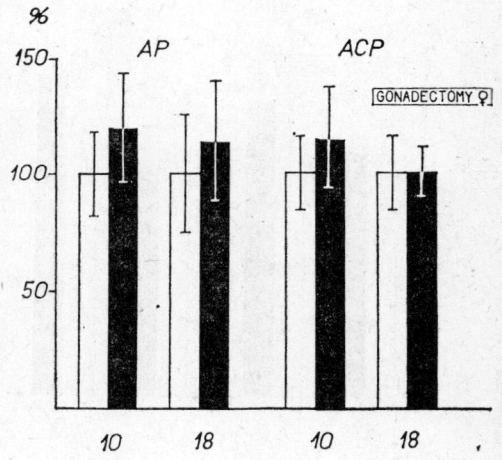

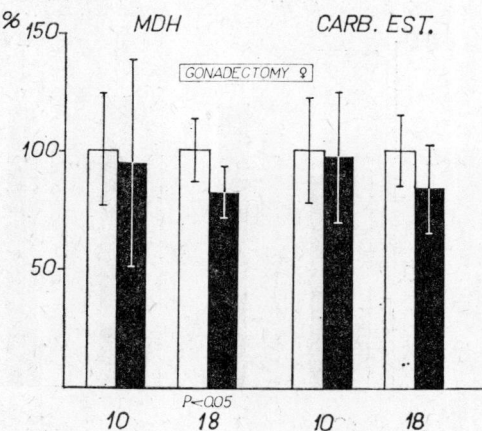

Fig. 21.—The activity of aortic malate dehydrogenase and non-specific carboxylesterase in the same animals as in Fig. 19.

activity in the aortas of the gonadectomized animals. However, in the aortas of animals killed 17 weeks after orchidectomy clear cut changes were apparent in some enzyme activities. The activity of alkaline and acid phosphatase and malate dehydrogenase was unequivocally higher in the aortas of castrated male rats than in those of the control animals, and the activities of 5′-nucleotidase and carboxylesterase showed a tendency to increase in the aortas of castrated animals. No effect of castration could be demonstrated on aortic ATPase activity.

It appears that the changes of enzyme activities after orchidectomy do not depend only on the time interval between the operation and the killing of the animals. In another similar series of experiments, where male rats only 2 months of age were used, the aortas exhibited a definite tendency towards increased activity of phosphomonoesterases as early as 10 weeks following castration. In all probability, the age of the animals at the time of gonadectomy is one of the decisive factors that determines the onset of enzymatic changes in such experiments.

Results obtained in experiments with gonadectomized female rats are presented in Figs. 19–21.The activity of aortic 5′-nucleotidase tends to decline at 10 weeks ofter ovariectomy. Later, i.e. 18 weeks after ovariectomy, the activity of 5′-nucleotidase and malate dehydrogenase significantly decreases. The activities of non-specific carboxylesterase and aspartate aminotransferase (not shown in the figures) manifest a similar declining trend. No significant changes could be found in the activities of ATPase, alkaline and acid phosphatase either 10 or 18 weeks following ovariectomy.

From the above findings, it is clear that gonadectomy can lead to changes in activities of aortic enzymes and that some converse changes are found in male animals when compared with females. Thus, gonadectomy in female rats caused a drop and in males an elevation of some enzymatic activities. In this respect we have obtained pertinent data about aortic malate dehydrogenase, 5′-nucleotidase and, perhaps, non-specific carboxylesterase activities. One may assume that the activity of such enzymes is under the control of some antagonistic and well-balanced regulatory factors, probably of a hormonal character. Gonadectomy displaces the equilibrium of these factors and causes a shift towards the unaffected hormonal function.

As far as sexual hormones are concerned, these results are not unexpected because it is known that male and female organisms produce both androgens and oestrogens. The testes of the male horse are for instance very rich in oestradiol, while the ovary produces measurable amounts of androgens (quoted by Charvát, 1952).

It must be emphasized that the activities of the above-mentioned enzymes are either higher in female than in male rat aortas or their activities are equal in both sexes. The situation is quite different with the phosphomonoesterases. Alkaline phosphatase activity is higher in the male aorta and becomes even higher following gonadectomy, whereas in the female the activity does not change after such a procedure. These findings are more difficult to interpret but, at least in the male, they do not exclude the possibility of an antagonistic regulatory mechanism. (In this connection see Malinow's (1960) *in vitro* findings, which are mentioned below.) It is more likely, however, that changes in this enzyme activity reflect alterations in the connective tissue elements of the male rat's aorta, such as fibre formation and calcification. In fact some authors ascribe such functions to tissue alkaline phosphatase (see p. 126).

In animal experiments, the effect of sex, sexual hormones and gonadectomy on vascular enzymes has been very carefully investigated by Malinow and his co-workers. It was found that incubation of the chicken's aorta with oestradiol is accompanied by enhancement of its alkaline phosphatase activity (Malinow 1960). In a certain sense this finding corroborates the hypothesis that, in the castrated male, this enzyme activity increases under the influence of female sex hormones. However, it must be kept in mind that converse results are sometimes obtained with steroids when the *in vivo* situation is compared with that pertaining *in vitro* (Smith et al. 1953).

In their experiments concerned with the effect of gonadectomy on the rat's aorta, Malinow et al. (1962) investigated the activities of some enzymes other than those studied in our experiments. Following gonadectomy (60 to 90 days) they found decreased oxidation of fumarate in the male aortas, whereas increased activity of cytochrome oxidase and enhancement of oxidation of α-ketoglutarate were detected in the female aorta. The aortas of both males and females exhibited increased activity of succinate dehydrogenase. These findings do not seem to support the sex-linked converse changes in enzyme activity as observed in our investigations. However, in another study from Malinow's group (Lacuara et al. 1962), it was shown that, after gonadectomy, esterolytic activity against α-naphthylbutyrate as substrate decreases in male rat aortic extracts and increases in females. In the same laboratory Gerschenson et al. (1962) found in gonadectomized rats an increase of desoxypentose nucleic acid phosphorus in male aortas and a decrease in female aortas.

Unfortunately, it is difficult to compare our results with those of the latter studies, as not only were different enzymes estimated but the time interval between gonadectomy and killing were different. In addition, it must be stressed that we calculated our results on the basis of the protein content of the aortic extracts and not on the latter authors' wet weight basis.

Sex differences in enzyme activities of the rat's aorta have also been reported by Szendzikowski et al. (1961–1962), who claimed that the lipolytic activity of the female rat's aorta is about twice as high as that of the male. We shall return to these findings in part three of this book.

Although contradictory in some details, the above studies obviously indicate that the activity of arterial enzymes is under the influence of sex-linked factors.

When trying to extrapolate the results obtained in these animal experiments to the sphere of related human problems, much caution is needed. Thus, contrary to what happens in man, spontaneous arteriosclerosis and mediocalcinosis are more common in the female than male rat (Gillman and Hathorn, 1959 ; Wexler, 1964; Záhoř and Czabanová, 1965 and others). Cholesterolaemia is also higher in the female than male rat (Filios et al., 1958) and many other such differences could be quoted. Therefore, work along these lines that is directed towards human vessels is perhaps even more needed than in other fields where human and experimental atherosclerosis are being compared.

The results obtained in this connection by Kirk (1963*a*, 1964*a*) are very important. In the age group 18–54 years samples of human aortas, pulmonary arteries and coronary arteries were studied and the activities of many enzymes were compared in male and female vessels (Figs. 22 and 23). Male aortas exhibited significantly higher activity of two NADP-dependent enzymes, namely glucose 6-phosphate dehydrogenase and decarboxylating malate dehydrogenase; the activity of two further NADP-dependent enzymes (isocitrate dehydrogenase and decarboxylating phosphogluconate dehydrogenase) revealed a similar tendency ($P<0.1$). The same applied to two of the above four enzyme activities in the male coronary arteries. Two additional statistically significant sex differences in the aorta were observed in the activity of 5'-nucleotidase (higher in the female aortas—see similar results in animal experiments) and in the activity of purine nucleoside phosphorylase (higher in the male aortas). The significance of the sex differences in the latter two enzyme activities is not clear.

The increased activities of two and perhaps of all NADP-dependent enzymes in male aortas (and coronary arteries?), than in female vessels, is without doubt of great interest. The higher activities of enzymes producing reduced NADP is a distinctive feature of the male arterial wall, as pointed out by Kirk. This feature is particularly important because the reduced form of this coenzyme is intimately involved in the synthesis of cholesterol and fatty acids.

The effect of the above factors on vascular enzymes is, of course, not surprising and could be expected, because there are many reports that sex hormones or gonadectomy influence the activities of tissue enzymes. Valuable information about this topic is to be found in reviews by Kochakian (1947), Fishman (1951), Dorfman (1952, 1964) and especially in the text edited by Litwack and Kritchevsky (1964).

In connection with the well known anabolic effect of androgens, much attention has been given to the relationship between these substances and enzymes involved in protein biosynthesis. The activities of aspartate-glutamate transaminase, alanine-glutamate transaminase and tyrosine-glutamate transaminase substantially decrease in rat tissues (seminal vesicle, prostate, liver) following castration, and are restored by androgen treatment. However, in mouse and guinea-pig tissues no such changes can be observed (for details see the review by Frieden, 1964). Oestrogen administration causes transaminase activities to increase in the uterus of the intact rat.

In addition, recent evidence indicates that testosterone as well as oestradiol stimulate the DNA-dependent RNA-polymerase* contained in the cell nucleus of sensitive tissues (prostate or uterus, respectively). These hormones enhance the rate of messenger RNA synthesis, which plays a most important role in the incorporation of aminoacids into protein (see Chapter I). This is in line with the careful studies of Wilson (1962) which suggest that the most probable site of testosterone action on protein synthesis lies in the conversion of transport RNA-aminoacid complexes into microsomal ribonucleoprotein.

According to Villee and Hagerman (1958) oestrogens exhibit a specific action on transhydrogenases that effect the transfer of hydrogen between $NADPH_2$ and NAD in some tissues. However, good evidence seems to indicate that this transfer is mediated by specific "oestradiol dehydrogenases" that catalyse alternate oxidation and reduction of the hormone, which behaves in a coenzyme-like fashion (for details see the review by Williams-Ashman and Liao, 1964).

Sex hormones are also able to bring about alterations of the molecular

* RNA nucleotidyltransferase (E.C. 2.7.7.6.), see p. 3.

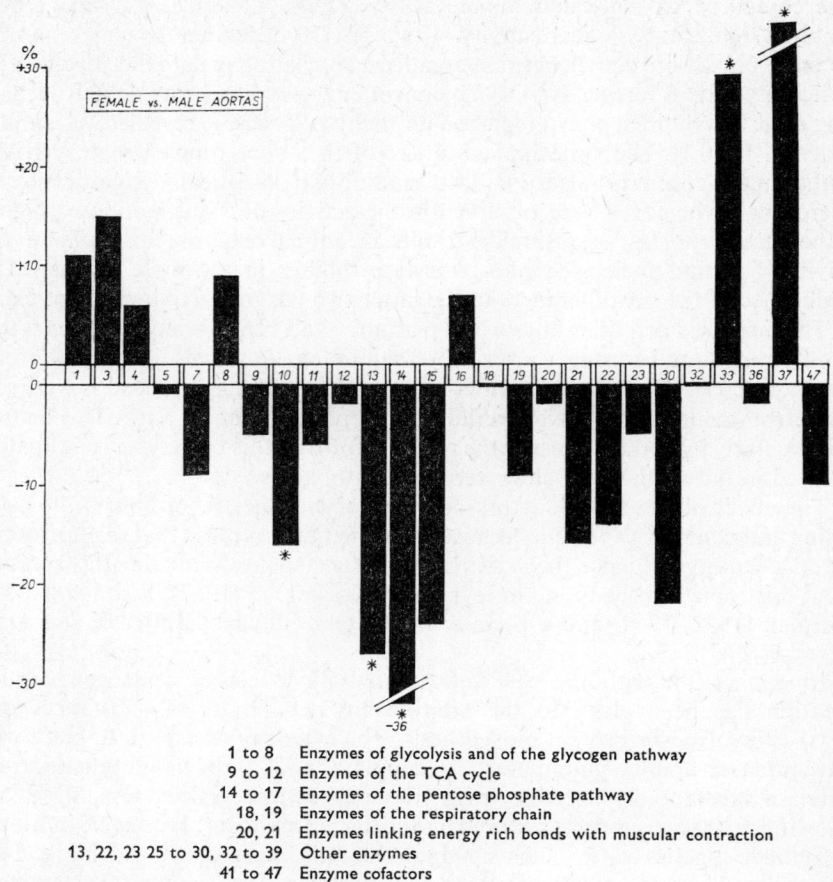

1 to 8	Enzymes of glycolysis and of the glycogen pathway
9 to 12	Enzymes of the TCA cycle
14 to 17	Enzymes of the pentose phosphate pathway
18, 19	Enzymes of the respiratory chain
20, 21	Enzymes linking energy rich bonds with muscular contraction
13, 22, 23 25 to 30, 32 to 39	Other enzymes
41 to 47	Enzyme cofactors

FIG. 22.—Differences in enzyme activities and cofactor levels between aortic specimens of sexually mature women and men. The enzyme activities and cofactor levels of male aortas, calculated on a wet weight basis = 100 per cent. (Constructed according to data from Kirk, 1963a, 1964a and some more recent publications by Kirk and co-workers.) For symbols see Fig. 9(a) and (b).

structure of some enzymes by binding their allosteric receptor sites.* This can modify the properties of the active site of the enzyme, but usually only at unphysiologically high concentrations *in vitro*. For example, such allosteric effects of sex hormones have been demonstrated with the succinoxidase system, with glucose 6-phosphate dehydrogenase and with a number of other oxidizing enzymes *in vitro*. Diethylstilboestrol and certain steroid hormones are even able to disaggregate glutamate dehydrogenase into monomers. Interestingly enough, they inhibit the glutamate dehydrogenase activity of the enzyme but stimulate its alanine dehydrogenase activity (Tomkins and Yielding, 1964).

Most of the information that is available in the literature, however, deals

* See p. 7.

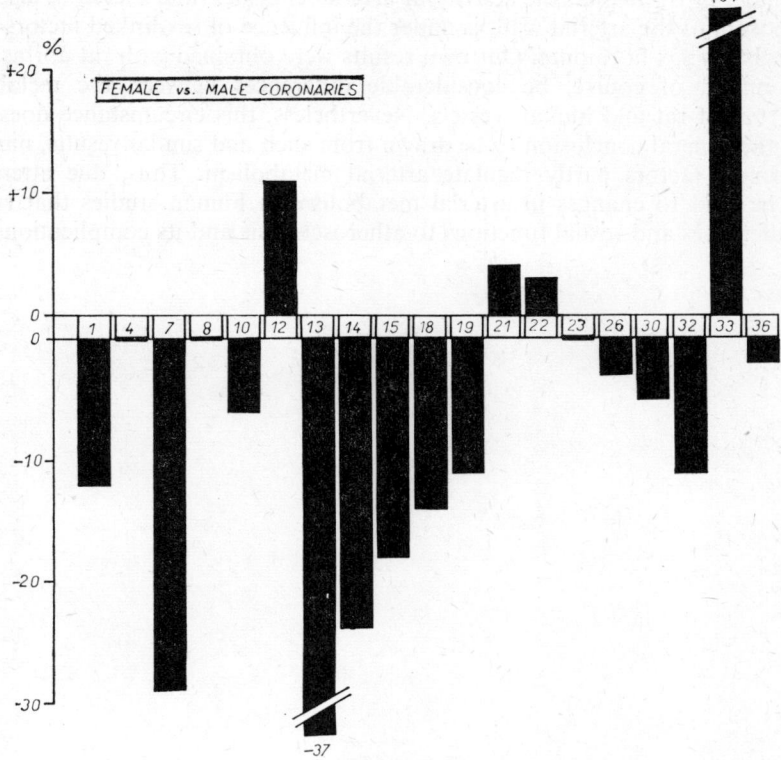

1 to 8	Enzymes of glycolysis and of the glycogen pathway
9 to 12	Enzymes of the TCA cycle
14 to 17	Enzymes of the pentose phosphate pathway
18, 19	Enzymes of the respiratory chain
20, 21	Enzymes linking energy rich bonds with muscular contraction
13, 22, 23, 25 to 30, 32 to 39	Other enzymes

FIG. 23.—Differences in enzyme activities between coronary artery specimens of sexually mature women and men. The enzyme activities of male coronaries, calculated on a wet weight basis = 100 per cent. (Constructed according to data from Kirk, 1963a, 1964a and some more recent publications by Kirk and co-workers.) For symbols see Fig. 9(a) and (b).

with enzymes contained in the tissues of reproductive organs and only in a few cases with enzymes of kidney, intestine or brain tissue. Androgens and oestrogens appear to control the formation of some enzymes in specific tissues, but the same enzyme activity may be catalysed in other tissues by somewhat differing chemical entities (see multiple forms of enzymes, Chapter I). It follows that the factors controlling enzymic biosynthetic rates are not identical (Dorfman 1952, 1964). Findings obtained with one tissue (e.g. prostate) cannot be simply extrapolated to the tissue of another organ such as, for instance, the vascular wall.

Returning now to the problems mentioned in the introduction of this chapter, one can conclude from the results obtained in several laboratories—

including our own—that the activity of arterial enzymes, and therefore also the metabolism of the arterial wall, is under the influence of sex-linked factors that are probably sex hormones. Our own results were obtained with rat aortas and there might, of course, be considerable differences between the metabolic properties of rat and human vessels. Nevertheless, this circumstance does not affect the general conclusion to be drawn from such and similar results, namely that sexual factors partly regulate arterial metabolism. Thus, due attention must be paid to changes in arterial metabolism in human studies that relate sex-differences and sexual functions to atherosclerosis and its complications.

Chapter XIII

ENZYMES OF THE VASCULAR WALL IN EARLY STAGES OF EXPERIMENTAL RABBIT ATHEROSCLEROSIS

FOLLOWING the early pioneer studies of Ignatowski (1908), Starokadomsky and Sobolew (1909), Stuckey (1912) and, in particular, the excellent studies of Anitschkow (1912, 1913, 1914), cholesterol-induced atherosclerosis in the rabbit has become the most commonly used experimental model of this disease.

In the presentation of our own enzyme results reference will often be made to the morphological picture of atherosclerosis. It will, therefore, be useful to summarize the morphological features of experimental cholesterol-induced rabbit atherosclerosis, based mainly on the classical description by Anitschkow (1933), and on our observations with Lojda (see Zemplényi et al., 1963c).

The onset of macroscopically discernible changes, as measured from the beginning of cholesterol feeding, varies considerably, but such changes are seldom to be seen before the end of the first month. These changes are in the form of yellowish spots or streaks that appear first in the aortic ring and arch near the orifices of the principal lateral branches. Afterwards similar changes develop in the ascending part of the aortic arch. The initial lesions grow large and become more prominent. The yellowish spots are transformed into elevated plaques of a more whitish colour, and they also appear in the upper part of the descending thoracic aorta. Here they are localized on the posterior wall and around the ostia of the upper and, later, also of the lower intercostal arteries. The lesions exhibit a tendency to fuse, so that large longitudinally oriented plaques are formed. Towards the end of approximately the fourth month the lesions are also seen in the caudal portion of the thoracic aorta and in the abdominal aorta; these lesions are usually less florid.

It can be seen that the progress of the aortic lesions generally follows a caudal direction.

In addition to the aorta, atherosclerotic changes often appear in the carotids, pulmonary arteries and the coronary arteries. The latter vessels develop lesions that appear as oblong streaks or as small spots localized at the orifices of lateral branches. Less frequently the cerebral and renal arteries also manifest distinct atherosclerotic changes. All the lesions are best detected after staining the vessels with Sudan III or other lipid soluble dyes. Early changes can usually be detected under the microscope at a time when they are as yet hardly perceptible to the naked eye.

During the earliest stages of atherosclerotic changes in the aorta, the subendothelial ground substance becomes swollen and changes in mucopolysaccharides can be detected histochemically. Special methods reveal diffuse or finely granulated lipids, including cholesterol, in those areas, and the lipid deposits occupy the somewhat enlarged interstitial spaces between the individual elastic lamellae or between bundles of muscle fibres in the innermost layers of the wall. Fine droplets of lipid material can also be detected in some macrophages (lipophages), which gradually increase in size and become typical foam cells. The origin of these cells has been much disputed and has so far not been definitely resolved. Recently, however, there seems to be some evidence, confirmed by the electron microscope, that they are derived from smooth muscle cells. Fine lipid droplets can also be detected in endothelial cells. Some initial lesions contain branching, stellate, or elongated cells of fibroblastic or muscular type, in which lipid droplets can be demonstrated, and foam cells may be absent. However, fibroblastic cells usually appear somewhat later in the course of the disease.

In later stages the lipophages become enlarged and more numerous, and some smaller cells of a monocytic character make their appearance. In the deeper part of large lesions the cells undergo necrosis and their lipid content is released. If cholesterol feeding lasts several months, the superficial layer frequently exhibits fibroblast multiplication so that a fibrous cap is formed. In some lesions stratification can be observed, characterized by alternation of layers of fibroblastic cells, muscle cells and foam cells.

Another characteristic process consists in the "reduplication" of fine elastic fibres around the lamina elastica interna. These fibres may form a dense network surrounding foam cells.

Some lesions are characterized by a considerable accumulation of an amorphous material that contains much cholesterol and is covered with endothelium or a fibrous cap; such changes acquire the character of atheroma. Atheromatous ulcers, however, do not make their appearance unless specific experimental techniques are used (see below). In some lesions newly-formed vessels can be observed entering from the media. Endothelial proliferation from the luminal surface also occurs in some places and may form small channels, but this picture is not very frequent. Deeper areas of the older plaques sometimes contain calcareous deposits.

Changes in the media are extremely variable and not infrequently no apparent changes can be seen at all. Often the internal elastic membrane becomes swollen and small gaps can be detected. In the more advanced stages of intimal lesions, oedema of the underlying media, lipid infiltration of the muscle cells and degenerative changes in the elastic membranes develop. These events are accompanied by degenerative fatty changes in the muscle cells and by accumulation of foam cells. Sometimes the muscle cells become reorientated so that a picture is seen that resembles migration into the plaque. In some places migration of macrophages can also be revealed. Focal necrosis of the media with calcification is common, but calcification may also make its appearance without apparent necrosis. At some places almost all elements of the media are substituted by foam cells. Vessels growing into the media from the adventitia are often detected. The detailed picture may differ under various experimental conditions and, even under identical condition, some variations may occur.

The lesions so far described mainly resemble the fatty streaks of human arteries. Rabbits fed cholesterol by the Anitschkow method do not manifest the complications of more advanced human atherosclerotic lesions, namely ulceration, haemorrhage and thrombosis (see p. 169).

However, Constantinides *et al.* (1960) succeeded in inducing rabbit disease that strongly resembles human lesions. If the animals are exposed to hypercholesterolaemia by feeding cholesterol for as long as is necessary to produce foam cell plaques and the feeding is then discontinued, the vascular lesions gradually turn into structures that, according to these authors, are very similar to advanced human changes. Under these conditions, most rabbit lesions develop a capsule, "fatty gruel" (i.e. a pasty mass of crystalline and amorphous lipid and protein, identical to its human counterpart) and calcification. A few lesions display capillarisation, haemorrhage or necrosis. The whole process takes at least six months or so to develop and is intensified with time (Constantinides, 1965).

An even more effective method of producing such advanced lesions consists in exposing the rabbits to prolonged intermittent lipaemia (Constantinides, 1961). A cholesterol-fat diet is fed for 2 or 3 months and alternated with normal diets for 2 or 3 months, and this is repeated for up to two years. The fat-free interval seems to enable the animals to recover and they can be kept alive for a relatively long time in a predominantly lipaemic condition.

In view of the above findings of Constantinides, it seems to be established that it is possible to induce in rabbits not only changes similar to fatty streaks but also the more complicated lesions encountered in human pathology.

Nevertheless, it is important to realize that morphological similarity does not necessarily imply a causal identity. There remain many important differences that justify a cautious attitude in extrapolating results obtained in the experimental rabbit model to the problems of atherogenesis in man.

To point out only a few of such important differences one has to keep in mind, first of all, that significant changes are also produced in other tissues by feeding rabbits cholesterol. Lipid is deposited in the viscera (Prior et al., 1961) and endocrine glands, as for example the thyroid (Bernick et al., 1962) and adrenals. These observations indicate that the syndrome induced in the rabbit is more comparable to a "lipid storage disease" and the arterial lesions are part of such an universal cholesterol thesaurismosis.

There exist fundamental differences in the structure of human and rabbit arteries. The intima of the human artery is thicker and has a subendothelial fibrous and musculoelastic layer which becomes gradually thicker with age. This feature of the human artery appears to be of special importance in the pathogenesis of human arterial lesions (see Chapter XIX). On the other hand rabbit arteries only exceptionally exhibit such a distinct subendothelial layer. Usually only a very thin layer of amorphous ground substance can be observed between the endothelium and inner elastic membrane and, rarely, some fibrocytes can also be detected. In view of this structural difference it is not likely that the details of the pathological process are quite the same.

The distribution of the atherosclerotic changes in rabbit experimental atherosclerosis, as outlined above, is also different from that in man. In the human aorta the extent and severity of atherosclerotic lesions is highest in the abdominal part, the proximal part being mainly affected only in syphilitic aortitis. In contrast, in experimental rabbit atherosclerosis the parts of the aorta mainly affected are the arch, the ascending and upper thoracic aorta.

Despite the above and other differences and evident limitations, we believe that the cholesterol-induced experimental rabbit model of atherosclerosis remains a useful "tool" for studies concerned with the mechanism of the effect of increased fat intake upon the morphology and metabolism of the arterial wall. General conclusions can be made, however, only after cautious evaluation and comparison of results obtained in other experimental models and in studies of human arteries as well.

In our own experiments, female albino rabbits about 5 months of age were fed daily 1 g. of cholesterol dissolved in 10 mg. of margarine in addition to a standard laboratory diet. Control rabbits of the same strain, age, and sex were fed the standard laboratory diet. The duration of the experiment was 2 weeks (16 rabbits), 4 weeks (33 rabbits), and 10 weeks (18 rabbits).

The aortas were removed immediately after killing the animals and prepared for biochemical and histochemical assays as described in Chapter XI. For histochemical studies other vascular segments were also used. In addition, the serum glucosamine level was determined by the method of Elson and Morgan (1933) as modified by Stary et al. (see Ledvina, 1958). The serum mucoprotein level was estimated by the method of Winzler et al. (1948). Serum cholesterol levels were determined by the method of Abell et al. (1952).

Figure 24 shows that as early as two weeks after the beginning of the experiment, the serum cholesterol level of the experimental animals was high and gradually increased with the duration of feeding. In spite of such cholesterolaemia, practically no pathological changes were observed macro- or microscopically in the aortas of the 4-week experimental series. Maybe this was connected with some features in the basic composition of the diet.*

The aortas of the animals of experimental series of 10-weeks duration exhibited atherosclerotic changes, mostly of degree 1–2·5 according to our grading system (see p. 107).

In some of the animals of the 4-week experimental series serum glucosamine and mucoprotein levels were determined. As indicated in Fig. 25 the serum glucosamine level unequivocally increased in cholesterol-fed animals, but no changes in mucoprotein could be detected.

(a) Phosphoric monoester hydrolases and ATPase

The results of alkaline and acid phosphatase determinations (Zemplényi et al., 1963a) are summarized in Fig. 26. No clear-cut changes in activity of alkaline phosphatase could be detected after 2 weeks of cholesterol-fat feeding. In the experimental series of 4-weeks duration, however, activity was significantly increased in the aortas of cholesterol-fed animals with a probable error of less than 0·001. The increase of enzyme activity in the aortas of rabbits from the 10-week experiment was also unequivocal as compared with the aortas of control animals.

It must be pointed out, however, that the activity of alkaline phosphatase is very low in the normal rabbit aorta. This seems to be one of the reasons why histochemical methods (see p. 125) cannot detect any activity in the intact intima and media of the rabbit's aorta.

* According to Horňáček et al. (1962) and Cookson et al. (1967) the alfalfa content of the diet significantly reduces the development of cholesterolaemia and the degree of atherosclerotic lesions in cholesterol-fed rabbits. Extracts from alfalfa also display this effect (Horňáček et al., 1966; Trčka et al., 1967). Similar observations concerning the basic composition of the diet were made in experimental chicken atherosclerosis by Reiniš et al., 1961a, b.

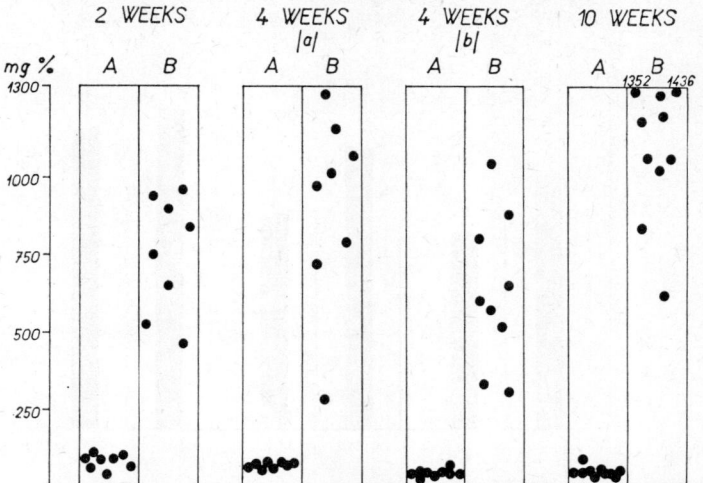

Fig. 24.—Serum cholesterol levels in control rabbits (A) and in cholesterol-fat fed rabbits (B).

In contradistinction to the results with alkaline phosphatase assays, acid phosphatase activity was only slightly increased in the aortas of rabbits fed cholesterol for 10 weeks with a borderline statistical significance of 0·05.

No significant differences in the activities of aortic 5′-nucleotidase and ATPase could be detected between cholesterol-fed and control rabbits in either of the experimental series of shorter or longer duration. This observation is summarized in Fig. 27.

For reasons mentioned in Chapter XI, histochemical findings will be presented in this section, as well as in the following parts dealing with experimental rabbit atherosclerosis. (For more details see Lojda, 1962; Lojda and Zemplényi, 1961; Zemplényi et al., 1963c).

Fig. 25.—Serum glucosamine and mucoprotein levels in control rabbits (A) and in cholesterol-fat fed rabbits (B).

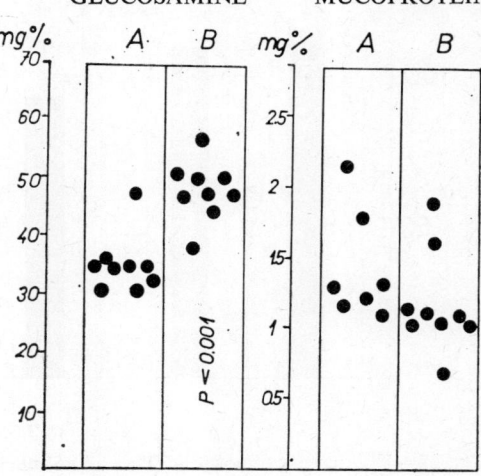

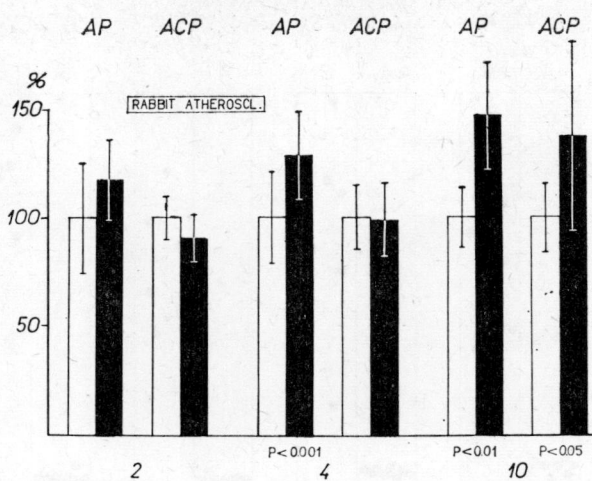

Fig. 26.—The activity of aortic alkaline and acid phosphatase activity in rabbits fed two, four or ten weeks on a cholesterol-fat diet. Empty columns = control rabbits. Black columns = cholesterol-fat fed rabbits. Upright line with bars ± S.D. Results expressed as percentage differences. (Data from Zemplényi et al., 1963a, c).

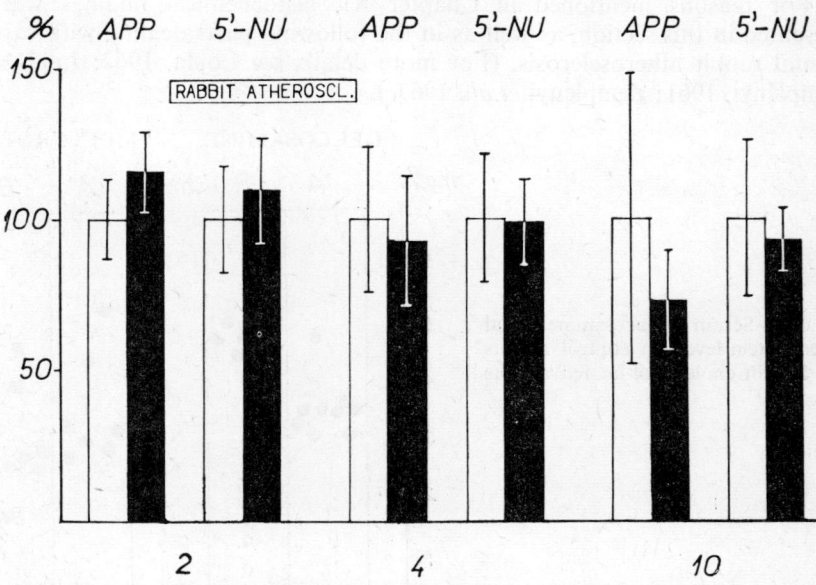

Fig. 27.—The activity of aortic ATPase and 5'-nucleotidase in the same animals as in Fig. 26.

In normal rabbit aortas low alkaline phosphatase activity could be observed histochemically in only some sections, and then in the adventitia, often exclusively in the endothelium of the *vasa vasorum*. In the branches of the pulmonary arteries the endothelium displayed comparatively high alkaline phosphatase activity, whereas the coronary arteries showed practically no reaction.

The aortas of rabbits in the 2- and 4-week experimental series did not show any change in intensity or localization of alkaline phosphatase activity. However, with the appearance of plaques, clear-cut changes in activity were observed particularly in certain parts of the intima (Plate II, 3). With some techniques certain plaques were negative, whereas other plaques displayed distinct activity. In the plaques with positive staining reactions, the activity was mainly located in the endothelium of newly-formed capillaries and was also noted in some macrophages. The endothelium covering the plaques did not show any activity at all. An important finding was the remarkably high activity accompanying calcification at some sites of the media (Plate II, 4).

Acid phosphatase activity was clearly demonstrated in the endothelium and also in the muscle cells of normal rabbit aortas, especially with Gomori's method. The coronary arteries exhibited lower activity.

With the appearance of atherosclerotic lesions increased activity was observed in the plaque components, particularly in the cytoplasm of young macrophages (lipophages). The endothelium that covered the smaller plaques and adjacent tissue gave more intense reactions than the endothelium that covered intact areas (Plate I, Plate II, 5). At sites of increased activity endothelial proliferation was always seen. Larger lipophages exhibited decreased activity and some of them—especially those in which necrobiotic changes or fibrous transformation had occurred—failed to react. Fibroblasts usually manifested only weak acid phosphatase activity.

The changes in the media were not always consistent. At sites with oedema the staining reactions for acid phosphatase activity were stronger in some sections, but others showed a rather weaker reaction. In the areas where calcification had occurred, a weak positive reaction was observed; the muscle cells surrounding calcified foci exhibited increased activity that was localized to small plasmatic granules.

In the atherosclerotic coronary arteries the picture was similar to that in the aorta. The smaller coronary branches with severe lesions exhibited high activity in the endothelium and a weaker one in the muscle cells.

ATPase activity was detected in the endothelium and in the muscle cells of the media of normal rabbits. The reaction in muscle cells was comparatively high and, in cryostat sections, it could be detected as early as 15 minutes after the beginning of incubation. In contrast, 5'-nucleotidase activity could be detected only after prolonged incubation and first appeared, when watched continuously, in the *vasa vasorum* of the adventitia (i.e. like alkaline phosphatase). Later, activity of this enzyme could also be demonstrated in the endothelium and muscle cells of the media.

In those aortas with manifest atherosclerotic changes, a slight or almost negative ATPase activity was observed in the plaque components. In some cases the reaction in the underlying media was weaker than at intact sites of the same vessel.

The histochemical findings as described above are similar to those described by Cavallero and Turolla (1960), Cavallero et al. (1963) and Gonzales (1963) in connection with phosphatase and/or ATPase activity.

Comparison of our biochemical and histochemical observations indicates that the respective results for acid phosphatase, ATPase and 5'-nucleotidase correspond fairly well. With alkaline phosphatase, however, some discrepancy exists between the results obtained by these methods in normal aortas. According to our histochemical evidence the activity of this last enzyme is located in the adventitia of the aorta and is absent from the inner layers of the normal vessel. On the other hand, aortic samples devoid of adventitia were used for biochemical assay and they always manifested distinct, although low, activity. We have not yet established to what extent this divergence is caused by the lower sensitivity of the histochemical methods, by partial or complete loss of extractable enzymes during the histochemical procedure or—on the contrary—by the presence of *vasa vasorum* in the external part of the media in the samples used for biochemical assay.

The most interesting of the biochemical findings is the unequivocal rise in alkaline phosphatase activity as early as 4 weeks after the beginning of the cholesterol-fat feeding experiment. On page 61 some of the functions ascribed to alkaline phosphatase have been mentioned. Nevertheless, so far there is very little definite knowledge about the *biological significance* of this enzyme. Aortas of calciferol-fed rats exhibit a very early increase in both alkaline and acid phosphatase activity (see p. 149), at a stage before manifest vascular calcification. In the present experiments with rabbits, the histochemical results again indicate, in addition to other findings, distinct alkaline phosphatase activity in both the media and intima at sites of calcification. Such findings seem to suggest that the rise in the activity of the enzyme could reflect a relationship between the process of calcification and alkaline phosphatase activity. Such a suggestion, however, contradicts most present-day theories of calcification, which deny any causal relationship between this process and alkaline phosphatase activity (see Glimcher, 1959). As 5'-nucleotidase is supposed by some workers (see p. 63) to be a regulatory factor in tissue calcification, it is of some interest that we could not observe any change in 5'-nucleotidase activity in the present experiments.

Another much disputed question is the relationship between alkaline phosphatase and intercellular ground substance. For example, administration of androgens causes an increase in ground substance as well as a local rise in alkaline phosphatase activity (Charvát, 1952). It has also been suggested that the enzyme participates in the synthesis of collagen and other fibrous proteins (see review by Gould, 1960). It appears that the enzyme may play this role in some tissues but not in others. As mentioned earlier on page 62, Schlief et al. (1954) observed a rise in the activity of alkaline phosphatase at those sites in human atherosclerotic vessels that exhibited enhanced fibre formation.

Romanul and Bannister (1962) suggested that the function of alkaline phosphatase at small branches of the arterial tree is that of active transport similarly as in the proximal tubules of the kidney and in the mucosa of the small intestine. One could speculate that the rise in alkaline phosphatase activity in the endothelium is the result of an overload of a normal transfer mechanism (Hess and Stäubli, 1963).

It must be mentioned that Paterson et al. (1957) consider that the presence of alkaline phosphatase in atherosclerotic lesions is evidence of early vascularization of the affected intima. This view, however, conflicts with the above histochemical findings that show, in addition to positively reacting capillaries, distinct activity in macrophages and calcified foci.

As with alkaline phosphatase, the actual *biological role of acid phosphatase* still remains unresolved (see p. 61). Our histochemical and biochemical data indicate that acid phosphatase activity increases simultaneously with the formation of plaques, the activity being located in the macrophages (lipophages) and endothelium, as well as in the damaged muscle cells surrounding calcified foci. It is likely, therefore, that the increased activity of the enzyme in these experiments reflects increased phagocytosis and tissue injury, conditions that are usually associated with corresponding changes of lysosomal enzymes (see pp. 61 and 159).

Finally, it has to be mentioned that, in contrast to our negative results with 5'-nucleotidase and ATPase activity, Fouquet (1961) reported positive results in this respect. He observed a definite relationship between pH and histochemically demonstrable aortic ATPase activities. He noticed no activity at pH 6·3, weak activity in the intima at pH 7·4 and strong enzyme activity of muscle cells in the media at pH 8·5 and pH 9·4. The aortas of cholesterol-fed rabbits revealed increased ATPase activity after 16 days of feeding but, after 30 days, the enhanced activity tended to decrease.

(b) Non-specific carboxylesterase

The biochemical results concerned with carboxylesterase activity, using β-naphthol acetate as substrate, did not reveal any significant difference between the cholesterol-fat fed and control rabbits in either of the series. As lipolytic and non-specific carboxylesterase activity seem to run parallel in the aorta, the above negative finding is not unexpected. In other experiments, to be described later (Chapter XXIII) we found *increased lipolytic activity* only in those rabbit aortas that displayed *more advanced* atherosclerotic lesions. However, in the present experimental series, definite atherosclerotic changes were found only in aortas of animals fed the cholesterol diet for 10 weeks. The atherosclerotic lesions in the aortas of most animals were of a minimum degree (1–2·5 in our grading system), in one animal of medium degree (grade 3) and in only one animal were they severe (grade 4·5). Interestingly enough, the latter aorta exhibited the highest carboxylesterase activity (323 per cent) of all the aortas of the 10-week experimental series.

Histochemically, there was a considerable variation in carboxylesterase activity according to the substrate used. With Tweens as substrate no regular reaction could be detected in normal rabbit arteries, whereas the azo-dye methods showed activity located chiefly in the muscle cells of the aortic media and increasing in a radial direction towards the adventitia. It was not possible to decide whether the surfaces of elastic fibres and lamellae were involved, since following protracted incubation a very strong positive reaction occurred between the elastic lamellae and exact localization was not possible (Lojda and Zemplényi, 1961). The arteries of rabbits kept for 2 to 4 weeks on the cholesterol-fat diet did not reveal apparent changes in the intensity and localization of non-specific carboxylesterase activity. The first changes could be detected as soon as atherosclerotic lesions appeared, i.e. at 10 weeks on diet. The endothelium covering most of the early plaques reacted more intensely than the endothelium of the unaffected portions of the vessel. The intensity of the reactions in the plaques depended upon their cellular composition. Whereas small macrophages (lipophages) reacted very strongly, larger cells showed reduced activity with all staining reactions. Extracellular reaction could be clearly seen at sites where disintegrated lipophages had accumulated, especially in the deeper layers of the plaques. A strong reaction was observed in transitional and muscle cells, whereas fibrocytes showed less staining.

In larger plaques some stratification could be observed, loci of higher activity alternating with those of lesser or no activity (Plate II, *1*, *2*). The reaction was very weak where fibrous transformation had occurred. Positively reacting macrophages could sometimes also be found in the tunica media. Medial muscle cells in the vicinity of the plaques sometimes stained variably but, in most cases, remained unchanged. In areas with calcification the histochemically detectable enzyme activity completely disappeared.

The plaques in other arteries, including the coronaries, displayed practically the same changes in non-specific carboxylesterase activity as those in the aorta.

Comparison of biochemical and histochemical results shows the converse pattern to that seen in the case of alkaline phosphatase. The histochemically detectable increase in carboxylesterase activity of the few small plaques in the early stages of atherosclerosis seems not to be large enough to influence the overall activity of the arterial wall. In addition, the possibility cannot be excluded that decreased mean activity in the media counteracts increased carboxylesterase activity in the newly-formed plaques.

(*c*) **Dehydrogenases**

In our biochemical and histochemical studies on dehydrogenase "systems" in experimental rabbit atherosclerosis, we only used tetrazolium salts as hydrogen acceptors; phenazine methosulphate was not employed as an intermediate carrier (see Chapter XI).

In the normal rabbit aorta the activity of the lactate dehydrogenase system was found to be about 90 per cent higher than the activity of the succinate dehydrogenase (SDH), glycerol-3-phosphate dehydrogenase (AGPDH) and malate dehydrogenase (MDH) systems; the activity of these last three enzyme systems was approximately of the same order.

The activity of the aortic AGPDH system was found to be unequivocally decreased in about half of the animals of the 4-week and 10-week experiments, when compared with the control series. Statistical analysis, however, revealed borderline significance ($P < 0.05$) in only the former group of animals (Fig. 28). The lack of statistical significance in the 10-week experiment was probably caused by the wide scatter in the results (45–120 per cent).

Similar results were obtained in the assays for LDH activity. In this case, however, statistical analysis showed significant difference only in the 10-week experiment (Fig. 29). In view of the borderline significance of the above results, it seems wisest only to conclude that there is a tendency toward decreased activity of these enzyme systems in the aortas from both longer feeding experiments, while 2-week's cholesterol feeding does not change the activity of either system.

Study of two enzymes in the tricarboxylic acid cycle revealed a significant decrease ($P < 0.01$) of the aortic SDH system in both the 4-week and 10-week experiments (Fig. 30). The activity of the MDH system did not change significantly in either of the experimental series, although the activity found in a few aortas at 10 weeks was much lower than the activities of control animals (Fig. 31).

Estimations of LDH activity by the spectrophotometric optical test showed an average activity of 163·0 Wróblewski units/100 μg. aortic extract N_2 in the 2-week and 4-week series of control animals. The aortas of the experimental animals did not exhibit any change in activity. This agrees with the above-mentioned findings with the tetrazolium method. No data are available with this method for the 10-week experimental series (Mrhová *et al.* 1963*a*).

The description of the histochemical findings in the aortas of normal and cholesterol-fat fed rabbits is based on Lojda's observations (see Lojda and Zemplényi, 1958; Lojda and Felt, 1960; Lojda, 1962; Zemplényi *et al.*, 1963*c*), and

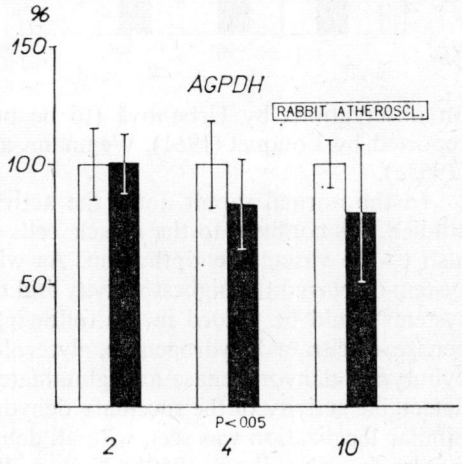

Fig. 28.—The activity of aortic glycerolphosphate dehydrogenase in the same animals as in Fig. 26.

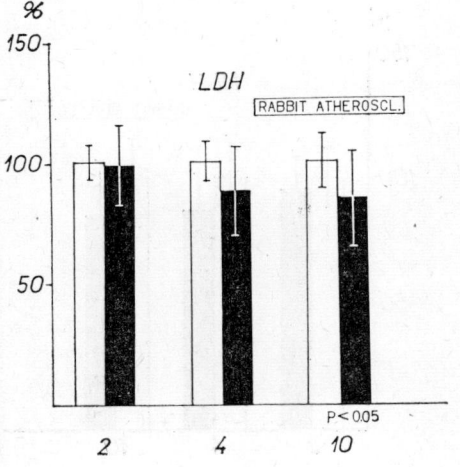

Fig. 29.—The activity of aortic lactate dehydrogenase in the same animals as in Fig. 26.

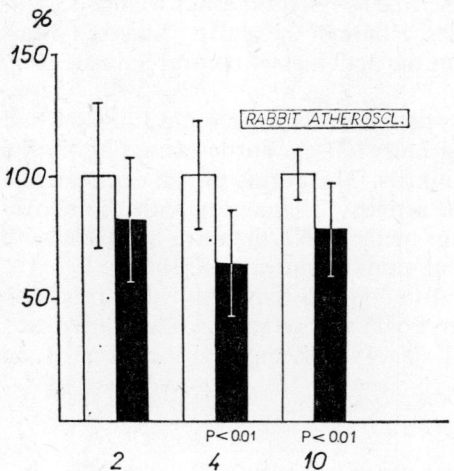

Fig. 30.—The activity of aortic succinate dehydrogenase in the same animals as in Fig. 26.

on observations by Urbanová (to be published). Similar findings were also reported by Fouquet (1961), Wegmann and Fouquet (1961) and Adams *et al.* (1963*a*).

In the normal rabbit aorta the activity of all the dehydrogenase systems studied was confined to the muscle cells of the media and to a varying degree also to the vascular endothelium. As with the biochemical assays, the LDH system displayed the highest activity and the activity of the other dehydrogenase systems could be graded in the following descending order: malate dehydrogenase, isocitrate dehydrogenase, glycerol-3-phosphate dehydrogenase, β-hydroxybutyrate dehydrogenase and glutamate dehydrogenase. The histochemically detectable activity of the succinate dehydrogenase system was again very low. Similar localization was seen with all dehydrogenase systems. This is explained by the fact (Novikoff, 1960*b*; Farber, 1962) that the histochemical methods

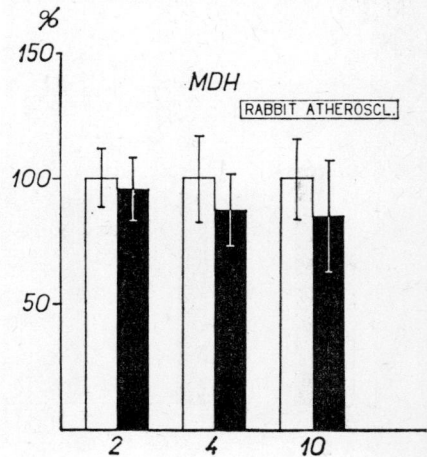

Fig. 31.—The activity of aortic malate dehydrogenase in the same animals as in Fig. 26.

used only localize the corresponding $NADH_2$- or $NADPH_2$-dependent "diaphorase".

In the coronary arteries the activities of all dehydrogenase systems were higher than in the aortas.

The aortas of rabbits from the 4-week experimental series displayed in some cases slightly decreased activity of the succinate dehydrogenase system, whereas no change in activities or localization of the other enzyme systems could be detected in this series. However, with the appearance of atherosclerotic lesions (i.e. after 10-weeks' cholesterol feeding) more definite changes in enzyme activity could be detected. The endothelium covering the initial lesions usually, but not always, showed a more intense reaction than the endothelium of the unaffected parts (Plate I, Plate II, 6, 7).

The overall intensity of the reactions in the plaques was very much dependent on their cellular composition. Macrophages and smaller lipophages reacted very strongly. However, with increasing size and increasing amount of histochemically-detectable cholesterol, activity decreased and often no reaction could be observed at all. Fibrous parts of older plaques did not display any positive reaction. Nevertheless fibrocytes themselves showed slight activity.

In the media underlying atherosclerotic plaques, some muscle cells exhibited decreased or increased activity, while at calcified sites activity disappeared.

Similar changes to those in the aortas were seen in the plaques of the larger branches of coronary arteries, especially in longer term experiments. In the endothelium of the affected coronaries of such rabbits Lojda observed increased reactions for all the dehydrogenase systems investigated. However, the plaques did not usually show a positive reaction. Activity was clearly decreased in sites of degenerative change in the tunica media, a finding that was quite the reverse of that obtained with acid phosphatase (see p. 125).

Returning now to our biochemical findings, cholesterol-fat feeding in rabbits seems to impair the tricarboxylic acid cycle in the aortas as early as 4 weeks after the onset of the experiment. Activity of the SDH system is unequivocally diminished. At this stage of the experiment distinct morphological lesions are absent, although the serum cholesterol levels are high (average 912 mg./100 ml.) and the livers of most animals manifest considerable fatty change. We shall see later in Chapters XVII–XIX that decreased activity of dehydrogenases of the TCA cycle, and probably of all enzymes associated with this metabolic pathway, seems to be an important factor connected with atherogenesis. The reason for decreased activity of these enzymes in spontaneous human atherosclerosis is of course different from that in experimental rabbit atherosclerosis. In the animal disease there is some possibility that metabolic hypoxic damage to the arterial wall results from the formation of a lipid film on the intimal surface, and that this is a prerequisite for the subsequent development of atherosclerotic changes (see Hueper, 1944, 1945). The decline in the overall activity of some enzymes, particularly certain dehydrogenases, might be the result of such a metabolic injury. It is more likely, however, that the decline is the result of gross overloading and injury of arterial tissue by cholesterol itself (see Chapter XX).

It is interesting that with the appearance of streaks and plaques, the focal increase of histochemical reactions for most dehydrogenases in the endothelium and young lipophages is obviously too small to be detected biochemically against the background of decreased overall activity. The fall in overall activity is reflected not only in the biochemical findings but also not infrequently in the overall staining intensity (Plate I, *3* and *4*). Furthermore, larger and more advanced plaques show decreased activity of all dehydrogenase systems.

(*d*) Other Enzymes

As shown in Fig. 32 we could detect no changes in the activity of aortic *β-glucuronidase* after 2 weeks of cholesterol-fat feeding. In the 4 and 10 weeks experimental series, however, such activity increased, this increase being of a

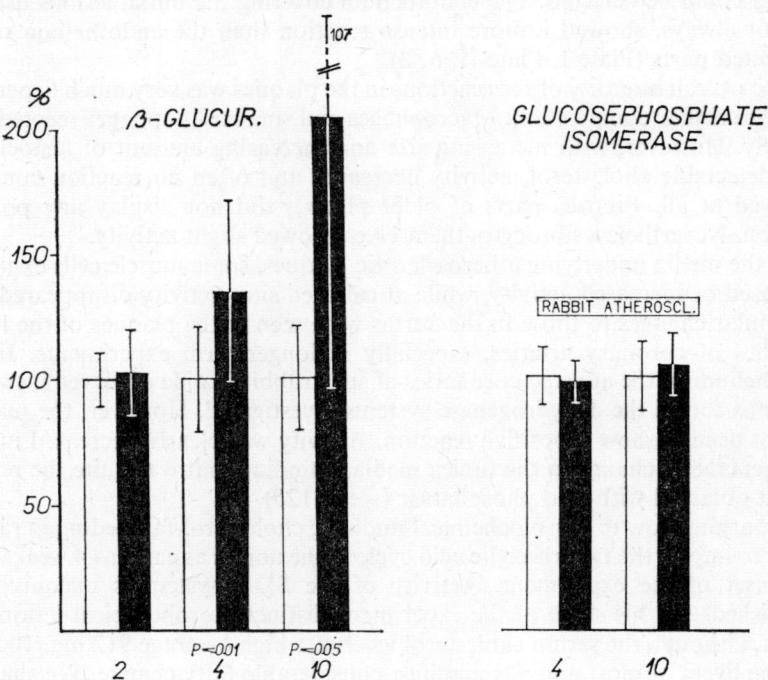

FIG. 32.—The activity of aortic β-glucuronidase and glucose phosphate isomerase in the same animals as in Fig. 26.

somewhat lower statistical significance in the 10-week experiment. Practically the same results were obtained when activity was referred to the protein or nitrogen content of the extracts (Mrhová *et al.* 1963*b*).

Histochemical activity was only studied in the later stages of experimental atherosclerosis. Some plaques exhibited focally increased activity, especially at sites where alcianophilic acid mucopolysaccharides had accumulated.

The possible biological significance of β-glucuronidase was discussed on page 65. Apart from other functions, β-glucuronidase appears to play a part in mucopolysaccharide metabolism and perhaps also in tissue proliferation. If so, our findings of raised activity in the morphologically normal aortas of rabbits fed cholesterol for only 4 weeks suggest that, even in cholesterol-fed animals, changes in the ground substance and cell proliferation are a prerequisite for the subsequent development of atherosclerotic changes. Another possible interpretation is, of course, that the raised activity of this "lysosomal" enzyme is a sign of increased activation in response to tissue damage, as is probably the case with acid phosphatase.

In connection with the possible role of this enzyme in disorders of the vascular ground substance, it is of some interest that serum glucosamine levels are significantly elevated ($P < 0.001$) in cholesterol-fed rabbits (47.8 ± 4.8 mg./100 ml.), when compared with control animals (34.8 ± 5.2 mg./100 ml.) (Fig. 24). This observation agrees with the findings of Gerö et al. (1962a, b) and others.

We investigated biochemically the activity of aortic *glucose phosphate isomerase* (Fig. 32) in the early stages of experimental rabbit atherosclerosis and, in co-operation with Grafnetter (unpublished), the activity of *aspartate aminotransferase*. No significant differences could be detected between cholesterol-fed and control animals. In the case of aminotransferases our findings appear to disagree to some extent with those of Alekseeva and Nekrasova (see p. 57). However, these authors studied the activities of aortic alanine aspartate aminotransferases either much earlier (5 days) or much later (100 days) after the onset of cholesterol feeding than in our experiments. They reported an early increase followed by a late decrease in the activity of alanine and perhaps also of aspartate transaminase.

From the evidence presented it is clear that increased fat and cholesterol intake results in an early local metabolic response by the rabbit arterial wall. In the early stages, where no apparent morphological alterations can be observed, the overall activity of some dehydrogenase systems—in particular succinate dehydrogenase—tends to decline, whereas the opposite trend is exhibited by phosphatase and β-glucuronidase activities. With the appearance of streaks and plaques the activities of many enzymes *focally* change, but the picture varies very much according to the particular cellular population of the lesion. The early changes in enzyme activities that we observed lend support to the view that local metabolic factors are important in the development of vascular lesions, even in the cholesterol-fed animal.

TABLE I

Average enzyme activities in normal female rabbit aortas

Enzyme	Av. activity	Unit
Alkaline phosphatase	0.25	μM phenol/1 mg. extract N_2/hour
Acid phosphatase	0.86	Same as above
ATPase	12.4	μg phosphate /100 μg extr. N_2/hour
5'-Nucleotidase	21.4	Same as above
Malate dehydrogenase (syst.)	53.5	μg diformazan/100 μg DNA/45 min.
Lactate dehydrogenase (syst.)	107.5	Same as above
Succinate dehydrogenase (syst.)	68.5	Same as above
Lactate dehydrogenase (opt. test)	163.0	Wróblewski u./100 μg extract N_2
Non-specific carboxylesterase	56.6	μg β-naphthol/1 mg. extract N_2/hour
β-Glucuronidase	110.5	μg phenolphthalein/100 μg extr. N_2/16 h.

We shall see in subsequent chapters that similar changes of enzyme activity are characteristic of situations associated with vascular wall injury and with connective tissue reactions to injury. Does this mean that in cholesterol-induced experimental atheroma, where arterial lipid accumulation doubtlessly plays a primary role, the development of more advanced lesions depends on metabolic injury to the arterial wall as mentioned in connection with dehydrogenases on page 131 ? The topic of mechanical and metabolic injury to the arterial wall and its relationship to atherogenesis will be discussed later in more detail in connection with spontaneous atherosclerosis in man and other mammals and birds (Chapters XVI, XVII, XVIII and XIX).

Chapter XIV

PROBLEMS OF EXPERIMENTAL ATHEROSCLEROSIS IN THE RAT

IN some animal species, especially the rabbit, the experimental production of atherosclerotic or atherosclerotic-like lesions by simple cholesterol feeding is quite easy. On the other hand, the rat develops neither atherosclerosis nor convincing hypercholesterolaemia when fed on ordinary cholesterol-enriched diets.

In Chapter XV mention will be made of the mainly successful use of combined vascular injury and hyperlipaemia to produce atherosclerotic-like lesions in the rat. Over the last fifteen years many investigators have attempted to achieve this goal by other procedures.

For example Hartroft *et al.* (1952) observed atheromatous changes in the arteries of rats maintained for a long time on a low choline diet. The initial lesions consisted of lipid accumulations in the endothelial cells. Later, intimal cells proliferated and small plaques formed. Necrosis and eventual calcification of the subjacent media were observed in large vessels, the latter lesions somewhat resembled Mönckeberg's sclerosis of human femoral and radial arteries. It is possible that the underlying mechanism of this sort of lesion depends on the liver and renal damage that some animals develop when subjected to choline deficiency. In contrast to the above findings Wissler *et al.* (1954) produced atheromatous arterial lesions in rats fed a synthetic diet high in lard and choline and containing cholesterol. In these experiments choline served to increase the cholesterolaemia. So far the paradoxical effects of choline deficiency and excess remain unresolved.

As far as hypercholesterolaemia is concerned, Schönheimer (1924) and Friedman and Byers (1951) and others have shown that feeding bile acids increases the blood cholesterol level. Following these studies, Page and Brown (1952) produced high levels of blood cholesterol and lipoproteins by simultaneously feeding rats a diet that contained cholesterol and cholic acid and suppressing thyroid-function with radio-iodine. The prolonged lipaemia induced lipid infiltration of the arteries, but no tissue response, intimal proliferation or foam cells appeared. Subsequently, the basic dietary procedure has been modified by replacing radio-iodine with thiouracil (Wissler *et al.*, 1954; Fillios *et al.*, 1956; Hartroft and Thomas, 1957). In these experiments not only lipid infiltration but fibrotic lesions were induced.

The mortality rate of rats treated in such a way is usually high. Záhoř (1963) succeeded in substantially prolonging survival by allowing them several "recovery" periods during which thiouracil was stopped. Rats treated in this way displayed arterial lesions that were more similar to the spontaneous human disease.

Without going into great detail about experimental atherosclerosis in rats, Thomas and Hartroft's (1959*a, b*) findings must be discussed as they are relevant

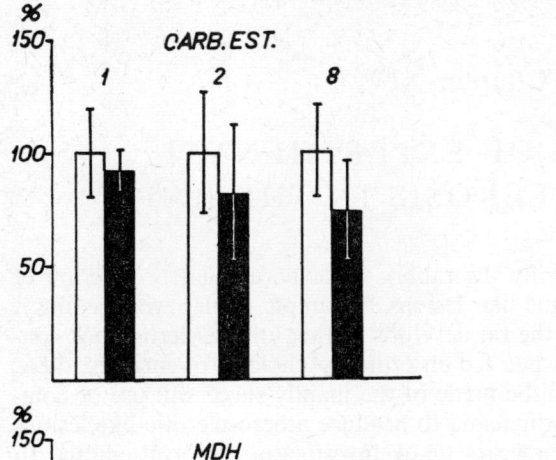
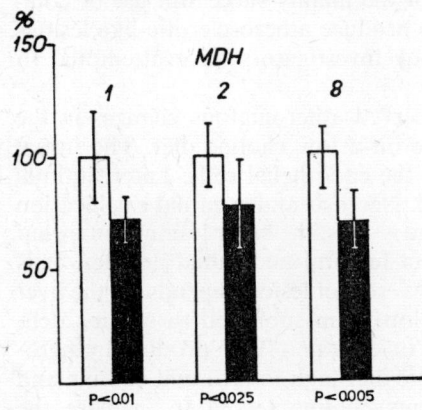
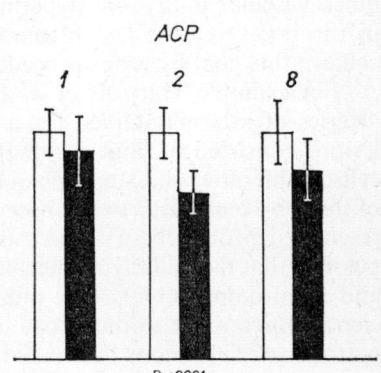

Fig. 33.—The activity of aortic non-specific carboxylesterase, malate dehydrogenase and acid phosphatase in rats fed for one, two or eight weeks on a semisynthetic diet containing cholesterol, sodium cholate, methylthiouracil and *butter*. Empty columns = 21 control rats fed the ordinary laboratory diet (Larsen's diet). Black columns = 23 experimental rats. Upright line with bars ± S.D. Results expressed as percentage differences. (Data from Zemplényi et al., 1965a.)

to our own work. These authors described a further modification of the dietary procedure just mentioned, in that a large excess (40 per cent) of butter was added to the diet already enriched with cholesterol, cholate and thiouracil. Those rats that survived for about 14 months on this diet developed arterial lesions which quite closely corresponded to the early and moderately advanced lesions in man (Hartroft and Thomas, 1963). However, many rats fed in this way, develop thromboses and multiple myocardial infarcts long before the development of clear-cut atherosclerotic lesions (hence the term "thrombogenic" or "infarctogenic" diet).

The type of fat used in the "Hartroft" diet seems to play a decisive role in determining the type of lesion produced, at least in the earlier stages of the experiment. Gresham and Howard (1960) confirmed the high incidence of myocardial infarcts in the rats fed this butter-containing diet. On average the animals survived about fourteen weeks and none of them lasted more than

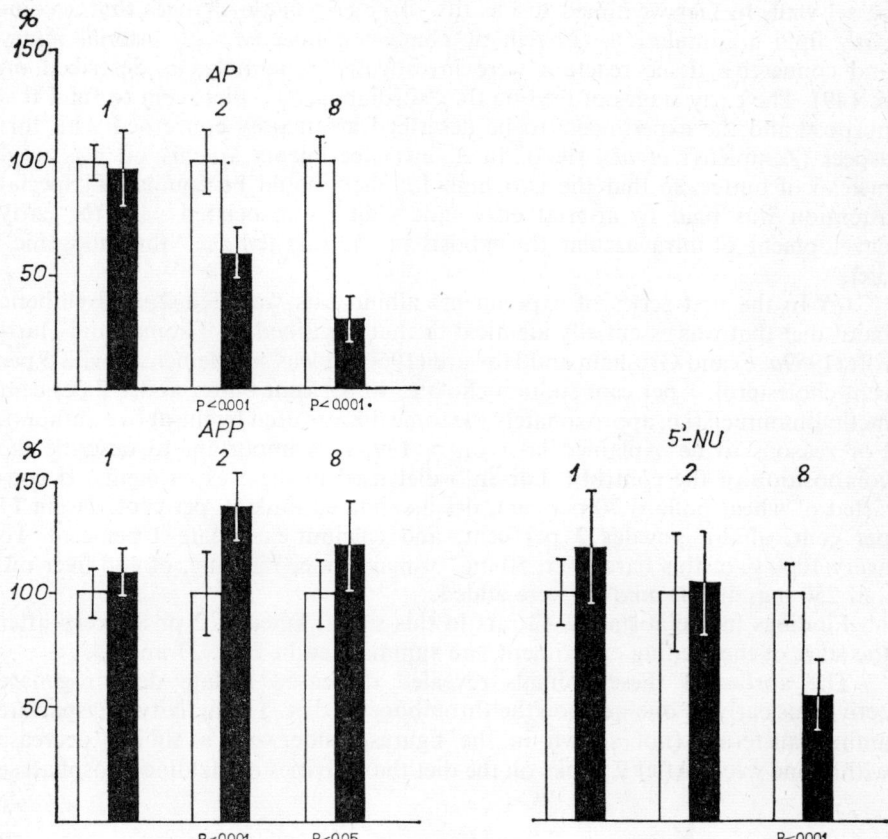

FIG. 34.—The activity of aortic alkaline phosphatase, ATPase and 5′-nucleotidase in the same animals as in Fig. 33.

twenty weeks. However, the mural arterial lesions mainly consisted of abundant endothelial lipid-filled macrophages. In a few animals small areas of intimal sudanophilia developed, but no tissue response was seen that amounted to atherosclerosis. In striking contrast, rats fed 40 per cent of arachis oil instead of butter, and who survived about the same time as the previous group, developed atherosclerotic plaques in the thoracic aorta and coronary arteries that closely resembled those of human atherosclerosis. The alterations were described as raised lesions capped by fibro-elastic tissue, beneath which amorphous debris was found that contained cholesterol crystals and other sudanophilic lipid; the underlying elastica was often fragmented. None of these rats developed arterial thrombosis (hence the term "atherogenic" diet). The omission of thiouracil or reduction of the butter content of the "thrombogenic" diet to 5 per cent prevented thrombosis.

In our own experiments we wanted to use an experimental model where only lipid accumulation and no connective tissue reaction would be present in the vessel wall. In fact we aimed to identify those enzymatic changes that accompany lipid accumulation. (Enzymatic changes connected with arterial injury and connective tissue reaction were investigated separately, as described on p. 149). The early stages of feeding the "thrombogenic" diet seem to fulfil this purpose and the experiments to be described are mainly concerned with this aspect (Zemplényi et al., 1965). In a few experiments arachis oil was used instead of butter, so that the two high fat diets could be compared. Special attention was paid to arterial enzymatic changes associated with the early development of intravascular thrombosis in the rats fed the "thrombogenic" diet.

(a) In the first series of experiments albino rats were fed a semisynthetic basal diet that was essentially identical to that described by Thomas and Hartroft (1959a, b) and Gresham and Howard (1960). It was supplemented with 5 per cent cholesterol, 1 per cent sodium cholate, 40 per cent butter and 1·2 per cent methylthiouracil (i.e. approximately the same dose as used by the above authors). For reasons to be explained later on p. 139, it is important to describe the composition of the control ("Larsen") diet used in these experiments. It consisted of wheat pollard 70 per cent, dried skimmed milk 10 per cent, casein 17 per cent, alfalfa powder 2 per cent, and calcium carbonate 1 per cent. To every 1000 g. of this basal diet, 50 mg. of margarine, 7500 mg. of cod liver oil, and 250 mg. of salt mixture were added.

Findings in the aortas of 43 rats in this series, killed 1, 2 or 8 weeks after the start of the feeding experiment, are summarized in Figs. 33 and 34.

The aortas of these animals revealed decreased malate dehydrogenase activity as early as one week on the thrombogenic diet. The activity of aspartate aminotransferase (not shown in the figures) underwent a similar decrease within one week. After 2 weeks on the diet the activities of alkaline phosphatase

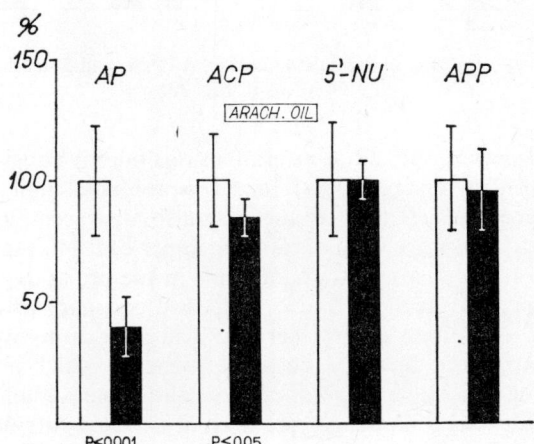

FIG. 35.—The activity of aortic alkaline and acid phosphatase, 5'-nucleotidase and ATPase in rats fed for two weeks on a semisynthetic diet containing cholesterol, Na cholate, methylthiouracil and *arachis oil*. Empty columns = 11 control rats fed the ordinary laboratory diet (Larsen's diet). Black columns = 11 experimental rats. Upright line with bars ± S.D. Results expressed as percentage differences. (Data from Zemplényi et al., 1966e.)

and acid phosphatase decreased, whereas aortic ATPase activity increased. It is important to note than in these early stages the aortas showed no accumulation of lipids and no microscopic changes attributable to atherosclerotic or thrombotic processes. In later stages, i.e. after 8 weeks on the diet, decreased activity of 5′-nucleotidase was also observed. Non-specific carboxylesterase activity declined in the 2- and 8-week experimental series, but this trend was not statistically significant.

(b) To investigate the effect of the "atherogenic" diet containing arachis oil instead of butter, we studied a further series of 22 rats; control animals were again fed the "Larsen" diet. The duration of this experiment was only two weeks. Fig. 35 shows that, as in the preceding experiments with the "thrombogenic" diet, alkaline and acid phosphatase activities decreased. The activity of malate dehydrogenase did not show any significant change, whereas the activity of lactate dehydrogenase, which was not investigated in the preceding series, increased at a borderline level of significance (Fig. 36). In contrast to the effect of the "thrombogenic diet", the diet that contained arachis oil did not affect aortic ATPase activity.

(c) In both preceding experimental series, dealing with the effects of diets that contain either butter or arachis oil, the experimental animals were fed a *basal semisynthetic diet*, whereas control animals were maintained on the *"Larsen" diet* (see above). This circumstance could induce a possible bias in the interpretation of the results. Therefore, a third experimental series was designed to compare the effects of the "Larsen" and basal semisynthetic diets on aortic activities. As can be seen from Figs. 37 and 38 the activities of alkaline phosphatase and carboxylesterase are significantly lower in the aortas of rats maintained on the basal semisynthetic diet; the activity of acid phosphatase also falls but this is not shown in the figures. No differences could be observed in the activities of the other enzymes investigated (Zemplényi, 1966).

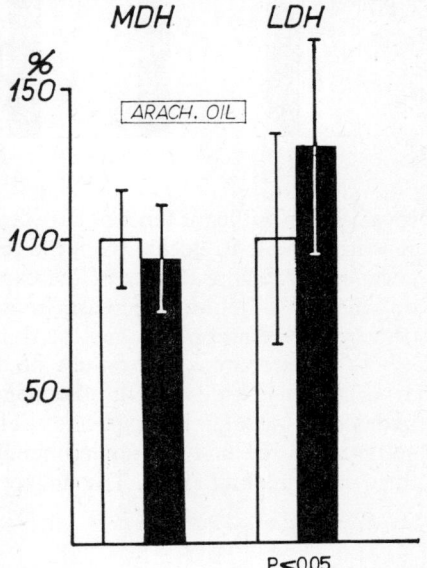

FIG. 36.—The activity of aortic malate dehydrogenase and lactate dehydrogenase in the same animals as in Fig. 35.

It is not clear why aortic phosphatase and non-specific carboxylesterase activities differ with these two basic diets. Possibly, the "Larsen" diet was deficient in some factor, as the semisynthetic diet used in these experiments was carefully checked from this aspect (Jelinek, unpublished observation).

Returning now to the results obtained in the first series of experiments, we see that an outstanding vascular effect of the thrombogenic diet is the very early decrease in aortic malate dehydrogenase activity. It appears that such decreased activity is in some way connected with the accumulation of lipids in the vessel wall. The aortas of rats maintained on the arachis oil diet did not

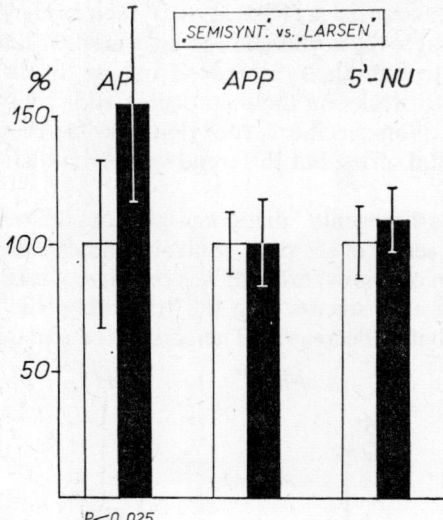

Fig. 37.—The activity of aortic alkaline phosphatase, ATPase and 5′-nucleotidase in rats fed the ordinary laboratory Larsen diet (black columns) and the basic semisynthetic diet (empty columns). Upright line with bars ± S.D. Results expressed as percentage differences. (Data from Zemplényi, 1966a.)

reveal any significant tendency towards reduced malate dehydrogenase activity. It is impossible to decide from the results whether this difference reflects differences in the final outcome of the experiments, as out-lined in the introduction, or whether a definite decrease in malate dehydrogenase activity would have occurred if we had prolonged the third experimental series for a few more days.

As aortic connective tissues do not proliferate in the early stages of the experiments just described, no change in the activities of phosphoric monoester hydrolases would be expected. The *paradoxical decrease* in phosphatase activities in the first two experimental series was clarified by the results of the third experimental series. The unexpected drop in phosphatase activities can be

Fig. 38.—The activity of aortic malate dehydrogenase, lactate dehydrogenase and nonspecific carboxylesterase in the same animals as in Fig. 37.

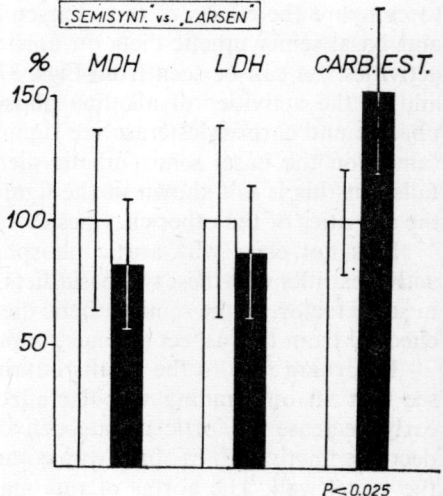

ascribed to differences in the basal diets *per se*, and not to the effects of the special experimental diets. The same conclusion applies to the tendency towards decreased non-specific carboxylesterase activity found in the experiments with the thrombogenic diet. (Zemplényi, 1966). In fact, recently, Kahn and Slocum (1967b) did not observe significant changes in acid or alkaline phosphatase activities in aortas of rats fed "atherogenic diets". When thiouracil was added to the diet, an increase of alkaline phosphatase activity in late stages (35–39 weeks) of the experiment could be observed.

An important feature of our experiments is the consistent *rise in ATPase* activity in the aortas of rats maintained on the *thrombogenic diet*; it was never observed when arachis oil was used instead of butter (Zemplényi *et al.*, 1966e). Since this finding will later be related to the increased incidence of arterial thrombosis in such rats (see p. 143), it is appropriate at this stage to mention some recent findings concerned with platelet aggregation and thrombogenesis.

Two steps seem to be essential in the initial stages of haemostasis and the formation of a white thrombus: (*a*) simultaneous adherence of platelets to damaged tissue and their aggregation to form a loose plug—this is a reversible process; (*b*) fusion of the aggregated platelets by so-called viscous metamorphosis.

Until recently relatively little was known about the first step. It was generally assumed that the platelet plug is produced by thrombin. However, adherence develops more rapidly than in any known mechanism for thrombin production in blood. Hugues (1960) and Bounameaux (1961) observed that blood platelets adhere to damaged mesenteric fibres *in vivo* as well as to aortic fragments *in vitro*, while Kjaerheim and Hovig (1962) identified the fibres by electron microscopy as collagen. At the same time Zucker and Borelli (1962) found that saline "extracts" (suspensions) of collagen and a variety of connective tissues cause platelets to aggregate. This effect was abolished by collagenase, thus indicating that collagen was probably the active material responsible for aggregation. "Extracts" from atherosclerotic human aortic plaques were only about half as active as those from grossly normal intima. However, the addition of such connective tissue "extracts" to platelet-rich plasma induces platelet aggregation in 50–60 seconds, whereas under physiological conditions platelets aggregate much more rapidly. This indicates that another independent or related mechanism may be present.

In 1960 Hellem extracted a water-soluble substance ("factor R") from red cells. *In vitro* it very effectively increased platelet adhesiveness to foreign surfaces, such as glass beads. The same substance also produced platelet aggregation within a few seconds, when added to platelet-rich plasma. Later Gaarder *et al.* (1961) showed that "factor R" is identical with adenosine diphosphate (ADP). Trace amounts of ADP caused almost instantaneous aggregation of platelets, and the effect of ADP on platelet adhesiveness was the same as that of "factor R".

If adenosine diphosphate plays the same role *in vivo* as it does *in vitro*, it need not necessarily originate from red blood cells. Born (1956a) found that platelets are extraordinarily rich in adenosine triphosphate (ATP), and that the latter disappears during both blood clotting (Born 1956b), and viscous metamorphosis of platelets. Since ADP is the first breakdown product of ATP,

Born (1962) suggested that ADP derived in this way might cause the increased adhesiveness of platelets and, furthermore, may underlie both the physiological mechanism responsible for the formation of haemostatic platelet plugs and the pathological mechanism responsible for thrombosis.

Hellem (1964) observed that the reaction between ADP and platelets could be inhibited by monoiodoacetic acid. This observation together with other related evidence suggests that an enzymatic process is present in which ADP takes part. On the basis of somewhat similar findings with iodoacetamide, Garvin (1961) concluded that this substance influences platelet adhesiveness either by blocking glycolysis or by interfering with an SH-linked enzyme reaction.

There are indeed good reasons to believe that the ADP mechanism also operates *in vivo*. When ADP is injected into rabbits (Hellem, 1960) or infused into cats (Born and Cross, 1963) the concentration of circulating platelets immediately falls. The same result was obtained with labelled platelets in human volunteers. However, the platelet count soon rises again, perhaps even during the infusion of ADP. This seems to suggest that ADP is broken down to derivatives, such as adenosine monophosphate and adenosine, which antagonize the aggregating effect of ADP (Born, 1965). Moreover plasma inactivation of ADP has been demonstrated by Ödegaard *et al.* (1964) and others.

Honour and Mitchell (1963) investigated the effect of some enzyme inhibitors on these events, in particular the *in vivo* formation of white thrombi after vascular injury. They observed that not only monoiodoacetate (see above) but also sodium azide and sodium fluoroacetate are inhibitory. The latter observation is of special interest, as fluoroacetate and fluorocitrate are well-known specific inhibitors of aconitate hydratase (aconitase) and thus of the whole tricarboxylic acid cycle (Morrison and Peters, 1954).

It appears that the aggregating effect of ADP and the above mentioned effect of collagen are in some way related.

Hovig (1963a) confirmed that the aggregation induced by a saline "extract" of rabbit tendon was due to a substance of collagenous nature and not to the presence of ADP. Such aggregation occurred within about a minute. However unequivocal evidence was presented by Hovig (1963b) that addition of tendon "extract" to platelet-rich plasma induces the release of ADP from the platelets themselves and the latter substance causes aggregation within 15 seconds. It could, therefore, be concluded that the reaction initiated by the collagenous substance consists of two steps: (*a*) release of ADP from the platelets, (*b*) aggregation caused by released ADP. It should be noted that some authors even assume that the aggregating effect of thrombin is mediated through release of ADP.

According to Mustard (1966) the reaction of platelets to surface stimuli, such as contact with collagen, has many characteristics similar to those of the white cell in inflammation, immunological injury and phagocytosis. Thus, under such conditions, the platelet also releases factors that increase vascular permeability and may produce focal injury of the vessel wall. Such factors may also be important in atherogenesis.

Most investigators seem to believe that the ADP that is responsible for these platelet changes originates from either red cells or platelets. On the basis of their above-mentioned observations, Honour and Mitchell (1963) suggested that white thrombus formation might be connected with the release of ADP and similar substances from damaged *vascular tissue itself*. In fact, Gaarder et al. (1961) in their paper demonstrating the importance of ADP as a factor in platelet adhesiveness also assumed that the release of ADP by cellular damage may play an important part in the initial stages of haemostasis and thrombosis.

Born (1965) recently expressed the view that the whole "chain reaction" that underlies the initial adhesion of platelets in the white thrombus can be initiated by the release of traces of ADP from the vessel wall. Afterwards it is assumed that a reaction takes place between ADP, calcium and protein on the surface of the adherent platelets and this in turn may induce the release of more ADP from the platelets.

Other possible schemes have been suggested by O'Brien (1963).

Without going into further details about this interesting topic let us now return to our findings in rats fed the "thrombogenic" and "atherogenic" diets. We have seen that there is a consistent increase in the activity of vascular ATPase in rats fed the butter-enriched "thrombogenic" diet but not in those given the "atherogenic" arachis oil diet. In view of the importance of ADP in thrombogenesis, the increased vascular ATPase activity may be related to the thrombogenic properties of the butter-enriched diet. It is reasonable to expect that ADP would accumulate in the vessel wall, as ADP is the product of vascular ATPase activity. As histochemical evidence shows that ATPase activity is present in both medial muscle cells and endothelial cells, an excess of ADP could become available even without the intervention of injury or increased permeability of the arterial wall. This conclusion is in line with the above mentioned views on the vascular origin of ADP, at least in the initial stages of platelet adherence. Unfortunately, these possibilities must be regarded as purely speculative until it is shown that the ADP content of the vascular wall is really increased under the conditions of the experiment.

The differences observed in vascular ATPase activity between animals fed the butter and arachis oil diets might be connected with known properties of ATPases. Pressman and Lardy (1956) observed that low concentrations of fatty acids can influence ATPase activity in mitochondria, while Skou (1964) reported similar findings in relation to membrane transport ATPase.

It is of interest in this connection that Shore and Alpers (1963) observed that stearate and longer-chain saturated fatty acids caused platelet aggregation. Haslam (1964) also found that stearate, behanate, palmitate and other saturated fatty acids in relatively high concentrations (0·1 mM) produced platelet aggregation. In an "adrenaline-ATP system" Ardlie et al. (1966) observed aggregation after addition of considerably lower concentrations of stearate but not after adding palmitate and the unsaturated fatty acids oleate, linoleate and linolenate.

It must be emphasized that the enhanced activity of vascular ATPase and the anticipated resulting accumulation of ADP cannot be assumed to be the only cause of the increased tendency to thrombus formation in the butter-fed rat. From the work of Davidson et al. (1961, 1962), Teitelbaum et al. (1962) and Naimi et al. (1962) it is evident that substantial changes occur in the blood clotting mechanism in rats fed the "thrombogenic" diet. Connor et al. (1965) summarized evidence that indicates that long chain saturated fatty acids induce

TABLE II

Average enzyme activities in normal male rat aortas

Enzyme	Av. activity	Unit
Alkaline phosphatase	19·83	µM phenol/100 µM extr. prot./hour
Acid phosphatase	1·14	Same as above
ATPase	6·34	µg phosphate/100 µg extr. prot./hour
5'-Nucleotidase	1·28	Same as above
Malate dehydrogenase	57·44	µg diformazan/100 µg extr. prot./hour
Lactate dehydrogenase	186·88	Same as above
Succinate dehydrogenase	66·8	µg diformazan/100 µg DNA/45 min.
Malate dehydrogenase (opt. test)	256·0	Wróblewski u./100 µg extr. prot.
Lactate dehydrogenase (opt. test)	47·9	Same as above
β-Glucuronidase	65·0	µg phenolphtaleine/100 µg extr. prot./16 hours
Non-spec. carboxylesterase	2·87	µg β-naphthol/100 µg extr. prot./hour

marked acceleration of blood clotting by several mechanisms, in particular by activation of the Hageman factor, whereas unsaturated fatty acids lack this effect. This increased coagulability is probably the main cause of the high incidence of thrombosis in animals fed the butter-containing diet. Nevertheless, it is important to realize that vascular ATPase activity also increases and, in this connection, the possibility of a cause-and-effect relationship has to be seriously considered.

Some other problems concerning experimental rat atherosclerosis will be discussed in Chapter XV and, in connection with vascular lipolytic activity, in part three of this volume.

Chapter XV

THE EFFECT OF INJURY ON VASCULAR ENZYMES

IN the middle of the last century Virchow (1856) and Rindfleisch (1872) suggested that local injury is an important factor in the genesis of arterial lesions (see also Aschoff, 1924). Since that time many different types of arterial damage have been studied with the purpose of elucidating the role of injury in the pathogenesis of atherosclerosis. In a recent publication Constantinides (1965) listed the large number of factors that have been found to injure the vascular wall. He called attention to the following agents and processes that have been reported to damage arteries: hypertension, catecholamines, renal damage, hypotension; mechanical insults such as stretching, air embolism, compression and injection of particulate suspensions; radiation, deficiencies of protein, choline, copper, magnesium, pyridoxine and tocopherol; excess of vitamin D or related sterols; diverse substances such as aminonitriles, allylamine, glucosamine, sodium caseinate, haematoporphyrin, alkaloids and various inorganic chemicals; endocrine dysfunction and calciphylaxis; finally, immune reactions, lipaemia, and nerve stimulation.

Experimentally, it has been established that injured animal arteries are selectively susceptible to lipid accumulation when exposed to some kind of lipaemia. This was reported in the older literature (Anitschkow, 1913*b*; Ssolowjew, 1929 and others) as well as in more recent studies in which different types of vascular injuries had been induced (Taylor *et al.*, 1950, 1954; Taylor, 1954; Oester *et al.*, 1955; Prior and Hartmann, 1956; Gutstein *et al.*, 1963; Constantinides, 1963, 1965; Courtice and Schmidt-Diedrichs, 1963; and many others). The increased sensitivity of the damaged arterial wall to experimental atherosclerosis is clearly demonstrated by the development of lesions within a much shorter time and at much lower levels of lipaemia than usual.

Gresham and Howard (1963) studied spontaneous and experimental atherosclerosis in many species, including man, and reached the conclusion that the early atherosclerotic lesion in man and other animals begins as a reparative response to injury, the cause of which is unknown. Likewise, Buck (1963) thinks that atherosclerosis represents the morphological response to some kind of arterial injury, although the nature of the injury is not completely understood. In 1957 Gillman expressed similar views on the basic process underlying atherogenesis. Waters (1965) maintains that "the atherosclerotic lesion is basically an inflammatory reaction of vascular tissues to injury", the reaction being "greatly modified by the anatomy and physiology of these specialized structures and by their contact with the circulating blood, as well as by the intensity and duration of the stimulus. A central feature of arterial inflammation, as related to atherosclerosis, is increased, transient, local permeability of the vessel wall to blood plasma".

At this point it is relevant to mention that the early experiments of Anitschkow (1921) and Petroff (1922) with colloidal dyes revealed that injury to the artery, however slight, promptly results in an increased arterial permeability. In fact Anitschkow (1933) very strongly emphasized the role played by arterial injury as a factor in the pathogenesis of atherosclerosis, not only by increasing permeability but also by promoting the development of intimal thickening. Such "predisposing" factors together with increased cholesterolaemia—the "exciting" cause—formed the basis of his "combination theory" of atherosclerosis.

Morphologically, injury to the arterial wall consists of a mixture of destructive and regenerative or repair processes.*

The destructive processes that result from arterial injury have been very clearly set out by Constantinides (1965) as follows:
(1) damage or necrosis of cells in the inner arterial lining,
(2) breakdown, splitting and eventual disappearance of certain intercellular materials that were maintained by the destroyed cells (notably the elastic lamellae),
(3) flooding of the injured area with varying amounts of fluid, protein and sulphated (acid) mucopolysaccharide.

In Constantinides's opinion, such "inundation" with polysaccharides is a common phenomenon widely encountered in the early stages of mesenchymal reaction to injury. Gillman et al. (1957), however, believe, that accumulation of mucopolysaccharides represents a normal phase in the repair of all injured connective tissues.

The regenerative processes consist of an attempt by arterial tissue to build a new wall to replace the destroyed parts (Hass, 1954). These processes lead to the formation of a new endothelium and the proliferation of modified smooth muscle cells ("myo-intimal cells") between the old internal elastic membrane and the new endothelium (Buck, 1963). Smooth muscle cells proliferate in from the media (but see p. 187) and enter the intima either through fenestrations or ruptures in the internal elastic membrane. These cells are longitudinally oriented in the vessel (Buck, 1963) and, according to Constantinides (1965), whenever we encounter cells oriented in this way in the inner media or in the intima we have to suspect that these sites have previously been injured. The repair processes that follow some types of arterial injury are characterized by connective tissue reactions culminating in fibrosis, derangement of collagen structure and production of "vascular scar tissue". This scarring narrows or dilates the lumen, and distorts the vascular wall and perivascular tissues (Gillman, 1964).

Modified smooth muscle cells are not only a characteristic feature of regeneration after arterial *injury*, but they appear to play an important role in both experimental and spontaneous *atherosclerosis*. These cells have been detected with the electron microscope in experimental atherosclerotic lesions in the rabbit's coronary artery (Parker, 1960) and aorta (Buck, 1963), in spontaneous lesions in animals (see Chapter XVIII) and also in human atherosclerotic lesions (Geer et al., 1961; McGill and Geer, 1963; Haust and More, 1963). Such findings

---

* As is well-known to general pathologists, Gillman (1964) recommends the consistent use of the term "repair" when referring to the replacement of injured tissues by new structures, including scar tissue, whereas "regeneration" should be reserved for those processes that replace the structure in its original form.

are further evidence of a connection between vascular injury and atherogenesis.

It is interesting to note that the newly-formed cells in regenerating vascular tissue appear transiently to possess a special affinity for lipids, but after a few weeks they lose the ability to accumulate these substances (Cox *et al.*, 1963; Taylor, *et al.*, 1963; Constantinides, 1965).

The findings discussed above clearly show the extreme importance of arterial damage (whatever its cause) in the pathogenesis of atherosclerosis. It is, therefore, relevant to see whether changes in arterial enzyme activities reflect vascular injury and related connective tissue changes and if so, whether such enzyme changes can be correlated with findings in atherosclerosis.

In the sections that follow we shall be concerned with these problems in connection with calciferol intoxication and experimental hypertension in rats.

Feeding Rats with Excess Vitamin D

Old rats are known to develop a kind of spontaneous arteriosclerosis with the characteristics of mediocalcinosis (Hummel and Barnes, 1938). This disease is most prominent in repeatedly bred female rats (Gillman and Hathorn, 1959; Záhoř, 1963; Wexler, 1964). The pathological arterial lesions are accentuated following ACTH administration (Wexler and Miller, 1959) or by the combined effect of unilateral nephrectomy and ACTH administration (Wexler and Miller, 1959; Wexler *et al.*, 1960). According to Záhoř and Czabanová (1964), lactation is connected with the appearance of such lesions, as they do not occur if female rats are prevented from suckling their offspring.

As early as 1930 Duguid demonstrated that feeding excess vitamin D causes mediocalcinosis in the arteries of rats. Subsequent work by Pfleiderer (1932) revealed that combination of excess vitamin D feeding with physical exercise and cholesterol feeding can induce atherosclerotic-like lesions even in the rat (see Chapter XIV).

Pfleiderer's work was for long forgotten and only recently has interest been re-awakened in producing experimental atherosclerosis, after first "preparing" the artery with calciferol. In 1957 and 1958 Wilgram described atherosclerotic changes in the aortas and coronaries of rats, which had been fed egg and cholic acid and had been intoxicated with vitamin D and thiouracil. Such rats developed coronary atherosclerotic changes (in 63 per cent), coronary occlusions (in 33 per cent) and myocardial infarctions (in 14 per cent).

However, a high mortality was encountered in rats treated in this way. Constantinides (1963) overcame this difficulty by modifying the dietary regime and, thus, induced atherosclerotic lesions in rats within 6 weeks. For the first two weeks he fed the animals a cholesterol-thiouracil-cholate diet, then he administered a large excess of vitamin D for 3 days in the third week and, thereafter, he fed them the original diet for a further three weeks before killing them. Mortality amongst these rats was minimal. The diet used in the experiments was essentially a butter-free variant of that described by Hartroft and Thomas (1957), and contained practically the same ingredients as Wilgram's diet.

It is now important to consider what kind of vascular enzyme changes develop in rats' aortas as a result of calciferol feeding and the resulting vascular injury. In this connection it must be emphasized that the rat's aorta reacts intensely to the acute injury caused by a short period of calciferol intoxication

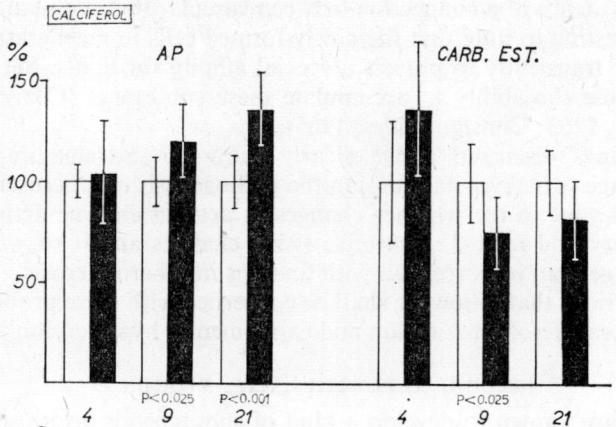

Fig. 39.—The activity of aortic alkaline phosphatase and non-specific carboxylesterase in calciferol-intoxicated rats and killed four, nine or twenty-one days after the beginning of calciferol feeding. Empty columns = control rats (24). Black columns = experimental rats (24). Upright line with bars ± S.D. Results expressed as percentage differences. (From Zemplényi and Mrhová, 1964b.)

(Gillman, 1964). The histological features of this reaction are those that might be expected to result from arterial injury and repair (see above), but with the addition of intense calcification. We shall later return to the morphological picture when the histochemical findings are discussed.

Accumulation or depolymerisation of vascular mucoproteins could play a role in the increased tissue avidity for calcium. Such changes are claimed to facilitate metastatic calcification (Sobel, 1955) and bind lipids (Gillman, 1958; Gerö et al., 1962a, b; Bihari-Varga et al., 1964).

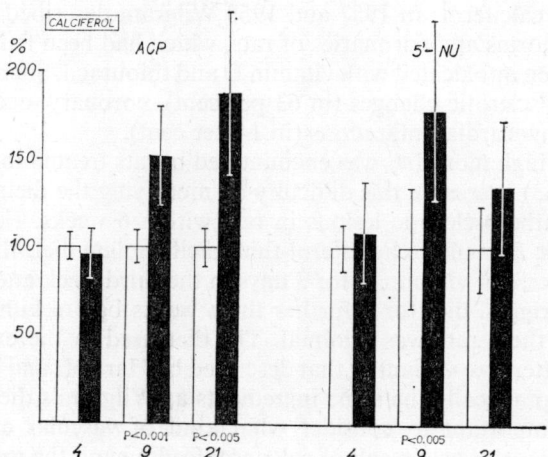

Fig. 40.—The activity of aortic acid phosphatase and 5'-nucleotidase in the same animals as in Fig. 39.

In the experiments to be described (Zemplényi and Mrhová, 1965a), male albino rats were fed an ordinary laboratory diet (see p. 138) and, for a period of 5 or 9 days, an oral dose of 30,000 units of vitamin D in oil was administered daily. The drug (Calciferol, Spofa) was fed to unanaesthetized animals by means of a syringe with a short length of attached polyethylene tubing. Control animals were fed the laboratory diet only. The animals were killed by decapitation, immediately after which the aortas were removed and processed as described in Chapter XI.

As seen from Figs. 39 and 40 the activity of the investigated enzymes did not significantly change at the fourth day of the experiment. However, in rats fed calciferol daily and killed on the ninth day of the experiment, aortic non-specific carboxylesterase activity significantly decreased, in contrast to the increased activities of 5'-nucleotidase and alkaline and acid phosphatase. The activity of the last two enzymes remained significantly elevated in rats fed for 9 days and killed on the 21st day, but the activities of 5'-nucleotidase and carboxylesterase were returning to more normal values at this time. No difference could be shown in aortic ATPase activity in experimental and control animals.

In another experimental series (Figs. 41–43), where animals were fed calciferol for only 5 days, increased activity of phosphoric monoester phosphohydrolases was again observed on the ninth day of the experiment. Alkaline phosphatase activity tended to increase as early as the fifth day. At this early stage of the experiment ATPase activity was slightly but significantly increased, but this was not demonstrable at the later stages.

No change could be observed in the activity of lactate and malate dehydrogenases at the very early stage of the experiment, either by the "optical test" based upon changes in NAD concentration or the combined phenazine methosulphate-neotetrazolium method. At later stages, however, there was a clear tendency for dehydrogenase activity to decrease, especially in the case of the succinate dehydrogenase system (Zemplényi et al., 1962b).

Before moving on to the histochemical findings, the morphology of calciferol-induced lesions must be outlined. The thorough investigations of Gillman and co-workers (Gillman et al., 1957, 1960; Gillman, 1964) have shown that the vascular injury caused by calciferol leads to a "wave of arterial necrosis" and subsequent calcification. The coronary arteries and other medium-sized visceral vessels are first afflicted; aortic changes are apparent somewhat later than in the coronaries. The abdominal aorta exhibits a typical reaction a few days later than the ascending aorta, arch and thoracic aorta. In addition to necrosis and calcification, Gillman et al. (1957) often found prominent cellular reactions in the intima, sometimes with heavy cellular infiltration at the coronary orifices. At later stages, large plaques of heavily metachromatic cartilage were a common finding in the aorta. Whereas coronary calcification rapidly resolves, the persistence of intense aortic calcification with cartilage formation is a striking feature of almost all cases and it becomes progressively more severe (Gillman et al., 1960; Gillman, 1964).

In our experiments with calciferol, Lojda and Urbanová histochemically investigated the same enzymes as had been studied in the arteries of atherosclerotic rabbits. The aortas of calciferol-fed rats were enzymatically different from those of control animals as early as the third day of the experiment, and

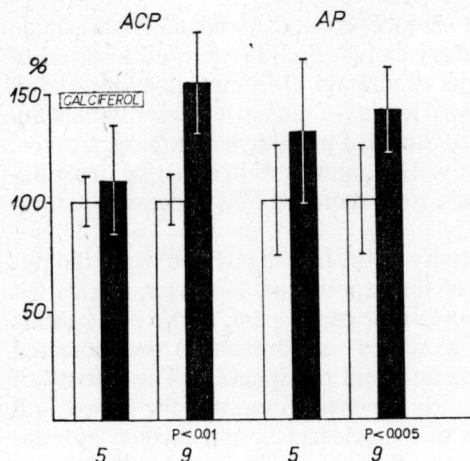

Fig. 41.—The activity of aortic acid phosphatase and alkaline phosphatase in another series of rats fed 5 days 30,000 I.U. of calciferol daily and killed the 5th or 9th day of the experiment. Empty columns = control rats. Black columns = experimental rats (12). Upright line with bars ± S.D. Results expressed as percentage differences. (Data from Zemplényi and Mrhová, 1965a.)

Fig. 42.—The activity of aortic ATPase, 5'-nucleotidase and non-specific carboxylesterase in the same animals as in Fig. 41.

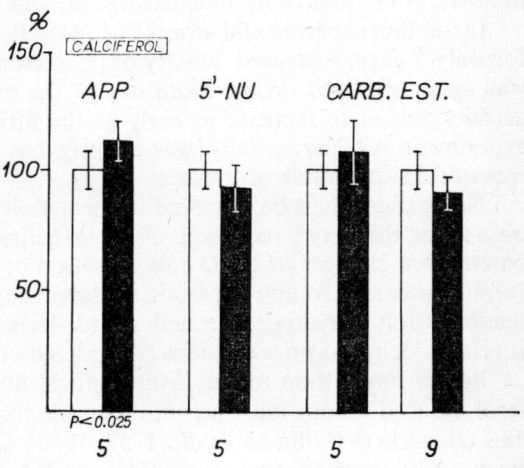

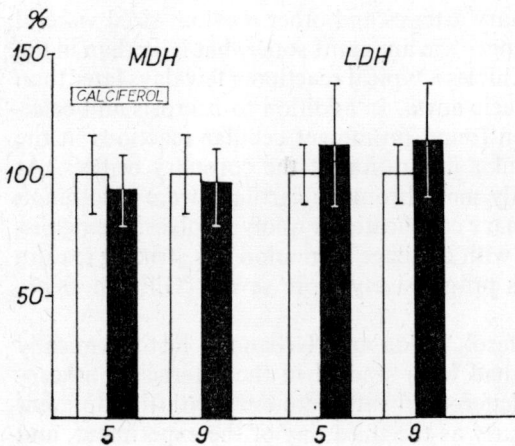

Fig. 43.—The activity of aortic malate dehydrogenase and lactate dehydrogenase in the same animals as in Fig. 41.

with the development of morphological lesions such differences became more pronounced (Plate III). The activities of all dehydrogenase systems and of non-specific esterase (together with cholinesterase) decreased and later even disappeared at foci of calcium phosphate deposits. Conversely, acid phosphatase activity increased in the muscle cells around these deposits, whilst alkaline phosphatase became more active both in the muscle cells and in the capillaries that invaded the lesions from the adventitia.

As seen from our biochemical and histochemical results, a salient feature of the aortas of calciferol-fed rats is an early rise in the activity of the phosphomonoester hydrolases (acid and alkaline phosphatases and 5'-nucleotidase). The increased activity of acid phosphatase can reasonably be attributed as a reaction to tissue injury (see p. 127). The rise in the activities of the other two phosphomonoester hydrolases might well reflect connective tissue reaction in the injured arterial wall. In Chapters VIII and XIII we mentioned that a function, ascribed by some authors to alkaline phosphatase, is the formation of fibrous proteins and in particular collagen fibres, whereas both alkaline phosphatase and 5'-nucleotidase seem to be connected in some ill-understood way with tissue calcification. Such views are consistent in particular with the histochemical findings described above.

The fall in the activity of carboxylesterase runs parallel with the decreased lipolytic activity displayed by the aortas of calciferol-intoxicated rats (see p. 219). This phenomenon might also be connected with the accumulation of abnormal mucopolysaccharides and calcium salts that is seen in the aortas of these animals, as under certain circumstances, both substances inhibit lipolytic activity in aortic tissue (see Chapter XXIII).

Our biochemical results show no overall change in the activity of aortic malate dehydrogenase in these early stages of calciferol feeding. Likewise, aortic succinate dehydrogenase activity was unaltered in another experimental series. Changes in the activities of tricarboxylic acid cycle (TCA) dehydrogenases appeared only in the late stages of the experiment (i.e. after 3 or more weeks). Therefore, we concluded that decreased activity of tricarboxylic acid cycle enzymes is not a characteristic feature of the early enzymatic changes in aortas damaged by calciferol intoxication (Zemplényi and Mrhová, 1965a). Nevertheless, the histochemical findings described above indicate that the activities of all dehydrogenase systems are focally decreased in the early stages. Presumably this change is initially too discrete to be reflected in overall enzyme activity and manifests itself "biochemically" only when the tissue is more grossly afflicted.

Returning now to the basic questions propounded in the introduction to this chapter, it can be stated that injury to the vessel wall, at least that caused by calciferol intoxication, causes clearcut early changes in the activities of the enzymes studied. These changes are characterized biochemically and histochemically by increased activity of phosphoric monoester hydrolases and decreased activity of non-specific carboxylesterase. The activities of some dehydrogenases, especially those of the tricarboxylic acid cycle, focally decrease in the early stages of intoxication but overall activity does not decline until later.

As injury consists of a mixture of destructive, regenerative and repair processes, it is very difficult to decide from our experiments which enzyme change is related to which of these events. However, in view of the functions previously ascribed to these enzymes, it is reasonable to assume that increased acid phosphatase activity and decreased TCA enzyme activity is connected with the tissue injury proper, whereas increased activity of alkaline phosphatase and 5'-nucleotidase probably reflects the onset of regenerative or repair processes.

Studies concerned with other forms of vascular injury will disclose whether or not these enzymatic changes are common denominators of arterial injury in general.

Arterial Wall Injury Caused by Experimental Hypertension

The relationship between atherosclerosis and elevated blood pressure appears to be a well-established aetiological factor in atherosclerosis.

Bell and Clawson (1928) found advanced atherosclerotic lesions of the left coronary artery in 38 per cent of hypertensive patients while only 4 per cent of normotensives manifested similar lesions. Davis and Klainer's (1940) study of 500 necropsies revealed markedly more severe coronary atherosclerosis in patients with essential hypertension than in normotensives, particularly in the age groups below 50 years. Wilkins et al. (1959) found that coronary and cerebral atherosclerosis and the occurrence of myocardial infarction are significantly associated with increased heart weight, which was taken as evidence of hypertension. From a study of more than 6,000 autopsies in Prague, Mentl (1966) recently reported that severe stenosing coronary atherosclerosis was present in 44·5 per cent of hypertensive males, whereas in patients of the same age group with chronic cor pulmonale the corresponding figure was only 6·4 per cent.

Young et al. (1960) observed a reasonably high correlation between *antemortem* blood-pressure levels and *post-mortem* quantitative assessments of atherosclerosis. The "Framingham study" (Kagan et al., 1962), based on eight years' prospective epidemiological observations, has revealed that the blood pressure level is an important factor in predicting coronary heart disease. Subjects with systolic blood pressure levels of 180 mm. Hg or over at the initial examination developed over twice as much coronary heart disease as would have been expected for the age/sex composition of the group. On the other hand, subjects with systolic blood pressure below 120 developed only a quarter of the expected coronary heart disease. Thus, there was an eight- to ninefold increase in the risk of developing coronary heart disease, when subjects in the upper blood pressure range were compared with those in the lower. Similar conclusions were reached when diastolic blood pressure was considered.

Turning now to experimental models of atherosclerosis, Wakerlin et al. (1951, 1957) and Moses (1954) demonstrated that renal hypertension enhances the degree of experimental atherosclerosis in dogs. Similarly, Nuzum et al. (1926), Wilens (1943), Wolkova (1953) and Heptinstall et al. (1958) have shown that the amount of aortic atheroma in the hypertensive rabbit correlates well with the blood pressure level. This was true not only in experiments with sustained blood pressure elevation, as even a short period of increased blood pressure exhibited this facilitatory effect (Heptinstall and Porter, 1957). The combination of hypertension and other atherogenic procedures has been successfully used in the rat by a number of workers (Malinow et al., 1954, Fillios, et al., 1956 and others). In the rhesus macaque, McGill et al. (1961) found that animals subjected to experimental renal hypertension for 19 to 49 months developed considerably more extensive and advanced aortic lesions than did normotensive animals. The lesions somewhat differed from those of human atherosclerosis and those induced in monkeys by dietary manipulation, but shared some of the characteristics of both.

All the above clinical, morphological and experimental evidence clearly demonstrates that there is a close relationship between intravascular pressure and atherosclerosis. However, the mechanism of this effect is not clear. Proponents of the filtration theory of atherogenesis maintain that increased transarterial filtration pressure in hypertension facilitates penetration of lipids (lipoproteins) into the vessel wall and thus promotes atherosclerosis. *In vitro* studies on arterial segments (Wilens, 1951a; Evans et al., 1952) seem to support this view. However, other considerations, including experimental and morphological evidence to be mentioned below, indicate that the mechanism is probably more complicated.

Waters (1954) has shown that even brief episodes of hypertension may damage the dog's aorta. Necrotizing arterial lesions can be produced by injection of catecholamines and renin preparations. The lesions occur predominantly at bifurcations or other sites of maximum arterial stress. Likewise, Magarey (1957) observed that experimentally hypertensive rats develop necrotic changes (hyalinization and vacuolation) in the aortic intima.

In rats with DOCA+salt-induced hypertension, Esterley and Glagov (1963) noted certain ultra-structural changes. In particular, they observed (*a*) degenerative and vacuolar changes in both endothelial cells and medial cells; (*b*) accumulation of extracellular osmophilic material in the subendothelial space and tunica media; and (*c*) the appearance of blood cells in the subendothelial space. They interpreted these findings as a reflection of altered permeability in the arterial wall. Such changes could again be regarded as the direct result of injury to the vessel wall.

From these collective observations it is reasonable to conclude that raised arterial pressure injures the vascular wall even if the lesions produced morphologically do not resemble those in any stage of atherosclerosis (Waters, 1965). With this in mind and because of the close correlation between hypertension and atherosclerosis, we investigated the enzymatic changes that develop in the arteries of hypertensive rats.

In collaboration with Drs. Nenov and Vranešič experimental hypertension was induced in rats by two independent methods (Zemplényi et al., 1967):

(*a*) Thirty unilaterally-nephrectomized young rats were administered a suspension of desoxycorticosterone acetate (DOCA) in Tween (1·1 mg./100 g. body weight, twice a week) and were given a 1 per cent NaCl solution to drink for a period of 42 days (Musilová et al., 1966). Control unilaterally nephrectomized rats were injected with either Tween or DOCA in Tween or were allowed to drink a 1 per cent NaCl solution *ad libitum*.

(*b*) One renal artery was clamped in 24 rats and the animals were killed 22 to 83 days later (Wilson and Byrom, 1939; Floyer, 1962). Rats treated in the same way who did not develop hypertension served as controls.

Arterial pressure was measured by a plethysmographic method (Vaněček and Trčka, 1959) at least once each fortnight and on the day before each animal was killed.

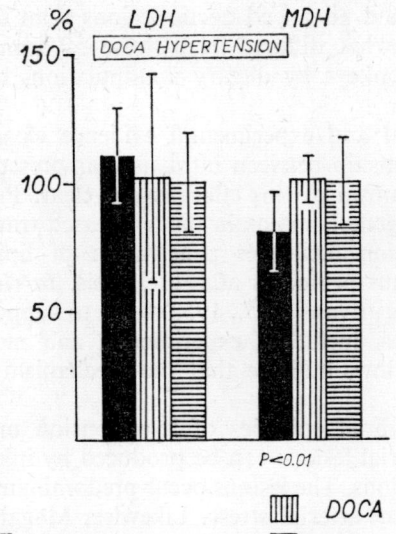

Fig. 44.—The activity of aortic lactate dehydrogenase and malate dehydrogenase in rats with hypertension induced by unilateral nephrectomy and DOCA + NaCl administration. Upright line with bars ± S.D. Results expressed as percentage differences. (Data from Zemplényi et al., 1966d, 1967.)

Figures 44 to 47 summarize some results obtained in experiments in which hypertension was induced with DOCA and NaCl. The aortas of rats that developed protracted arterial hypertension together with cardiac hypertrophy exhibited lower activities of malate dehydrogenase and non-specific carboxylesterase than those of control animals. The activities of aortic lactate dehydrogenase and β-glucuronidase remained unaltered. Alkaline and acid phosphatase activities increased while ATPase activity decreased. Aortic enzyme activities in

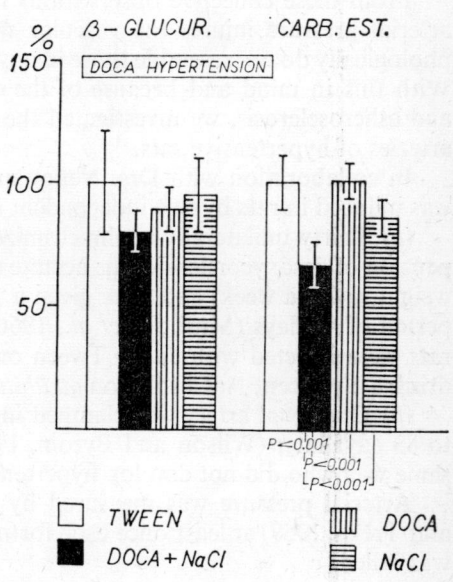

Fig. 45.—The activity of aortic β-glucuronidase and non-specific carboxylesterase in the same animals as in Fig. 44.

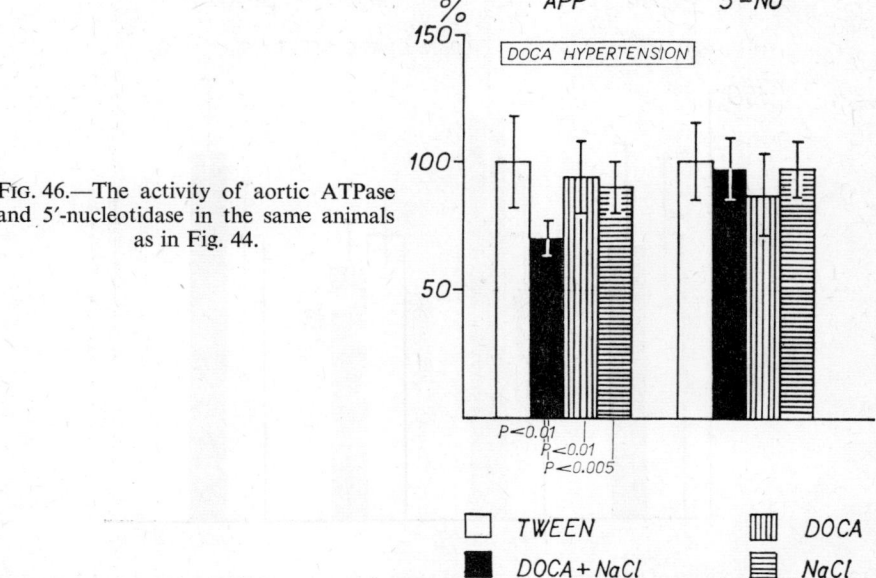

FIG. 46.—The activity of aortic ATPase and 5′-nucleotidase in the same animals as in Fig. 44.

all "control" groups remained unaltered, with the exception of rats treated with DOCA alone where aortic alkaline phosphatase tended to decline in activity.

Figures 48 and 49 summarize the changes in aortic enzyme activities in rats with unilateral renal artery stenosis accompanied by hypertension, as compared with similarly treated rats that did not develop raised arterial pressure. The

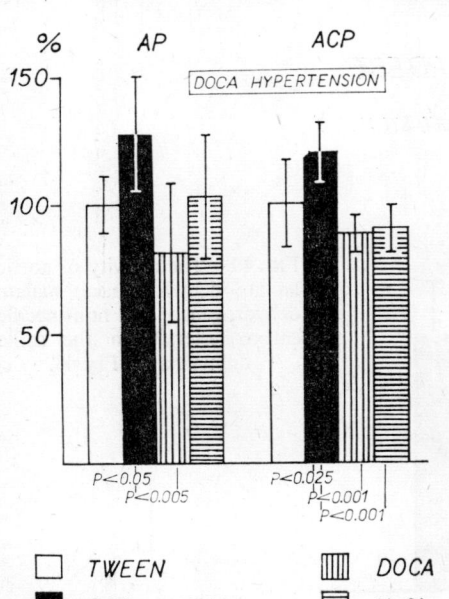

FIG. 47.—The activity of aortic alkaline phosphatase and acid phosphatase in the same animals as in Fig. 44.

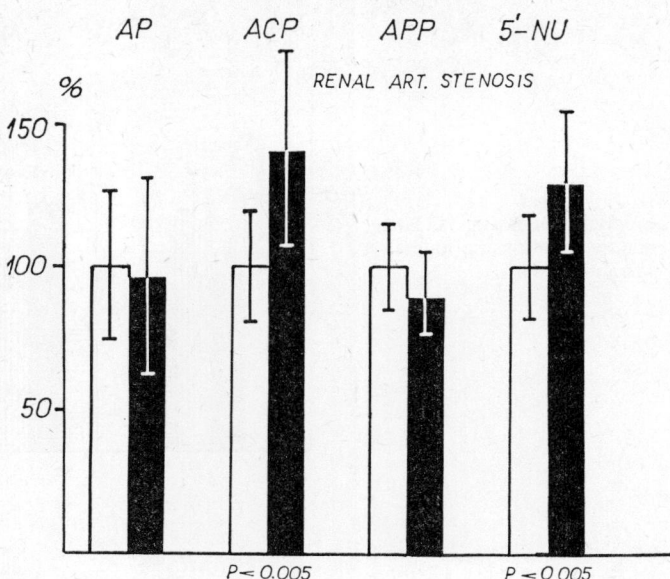

Fig. 48.—The activity of aortic alkaline phosphatase, acid phosphatase, ATPase and 5′-nucleotidase in rats with hypertension induced by unilateral renal artery clamping. Empty columns = operated rats which did not develop hypertension. Black columns = hypertensive rats. Upright line with bars ± S.D. Results expressed as percentage differences. (Data from Zemplényi et al., 1966d, 1967.)

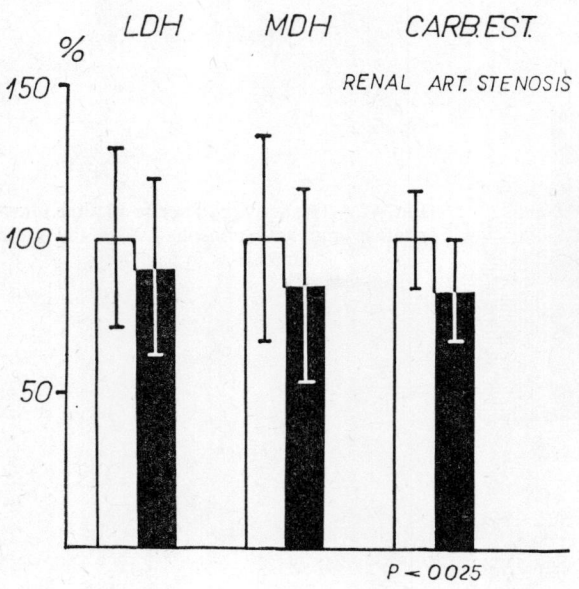

Fig. 49.—The activity of aortic lactate dehydrogenase, malate dehydrogenase and non-specific carboxylesterase in the same animals as in Fig. 48.

activities of acid phosphatase and 5'-nucleotidase increased, non-specific carboxylesterase activity fell, while the activity of malate dehydrogenase only tended to decline. It must be added that the rats in this experimental series did not form such a homogeneous group as the preceding one, because the time interval between arterial clamping and killing varied from 22 to 83 days. Nevertheless, in most instances the enzymatic changes showed the same trend as in rats with DOCA + NaCl hypertension, suggesting that the common denominator causing these changes was actually the raised arterial pressure.

Only aortas from the first series of experiments were available in sufficient numbers for histochemical studies. Independently from the biochemical results these studies revealed increased overall activity of acid phosphatase and decreased succinate dehydrogenase activity in the aortas of hypertensive rats. These findings accord with those of Postnov (1965), who also used histochemical methods to study arterial enzymes in rats with renal hypertension. He reported decreased activities of succinate dehydrogenase, non-specific esterase and cholinesterase, and increased alkaline and acid phosphatase activities, especially in the aortic media. It is surprising, however, that in *very small visceral arteries* of such rats Gardner and Laing (1965) could not detect by conventional histochemical methods any change in enzyme activity.

Finally it must be mentioned that there was no electrolyte imbalance in either of our experimental hypertension series (Zemplényi *et al.*, 1967).

Table III contrasts some of the aortic enzyme activity changes caused by the two types of experimental hypertension with those resulting from calciferol-induced injury.

In all three experiments acid phosphatase activity increased while non-specific carboxylesterase activity decreased. The activity of aortic Krebs cycle (succinate or malate) dehydrogenases decreased in both DOCA plus NaCl hypertension and in the later stages of calciferol injury. Furthermore, they tended to decline in rats with hypertension induced by renal artery clamping. However, the other enzymatic changes did not conform to this pattern in all three types of experiments.

Assuming that vascular injury is a common denominator in all these experiments, it is reasonable to conclude that increased acid phosphatase activity, together with decreased activities of carboxylesterase and Krebs cycle dehydrogenases constitute a common feature of the injured vascular wall.

At this point it is appropriate to modulate the theme of this discussion and consider some physicochemical, biochemical and ultrastructural events underlying tissue injury. (The reader is also referred to the 1964 *Ciba Foundation Symposium on Cellular Injury* and to the recent review by Trump and Ericsson, 1965). These problems have mainly been studied in connection with liver injury induced by certain specific poisons. Although it would be unwise to try to extrapolate all the findings in the intoxicated liver to vascular tissue, some of them are undoubtedly relevant to the main theme of this chapter.

The first changes in the damaged liver cell may be interpreted in general terms as altered permeability (Gallacher and Rees, 1960; Rees *et al.*, 1961; Epinosa and Insunza, 1962); they are characterized by leakage of larger molecules—including enzymes and coenzymes—from the cytoplasm into the blood plasma. Similar changes are supposed also to occur with inorganic ions

(McLean, 1960). In this connection Judah *et al.* (1964) consider that mobilization of membrane calcium is a primary feature of cellular damage; they argue that it allows sodium to enter the cell and, thus, exhausts the sodium transport system (see transport ATPase, p. 47) and depletes intracellular ATP.

Electron microscopic studies reveal early changes in the rough endoplasmic reticulum of the damaged liver cell, especially dilatation of cisternae, degranulation (Bassi, 1960; Emmelot and Benedetti, 1960; Smuckler *et al.*, 1961) and ribosomal alterations (Steiner and Baglio, 1963). In accord with these findings

TABLE III

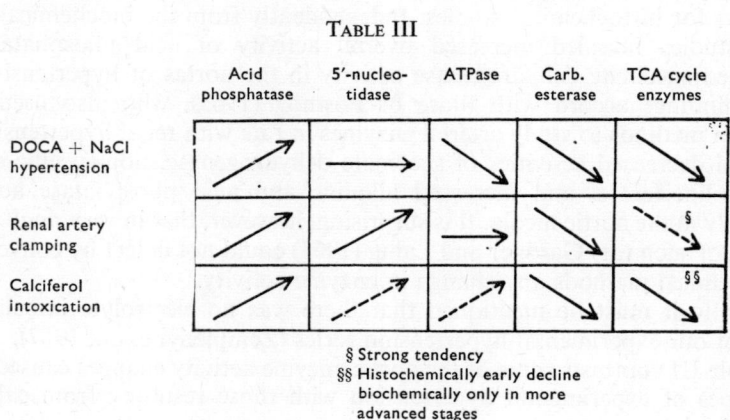

§ Strong tendency
§§ Histochemically early decline biochemically only in more advanced stages

is the altered soluble/bound RNA ratio (Richter, 1962) and the defective protein synthesis that results from decreased *in vivo* incorporation of amino acids into protein (Smuckler *et al.*, 1961, 1962).

The synthesis of the protein moiety of lipoproteins is also decreased and, as fat cannot then be secreted from the liver, fat accumulates in the organ and the serum lipoprotein levels fall (Harris and Robinson, 1961). This mechanism probably explains fatty change in the poisoned liver, but derangement of lipid metabolism can perhaps also be caused by alterations in the smooth endoplasmic reticulum, which in other tissues is believed to play a role in intracellular lipid transport (Palay and Karlin, 1959).

Somewhat later than the above changes, mitochondria become altered in response to various cell injuries, such as anoxia, lipid solvents and alterations in ionic balance (Zollinger, 1948). Electron microscopy of mitochondria in nearly all types of cell injury reveals changes in their membranes and matrix, the appearance of aggregates and especially swelling.

The functional significance of such changes is indicated by the fact that loss of soluble enzymes and cofactors seems to be a general feature of swollen mitochondria, together with a marked decrease in the organelles' ATP content (Dianzani, 1957). Liver injury is accompanied by uncoupling of oxidative phosphorylation (Dianzani, 1954) and, conversely, many substances that inhibit oxidative phosphorylation induce fatty change in the rodents' liver (Dianzani and Scuro, 1956). Mitochondrial alterations are accompanied by decreased oxidation of NAD-dependent Krebs cycle intermediates (Gallagher *et* al., 1956; Bailie and Christie, 1959 and others). As mentioned by Lehninger (1965), the ability to maintain oxidative phosphorylation (with a high P/O ratio) was for a time considered a criterion for the "intactness" of isolated mitochondria. (See also p. 10.)

It must be pointed out that in addition to the mitochondrial swelling that accompanies cell injury, isolated mitochondria display the phenomenon of respiration-dependent "active" uptake and extrusion of water. This swelling-contraction cycle has been extensively investigated by Lehninger and his co-workers (see Lehninger, 1962, 1965); its function may be to promote active intracellular movement of water in some tissues (e.g. the renal tubule). As contraction is generally accompanied by decreased permeability and swelling by increased permeability of the mitochondrial membrane, the cycle probably facilitates exchange of extra- and intra-mitochondrial metabolites—including ATP—according to metabolic needs.

Of considerable relevance to cell injury is the postulated role of lysosomes (see p. 61). De Duve (1964) remarked that lysosomes are ". . . identified biochemically as cytoplasmic particles containing a variety of acid hydrolases and surrounded by a membrane acting as a permeability barrier between the internal enzymes and susceptible substrates present on the outside (including the cell's own constituents)". It is assumed that a decrease in the bound/free ratio of a lysosomal enzyme indicates that it has been released as a result of injury to the fragile lysosomal membrane. Slater et al.'s (1963) results, based on the determination of such ratios for β-glucuronidase, acid phosphatase and acid ribonuclease, reveal lysosomal damage in the livers of poisoned rats. Since lysosomal changes in the poisoned hepatic cell develop later than the above-mentioned changes in permeability and damage to endoplasmic reticulum and mitochondria, it is highly improbable that lysosomal disruption plays an important part in initiating liver necrosis. Rees (1964) considers lysosomal disruption as a secondary event that is probably caused by such factors as altered ionic composition or increasing acidity due to the formation of organic acids.

Increased activity of lysosomal enzymes has been correlated with destructive changes in other tissues. For example, in murine muscular dystrophy the elevated activity of lysosomal enzymes can be partly attributed to the breakdown of muscle fibres (Tappel et al., 1963).

As pointed out above, it is hazardous unreservedly to extrapolate from special kinds of experimental tissue injury to the very different kinds of injury in vascular tissue. The *general conclusion* that can be derived from the experimental evidence is that acute tissue injury causes the concentration and activity of many cytoplasmic and mitochondrial enzymes to decline, probably as a result of leakage accompanying permeability changes and damage to intracellular structures. On the other hand, the increased intracellular level of some enzymes, as for example acid phosphatase, probably arises from damage and disruption of lysosomes and is indicative of cell injury.

It must be emphasized that the changes we have discussed apply mainly to acute injury. In more protracted sorts of tissue damage—as in our experiments—account must be taken of the interplay between destructive, regenerative and repair processes. The role tentatively ascribed to alkaline phosphatase in connective-tissue proliferation (see p. 126) would lead us to predict that the activity of this enzyme would increase during tissue repair by organization. However, some of the enzymatic changes that would be expected in tissue injury, such as decreased non-specific carboxylesterase activity, may become "masked" by the conflicting effect of subsequent tissue reactions (see part three of this volume).

Our findings in all three types of vascular injury essentially conform to the predictions derived from the above-mentioned investigations on cellular injury. We shall see later that the occurrence of increased acid phosphatase activity and decreased Krebs cycle enzyme activity is a hallmark of that form of protracted vascular damage that appears to prepare the ground for the development of atherosclerotic lesions in mammalian and avian arteries.

Chapter XVI

ARTERIAL HYPOXIA AND LACTATE DEHYDROGENASE ISOZYMES AS RELATED TO ATHEROSCLEROSIS

In several chapters the possibility has already been raised that arterial hypoxia may in some way be related to the pathogenesis of atherosclerosis.

Evidence to be discussed shows that the intima and inner third of the media in larger arteries obtain their oxygen and nutrition by diffusion through the intima from the blood. However, the outer part of the artery is nourished by the *vasa vasorum*, so that the mid zone constitutes a borderline between the layers supplied by these two routes.

The avascular inner third of the tunica media increases from about 0·16 mm. in the newborn human thoracic aorta (Wolinsky and Glagov, 1967) to approximately 0·5 mm. in the adult thoracic aorta (Geiringer, 1951). The *vasa vasorum* do not penetrate any previously avascular aortic tissue (Wolinsky and Glagov, 1967). Adams *et al.* (1962, 1963a) emphasized that the region of the watershed would be expected to be the first part of the media to suffer from anoxia. This could be the result either of intimal thickening (see p. 121), leading to impaired diffusion from the lumen, or the result of degenerative and other changes in the *vasa vasorum* (see Winternitz *et al.*, 1938) with impairment of blood supply from this source. But thickening is probably the more important mechanism (Adams, 1967a). Hueper (1944, 1945) suggested that arterial hypoxia resulting from the formation of a lipid film on the intimal surface might play a role in the genesis of atherosclerotic lesions (see Chapter XIII). Dixon (1961), Lazzarini Robertson (1963, 1966) and others also consider that arterial hypoxia is an important factor in the development of atherosclerotic lesions.

An important metabolic feature of the arterial wall is that it exhibits the phenomenon of aerobic glycolysis. As outlined in Chapter I, it is reasonable to expect that under hypoxic conditions the poorly nourished cells will partly derive their energy needs from glycolysis with the production of excess lactate. The question arises whether this adaptation of the mid-zone layers to hypoxia is in some way reflected in the behaviour of certain enzyme activities here.

In a series of publications Adams and co-workers demonstrated that with advancing age the histochemically detectable activity of some enzymes in the mid-zone layers of human aorta decreases (Adams *et al.*, 1962, 1963a, Adams, 1964a, b). The enzymes investigated were adenosine triphosphatase, 5'-nucleotidase, cytidine triphosphatase, "$NADH_2$-tetrazolium reductase" and the lactate dehydrogenase system (see Chapter XI). However, subsequent quantitative studies (Saudek *et al.*, 1966; Adams *et al.*, 1966), in which phenazine methosulphate was used as an intermediate electron acceptor (see p. 98), revealed that lactate dehydrogenase activity in the middle layers of the senescent human aortic media increases or remains unchanged, whereas the activities of the lactate dehydrogenase *system* (i.e. *without phenazine methosulphate*) and the $NADH_2$-tetrazolium reductase were decreased as compared with the other layers. The activity of malate dehydrogenase also tended to decline, but the activity under the experimental conditions used was so low that little confidence could be placed in comparative results from different aortic layers.

The increased lactate dehydrogenase activity of the middle aortic layers, as demonstrated by Adams and co-workers, appears strongly to support the hypothesis that these layers become hypoxic with advancing age, as it is reasonable to expect such an adaptive enzyme response in line with the greater dependence on aerobic glycolysis under hypoxic conditions (see Chapter I).

It is well known that lactate dehydrogenase is present in most tissues in multiple forms (see Chapter I). Isozymes of lactate dehydrogenase have recently been demonstrated in aortic tissue (Lojda and Frič, 1966a, b). In view of the different functions ascribed to these isozymes (see below), it seemed pertinent to investigate the relationship between hypoxia and the lactate isozyme pattern of the human aorta.

Some general problems concerned with multiple enzyme forms were mentioned in Chapter I. Before further discussion of vascular hypoxia, the special problems of lactate dehydrogenase isozymes must be considered.

The heterogeneity of lactate dehydrogenase in animal tissues (LDH) was established by Meister (1950), Nielands (1952) and by the work of Wieland and Pfleiderer and their co-workers (Wieland and Pfleiderer, 1957, 1961, 1962; Pfleiderer and Jeckel, 1957; Wieland et al., 1959). The latter authors showed that from one to five LDH components can be separated by high-voltage electrophoresis of extracts of liver, heart, kidney and other organs. LDH from heart and skeletal muscle differ in many respects, such as pH optimum, temperature coefficient, specific absorbance at 280 mμ., inhibition by sulphite and the action of pyruvate. Those LDH fractions from these two sources that had the same electrophoretic velocity responded in a similar way to sulphite and pyruvate. The fast moving (anodic) fractions were found to be much more sensitive to sulphite than the slower ones.

The heterogeneity of lactate dehydrogenase from different sources has also been intensively investigated by Vessell and Bearn (1957, 1961, 1962), Nisselbaum and Bodansky (1959, 1963), Plagemann et al. (1960a, b), Kaplan et al. (1960) and other workers. It is beyond the scope of this book to quote all the pioneer work performed in this field.

For the purposes of the present discussion it is relevant to mention the work of Appella and Markert (1961). They succeeded in dissociating the fast-moving bovine heart isozyme (LDH_1) into 4 subunits, and formulated the structure of the five isozymes as tetramers composed of two distinct subunit types (see Chapter I, p. 7). Elegant evidence to support the subunit hypothesis was subsequently provided by immunological techniques and, in particular, by recombination of the subunits (A and B) to form all five isozymes (Markert, 1963).

A *functional role for LDH isozymes* can be inferred from the subunit hypothesis as extended by the investigations of Kaplan and co-workers (Cahn et al., 1962; Kaplan and Cahn, 1962; Fine et al., 1963; Kaplan, 1964; Dawson

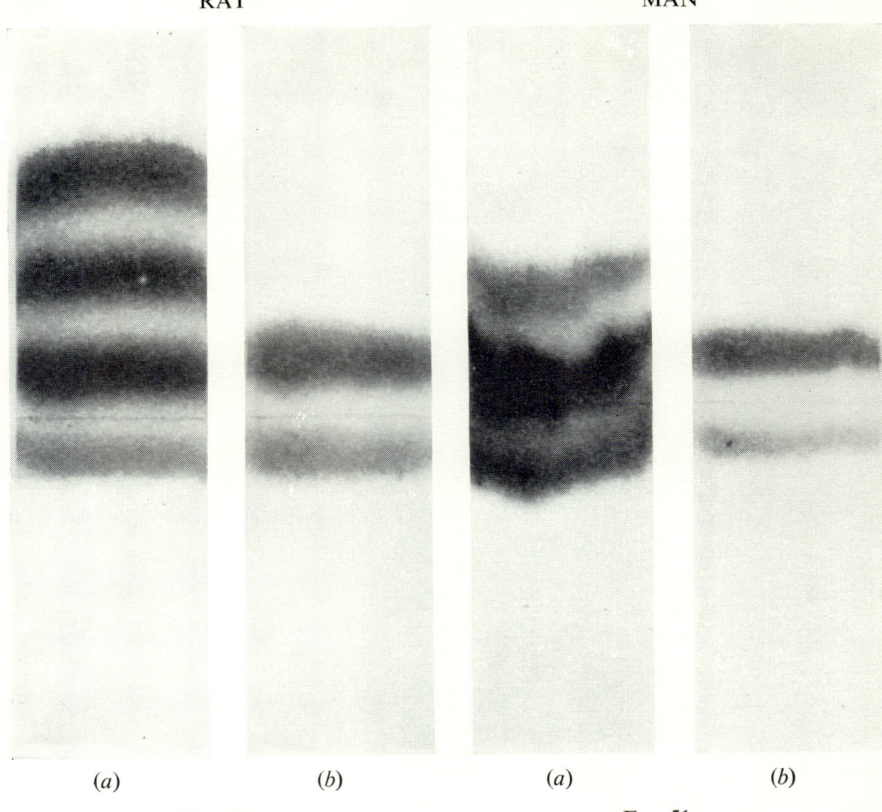

FIGS. 50 and 51.—Representative pictures showing electrophoretic separation of lactate dehydrogenase isozymes in rat aorta and human aorta. Untreated extracts (*a*) and supernatants of extracts after adsorption on DEAE-Sephadex (*b*).

et al., 1964). By analogy with Appella and Markert's suggestions, these authors maintain that LDH is a tetrameric molecule made up of four subunits of the two principal or "parent" forms of LDH. The authors termed the parent forms H (heart) and M (muscle), according to the organs in which they predominate. "Pure" H type ($= LDH_1$) is thus composed of four H subunits (H_4), "pure" M type ($= LDH_5$) is composed of four M subunits (M_4) while the other three forms of LDH are molecular hybrids consisting of mixtures of subunits: MH_3 ($= LDH_2$), M_2H_2 ($= LDH_3$) and M_3H ($= LDH_4$). This terminology is used to emphasize the postulated functional role of the two parent types of LDH. According to their concept the pure M type (LDH_5) is the principal isozyme in anaerobically metabolizing tissues, whereas the pure H type (LDH_1) is the most abundant isozyme in heart and other tissues where a steady supply of energy is required and maintained by the complete oxidation of pyruvate in mitochondria. The main basis for this theory is the different distribution of the two principal isozymes in metabolically different tissues and their different inhibition by high concentrations of pyruvate and lactate. In fact, there is a good deal of indirect evidence supporting this theory; the most interesting being that low oxygen tension apparently favours synthesis of the M type, which is the form assumed to be best suited for anaerobic metabolism. On the other hand high oxygen tension appears to favour synthesis of the H type, the form that is assumed to predominate in aerobic tissues (Dawson *et al.*, 1964). This popular theory has recently been challenged (Vesell, 1966*b*; Vesell and Pool, 1966; Stewart and Papaconstantinou, 1966), but, if it is true, the LDH isozyme pattern should be an ideal test for detecting hypoxia of insufficiently vascularized tissues.

Returning now to the topic of the present chapter, our study of arterial hypoxia was based on the above theory of Kaplan and co-workers. In the light of this theory, it was reasonable to expect that hypoxic layers of the aortic wall would display higher activity of isozymes with predominating anaerobic M subunits, i.e. higher activity of LDH_5 ($=M_4$) and LDH_4 ($=M_3H$). Lojda and Frič's (1966*a, b*) findings encouraged this expectation, because these authors observed higher activity of fast-moving LDH fractions in the human adventitia than in the intima or media. On the other hand, the activity of LDH_5 was highest in the media.

In our investigations human thoracic or abdominal aortas—obtained fresh at autopsy within 8–12 hours after death—were cleaned of all periaortic tissue and adventitia: multiple consecutive layers were prepared from grossly intact segments essentially as described by Saudek *et al.* (1966), but the sections were cut on a cryostat instead of the thermoelectric freezing microtome used by the latter authors. After removing the intima, which formed a separate layer, and after preparation of appropriate blocks, unfixed consecutive sections were cut at 25 μ in the circumferential plane of the aorta in such a way that the last sections comprised the outermost zone of the media. Between 5 and 7 consecutive sections were assembled into layers and usually 6 or 7 such layers were obtained from one aorta. From each layer 2 or 20 per cent homogenates were prepared (see Chapter XI) and, after centrifugation, the supernatants were used for isozyme determinations.

The isozyme pattern was either displayed by electrophoretic fractionation or was determined as the ratio of the principal isozymic fractions after selective adsorption on DEAE-Sephadex. The details of the techniques used are described in Chapter XI. In addition to many differences between the properties of the "heart" and "muscle" isozyme forms—such as pH optimum, substrate inhibition, sulphite inhibition etc.—the "heart" type is strongly bound by diethylaminoethyl (DEAE) cellulose and DEAE-Sephadex (see Chapter XI). This last property enables a satisfactory and sensitive estimation to be made of the ratio of isozymes with predominantly M subunits to those with predominantly H subunits, not only in blood serum but also in tissues (Hess, 1963; Wachsmuth and Pfleiderer, 1963).

Figure 50a illustrates that in rat aortic extracts 5 separate electrophoretic bands can be detected. The fast moving LDH_1 band is very faint, as is the usual

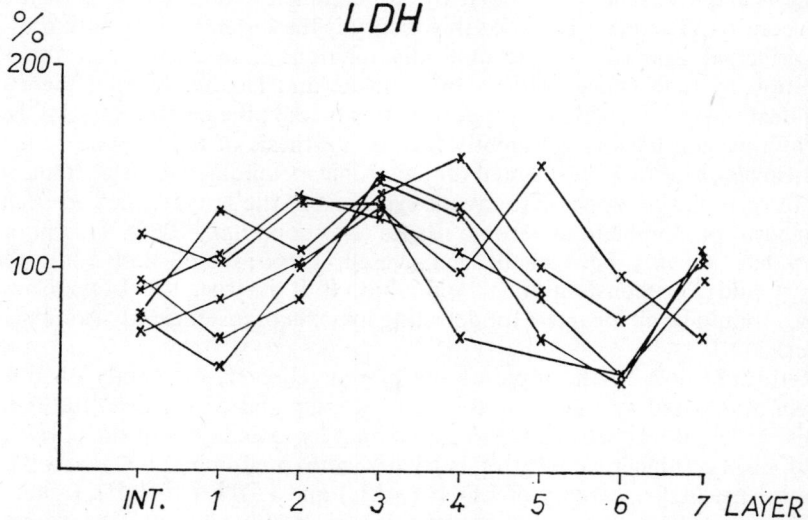

FIG. 52.—The activity of lactate dehydrogenase in multiple consecutive layers of human aortas (100 per cent = av. activity of all layers).

experience. Three bands corresponding to LDH_3, LDH_4, and LDH_5 can be clearly detected in human aortic extracts (Fig. 51a), but the fastest moving bands (LDH_1 and LDH_2) cannot usually be distinguished. After adsorption of the aortic supernatants on DEAE-Sephadex, only the fractions with predominating M subunits (LDH_5 and LDH_4) can be detected (Figs. 50b and 51b).

It is of interest that this method shows, for example, that the activity of the M types of LDH is higher in male than female rat aortas. Following the "aerobic-anaerobic" theory of LDH isozyme functions, this result suggests that the male rat aorta depends more on anaerobic metabolism than does the female vessel (Zemplényi and Mrhová, 1966).

Figure 52 demonstrates overall lactate dehydrogenase activity in the intima and consecutive layers of the media in grossly intact segments of six human aortas, from the inside to the outside of the vessel. In spite of a certain degree of variation, it is clear that LDH activity increases towards the middle layers of the

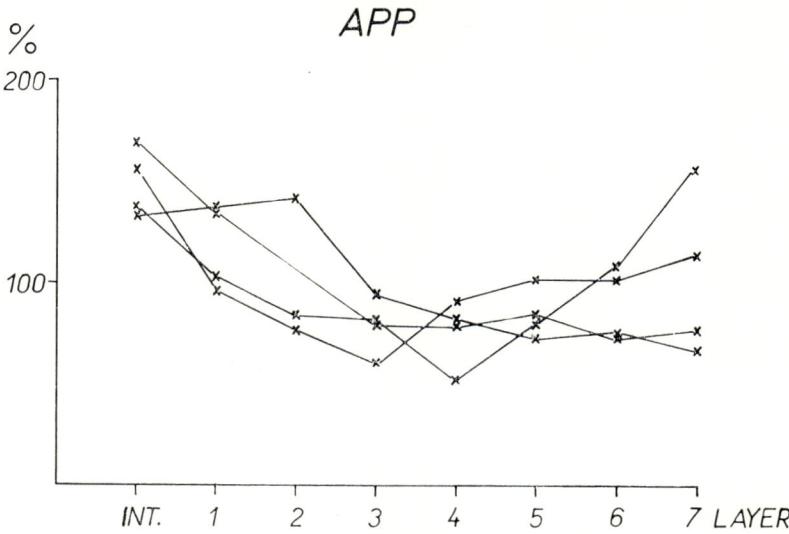

Fig. 53.—The activity of ATPase in multiple consecutive layers of human aortas (100 per cent = av. activity of all layers).

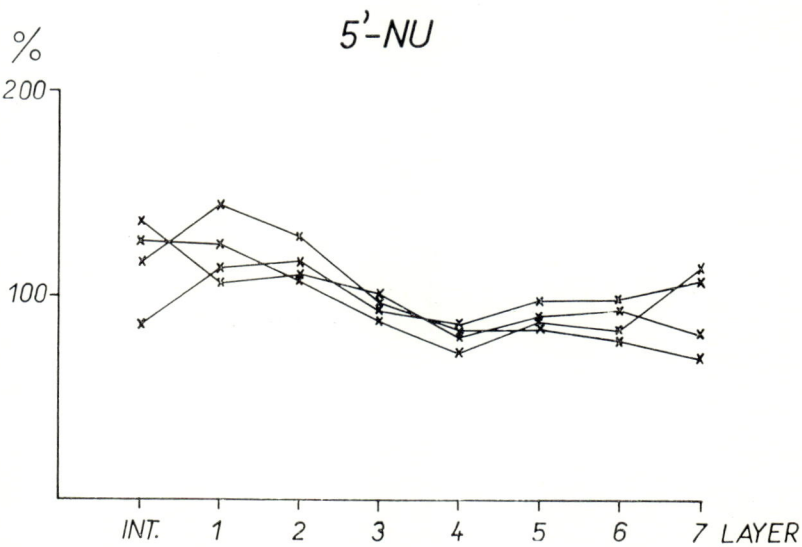

Fig. 54.—The activity of 5′-nucleotidase in multiple consecutive layers of human aortas (100 per cent = av. activity of all layers).

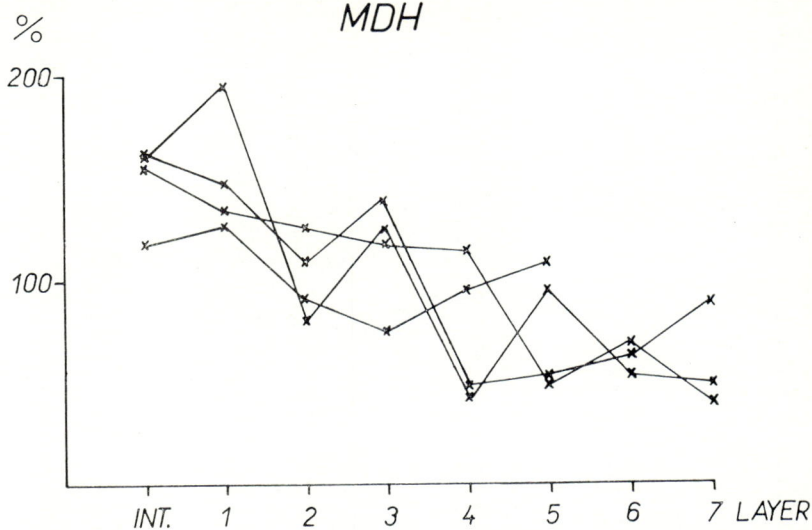

Fig. 55.—The activity of malate dehydrogenase in multiple consecutive layers of human aortas (100 per cent = av. activity of all layers).

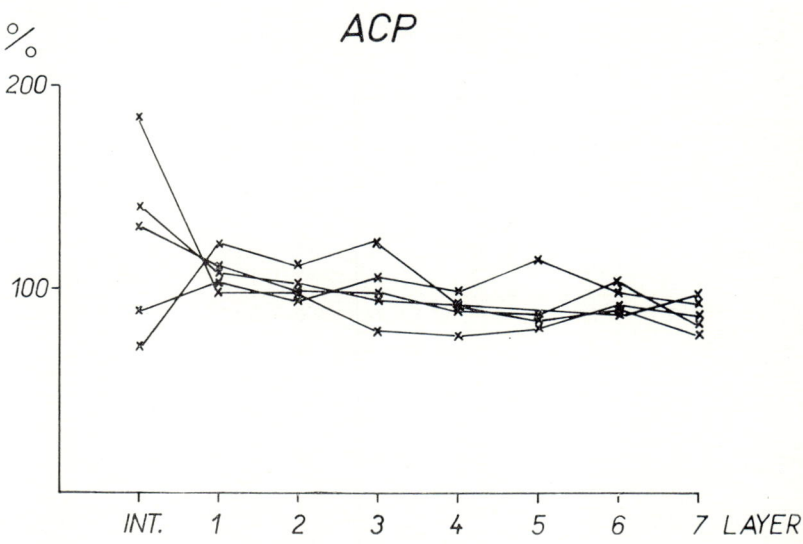

Fig. 56.—The activity of acid phosphatase in multiple consecutive layers of human aortas (100 per cent = av. activity of all layers).

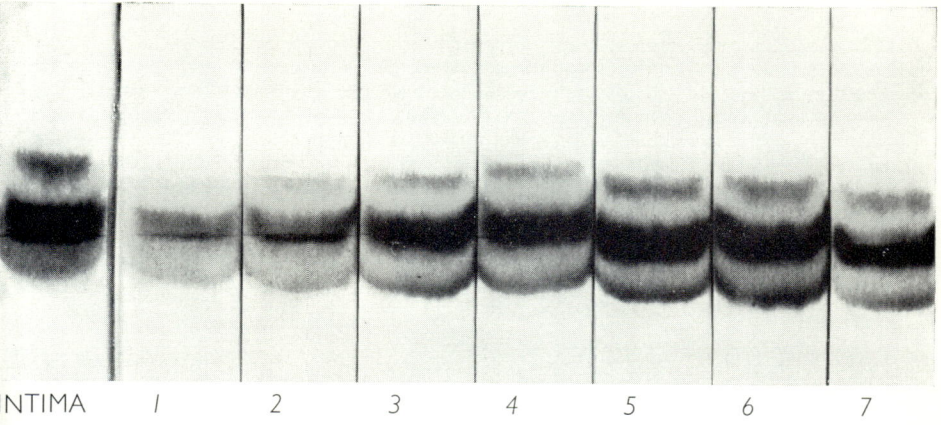

Fig. 57.—Representative picture showing electrophoretic separation of lactate dehydrogenase isozymes in multiple consecutive layers of a human aorta.

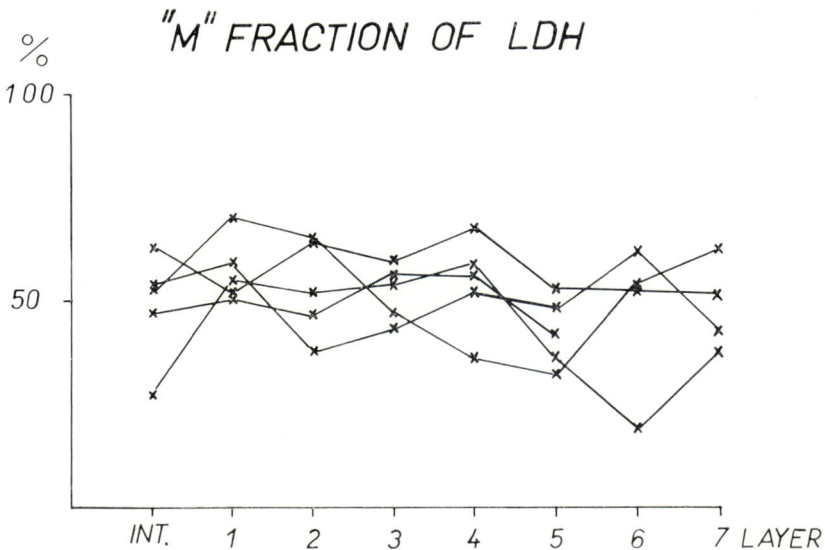

Fig. 58.—The activity of lactate dehydrogenase in multiple consecutive layers of human aortas. Supernatants of extracts after adsorption on DEAE-Sephadex. (Represents the activity of "M" subunits.) 100 per cent = total LDH activity of the layer.

PLATE I

Histochemistry of Aortic Enzymes in Cholesterol-fat fed Rabbits

1. Acid phosphatase in a normal rabbit aorta ($\times$ 120). *2.* Acid phosphatase in the aorta of a rabbit fed a cholesterol-fat diet for 10 weeks. The staining reaction is very intense in the thickened intima and the reaction in the media is also stronger than that of the control rabbit ($\times$ 120). *3.* Malate dehydrogenase in a normal rabbit aorta ($\times$ 120). *4.* Malate dehydrogenase in the aorta of a rabbit fed a cholesterol-fat diet for 10 weeks. With the exception of the surface of the thickened intima, the staining reaction is weaker both in the intima and media than in the aorta of the control animal ($\times$ 120). *5.* Lactate dehydrogenase in a normal rabbit aorta. *6.* Lactate dehydrogenase in the aorta of a rabbit fed a cholesterol-fat diet for 10 weeks. The staining reaction is slightly weaker in the thickened intima than media. No substantial difference in the staining reaction of the media between the experimental and control rabbit ($\times$ 120). (Microphotographs by Dr. D. Urbanová.)

PLATE I

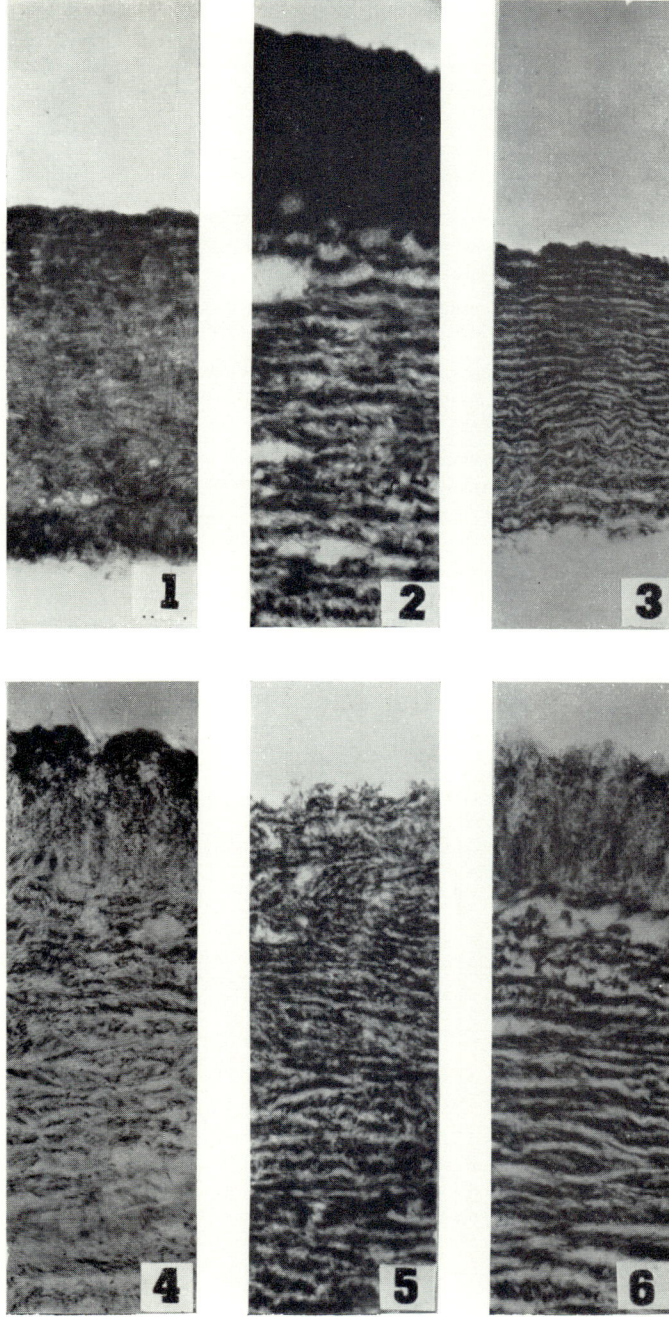

PLATE II
Histochemistry of Aortic Enzymes in Cholesterol-fat fed Rabbits

1. Non-specific carboxylesterase in an atherosclerotic rabbit aorta. Very strong reaction in the endothelium, macrophages and lipophages (upper part of the plaque) and in muscle cells (deeper part of the plaque). The reaction of the muscle cells of the media is much weaker ($\times 56$).

2. Non-specific carboxylesterase in a more advanced stage of atherosclerosis. Strong reaction in the endothelium and smaller lipophages. Somewhat weaker reaction in the muscle cells of the media. Very weak reaction in the fibrocytes of the plaque ($\times 56$).

3. Alkaline phosphatase in a small plaque of an atherosclerotic rabbit aorta. Plaque (on the left) exhibits a very intense reaction, a more precise localization cannot be unequivocally determined ($\times 105$).

4. Alkaline phosphatase in an atheromatous rabbit aorta. Positive reaction in the media at sites with calcification ($\times 140$).

5. Acid phosphatase in an atherosclerotic rabbit aorta. Early stage. The endothelium and the cellular elements of the small plaque reveal a stronger activity than the muscle cells of the media ($\times 56$).

6. Lactate dehydrogenase in a plaque with prevailing foam cells and with necrobiotic changes in the deeper parts. Strong staining reaction in the endothelium, macrophages, and muscle cells; weaker reaction in larger foam cells; necrobiotic areas practically devoid of activity ($\times 140$).

7. More advanced stage, plaque with prevailing fibrocytes. In comparison with the preceding microphotograph the overall staining reaction for LDH is less intense. Fibrocytes reveal a weak but distinct reaction ($\times 140$).

8. Malate dehydrogenase in an atherosclerotic rabbit aorta. Weak reaction in the cells of the plaque, strong reaction in the muscle cells of the media ($\times 56$). (From Lojda and Zemplényi, 1961 and Zemplényi et al., 1963c.)

PLATE II

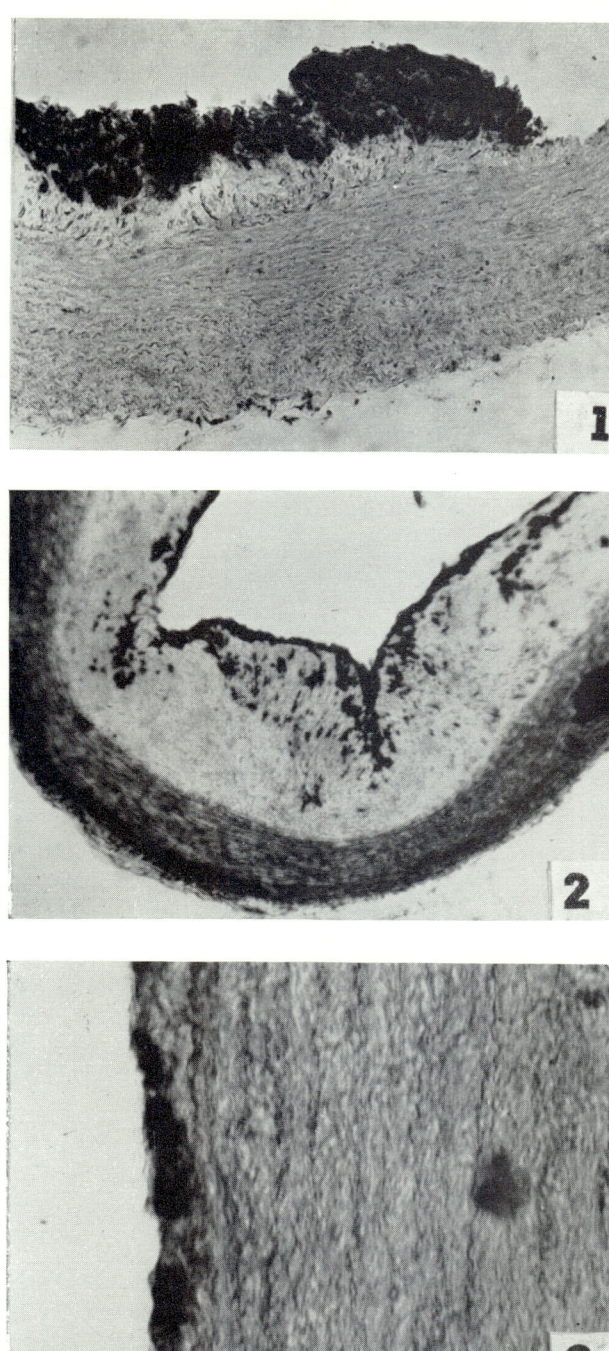

PLATE II (cont.)

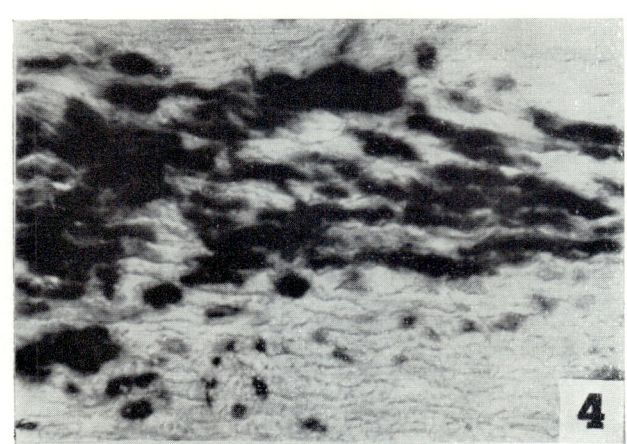

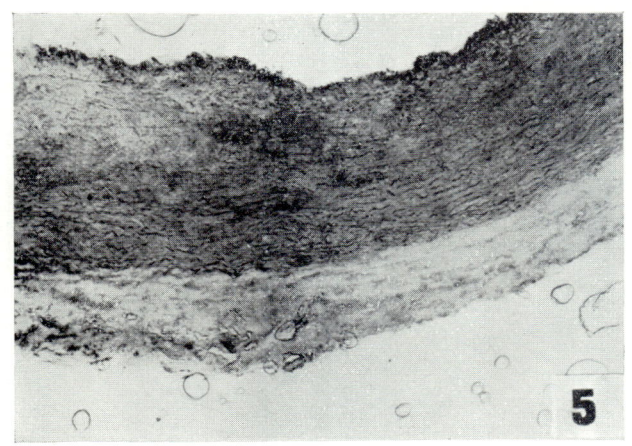

PLATE II (*cont.*)

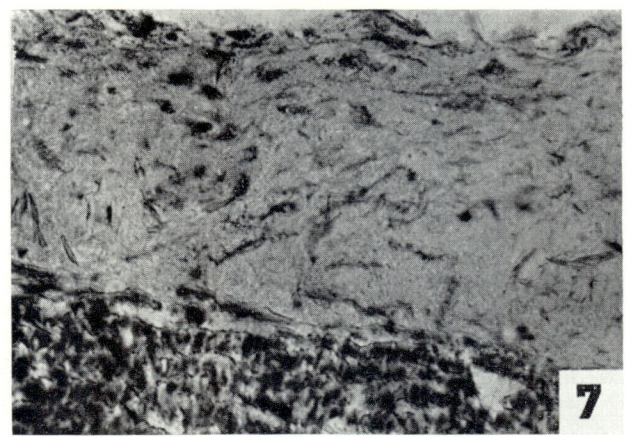

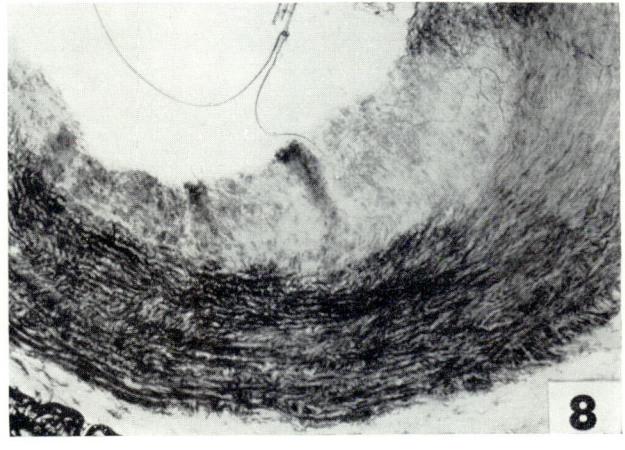

PLATE III

1. Acid phosphatase in a normal rat aorta ($\times 250$). *2.* Acid phosphatase in the aorta of a calciferol-intoxicated rat. Very intensive staining reaction around the calcified focus, no reaction within the focus ($\times 250$). *3.* Malate dehydrogenase in a normal rat aorta ($\times 250$). *4.* Malate dehydrogenase in the aorta of a calciferol-intoxicated rat. Weak activity around the calcified focus, as compared with the control animal. No staining reaction within the focus ($\times 250$). *5.* Lactate dehydrogenase in a normal rat aorta ($\times 250$). *6.* Lactate dehydrogenase in the aorta of a calciferol-intoxicated rat. No staining reaction in the calcified focus ($\times 250$). (Microphotographs by Dr. D. Urbanová.)

PLATE III

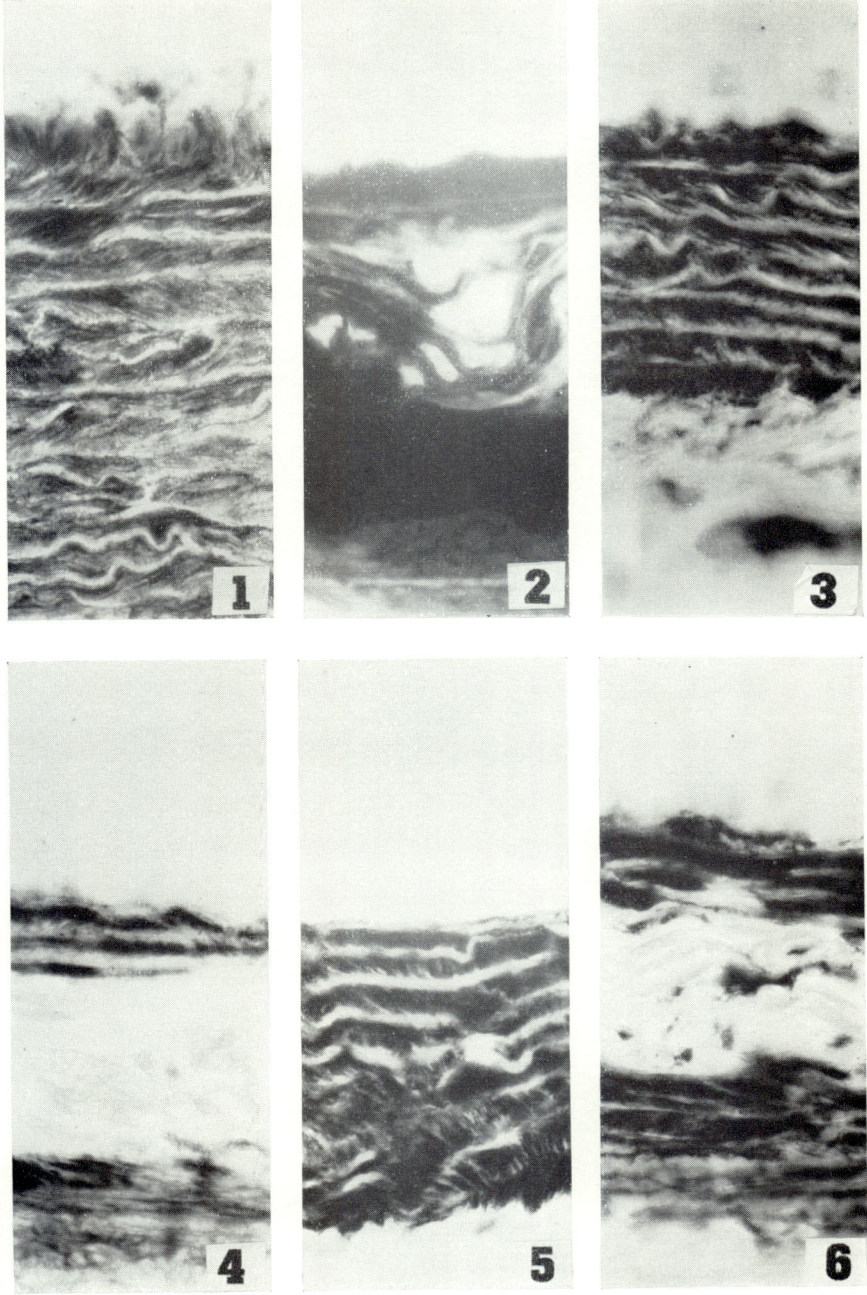

PLATE IV

1, 2, 3. Acid phosphatase in the ascending aorta, abdominal aorta and pulmonary artery of a boy, 3 months of age. Stronger staining reaction in the abdominal than ascending aorta ($\times 120$).

4, 5, 6. Malate dehydrogenase activity in the same vessels as above. Stronger staining reaction in the abdominal than ascending aorta ($\times 120$). (Microphotographs by Dr. D. Urbanová.)

PLATE IV

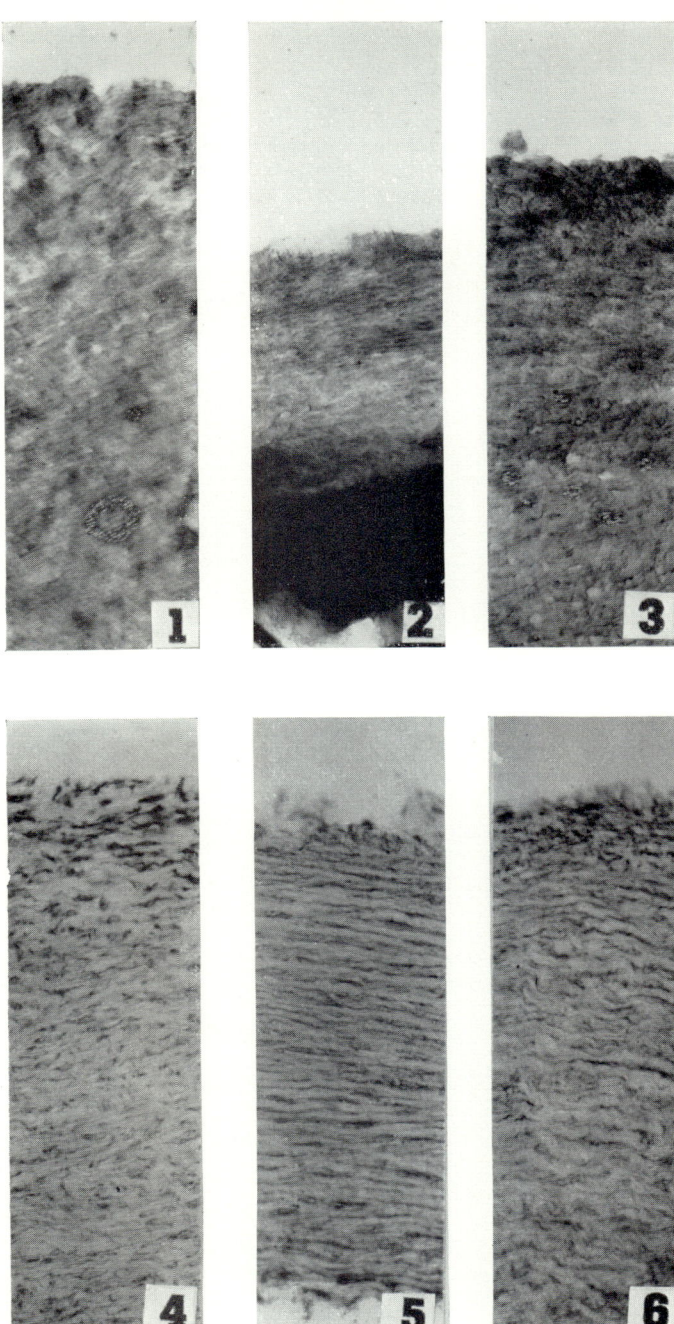

PLATE V

1, 2, 3. Lactate dehydrogenase in the ascending aorta, abdominal aorta and pulmonary artery of a boy, 3 months of age. The strongest staining reaction is in the abdominal aorta, the weakest reaction is in the pulmonary artery ($\times 120$).

4, 5, 6. Lactate dehydrogenase in the ascending aorta, abdominal aorta and pulmonary artery of a 40-year-old man. Weak staining reaction in the abdominal aorta with stronger staining in the macrophages of the thickened intima ($\times 120$). (Microphotographs by Dr. D. Urbanová.)

PLATE V

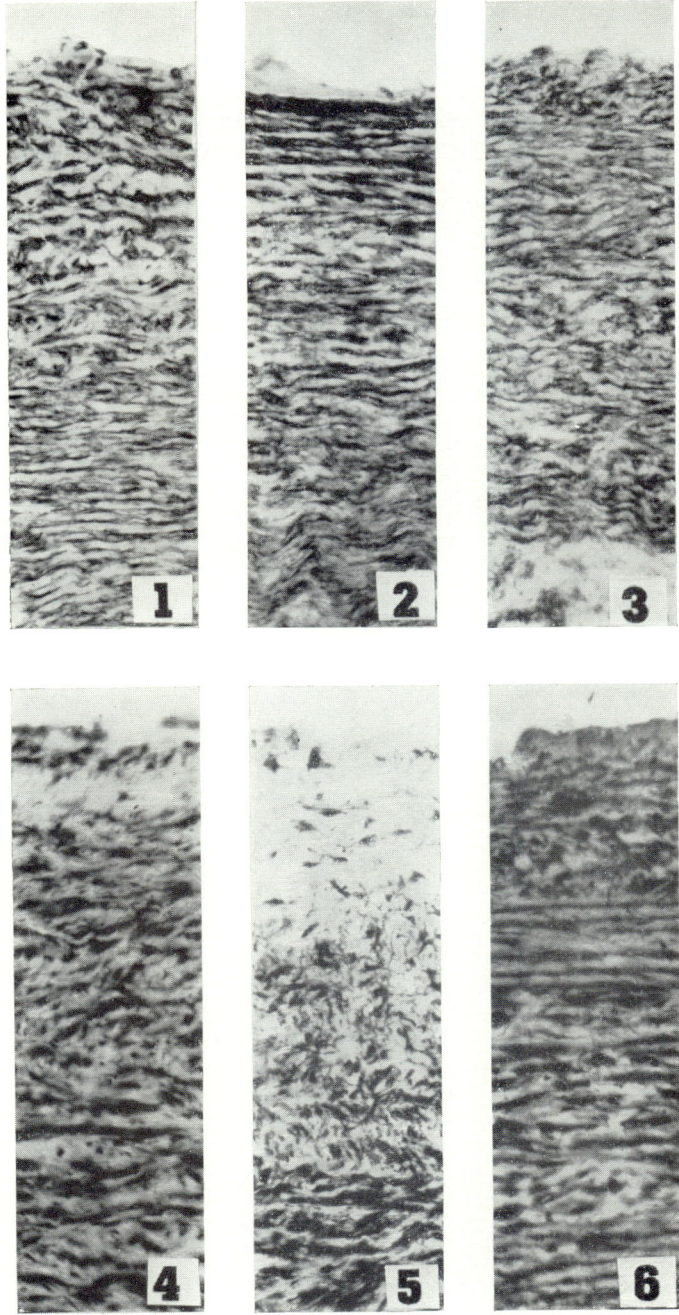

PLATE VI

1, 2, 3. Acid phosphatase in the ascending aorta, adbominal aorta and pulmonary artery of a 40-year-old man. Strongest staining reaction in the abdominal aorta, in the media and in the macrophages of the thickened intima. Weak staining in the pulmonary artery ($\times 120$).

4, 5, 6. Malate dehydrogenase in the ascending aorta, abdominal aorta and pulmonary artery in the same vessels as above. Weak staining reaction in the abdominal aorta, stronger only in the macrophages of the thickened intima ($\times 120$). (Microphotographs by Dr. D. Urbanová.)

PLATE VI

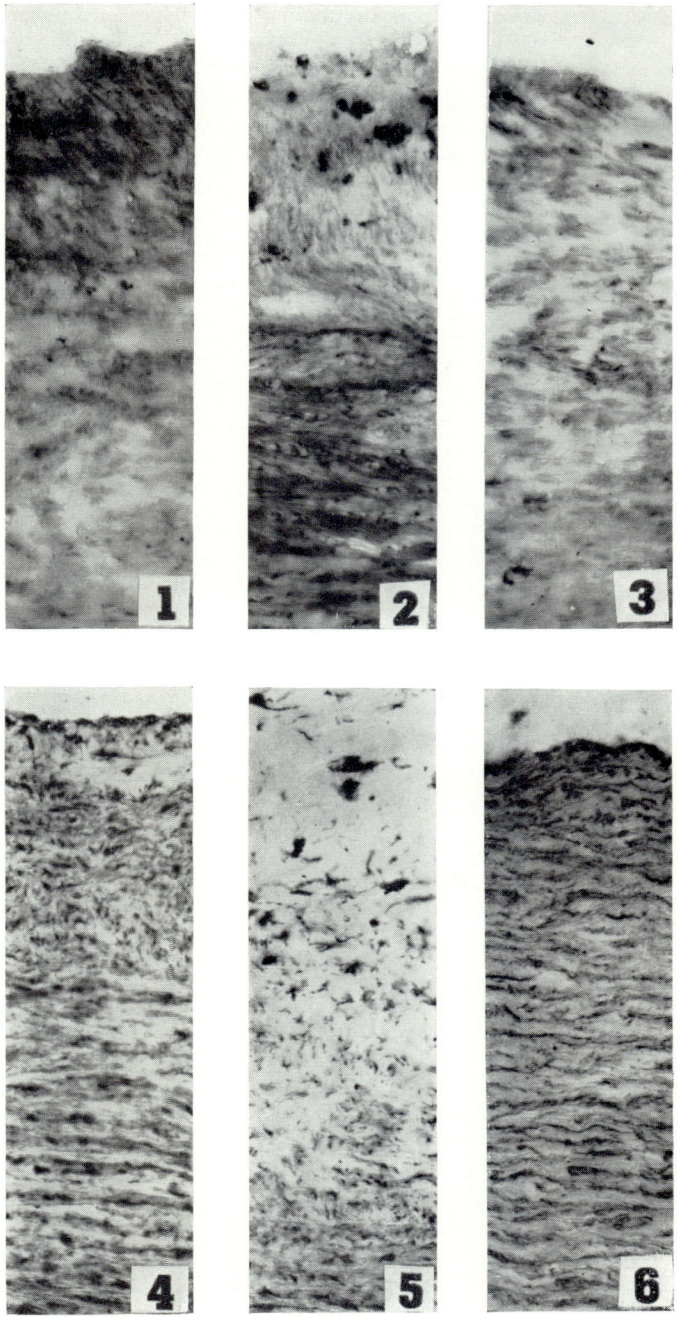

PLATE VII

1. 2. 3. Histological structure of the ascending aorta, abdominal aorta and pulmonary artery of a chicken. Transition between the elastic and muscular type of artery in the abdominal aorta. Hematoxylin-eosin and orcein ($\times 130$).

4. 5. 6. Histological structure of the ascending aorta, abdominal aorta and pulmonary artery of a duck. The abdominal aorta is a muscular artery. Hematoxylin-eosin and orcein ($\times 130$).
(From Zemplényi *et. al.*, 1965c.)

PLATE VII

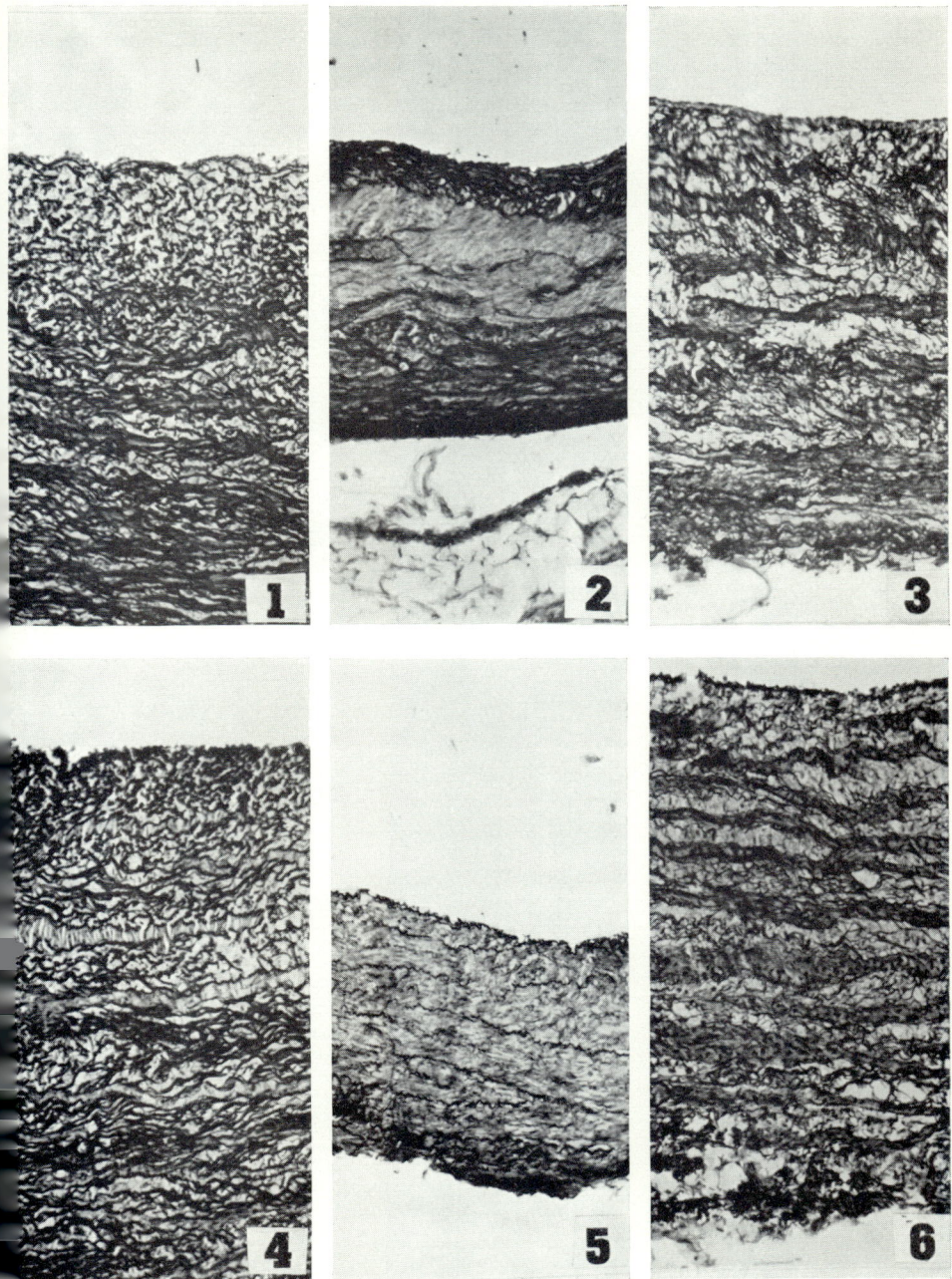

PLATE VIII

1 2. 3. Acid phosphatase in the ascending aorta, abdominal aorta and pulmonary artery of a 30-day-old chicken. Stronger staining reaction in the abdominal than ascending aorta. The staining intensity in the pulmonary artery is intermediate between the two. ($\times 120$).

4. 5. 6. Malate dehydrogenase in the same vessels as above. Stronger staining reaction in the abdominal than thoracic aorta ($\times 120$). (Microphotographs by Dr. D. Urbanová.)

PLATE VIII

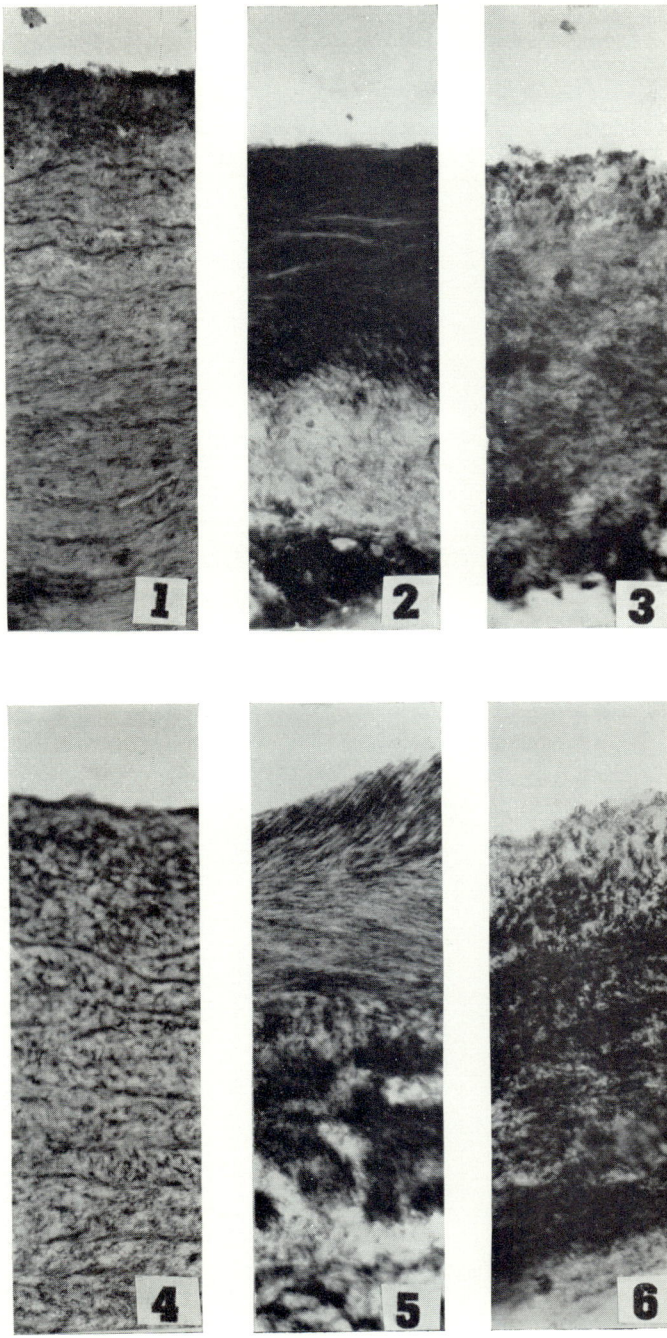

PLATE IX

1. 2. 3. Lactate dehydrogenase in the ascending aorta, abdominal aorta and pulmonary artery of a 30-day-old chicken. Stronger staining reaction in the abdominal aorta than ascending aorta. The reaction in the pulmonary artery is intermediate between the two ($\times 120$).

4. 5. 6. Lactate dehydrogenase in the ascending aorta, abdominal aorta and pulmonary artery of a 50-day-old duck. Strongest staining reaction in the abdominal aorta, lowest in the ascending aorta ($\times 120$). (Microphotographs by Dr. D. Urbanová.)

PLATE IX

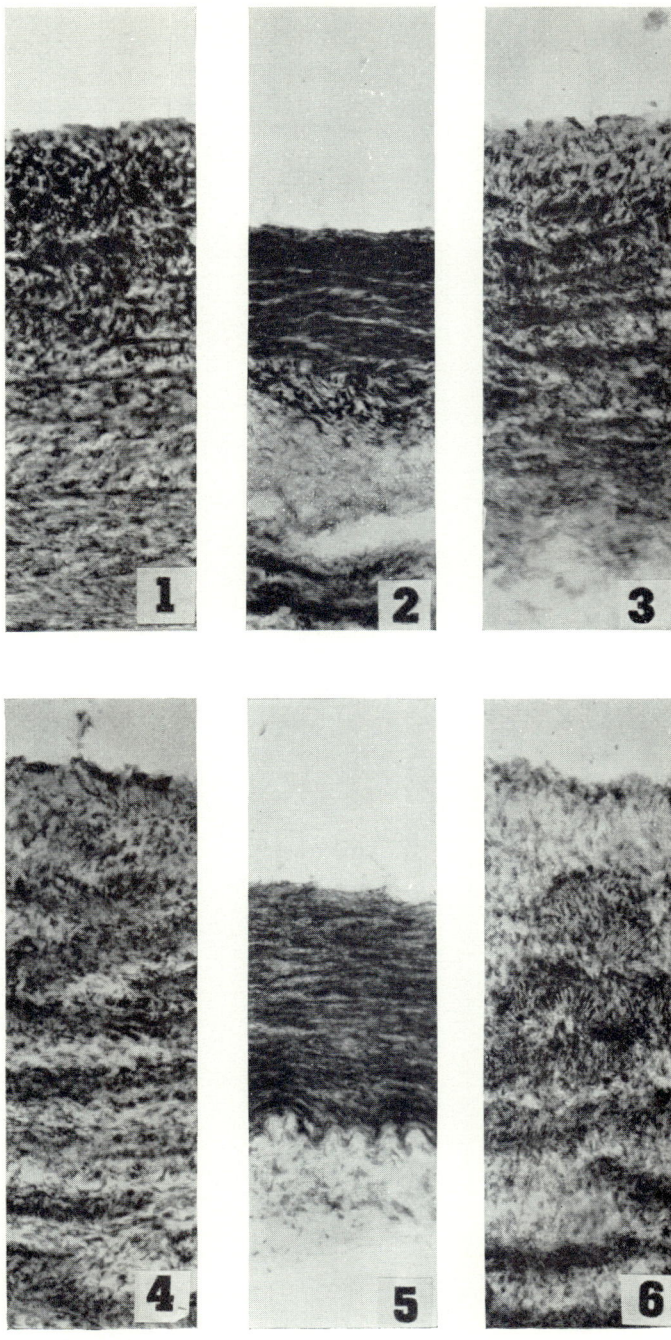

PLATE X

1. 2. 3. Acid phosphatase in the ascending aorta, abdominal aorta and pulmonary artery of a 50-day-old duck. Strongest staining reaction in the abdominal aorta, lowest in the ascending aorta ($\times 120$)

4. 5. 6. Malate dehydrogenase in the same vessels as above. Strongest staining reaction in the abdominal aorta ($\times 120$). (Microphotographs by Dr. D. Urbanová.)

PLATE X

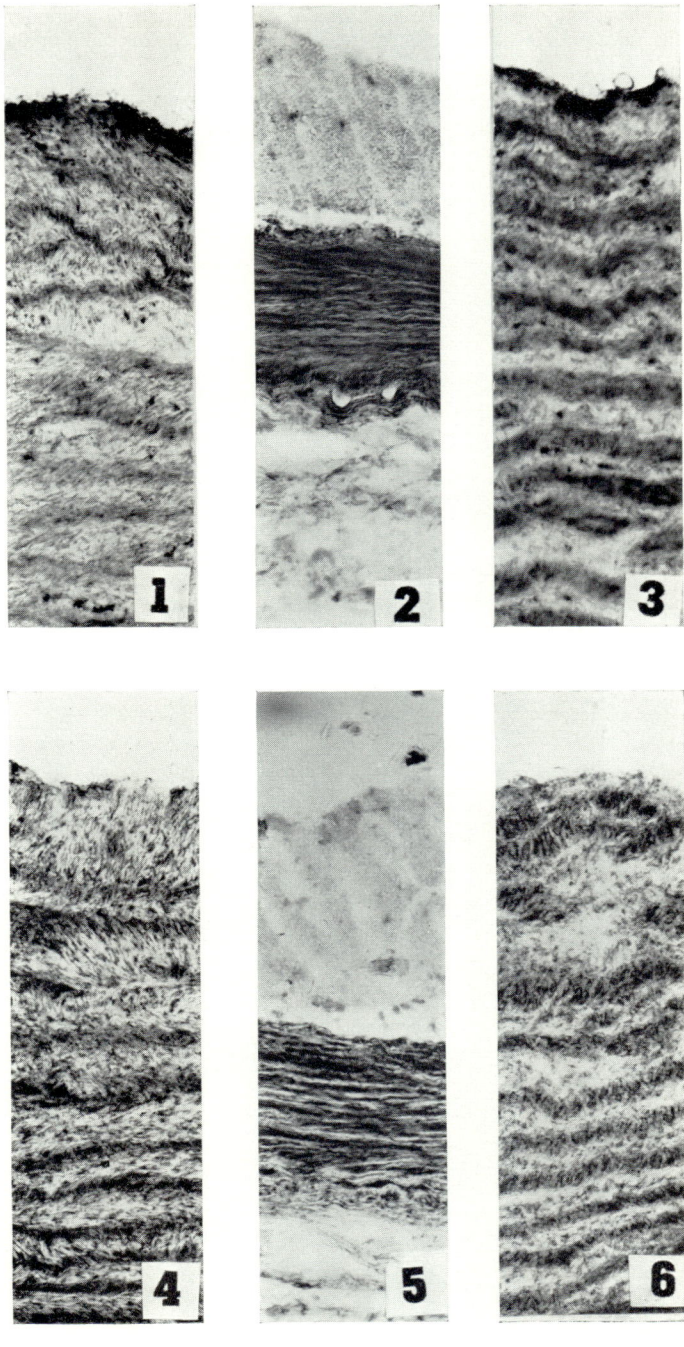

PLATE XI

1. 2. 3. Acid phosphatase in the ascending aorta, abdominal aorta and pulmonary artery of a young rhesus macaque. Intensity of staining reaction is practically the same in all three vessels ($\times 250$).

4. 5. 6. Malate dehydrogenase in the same vessels as above. Stronger staining reaction in the abdominal aorta than in the other vessels ($\times 250$). (Microphotographs by Dr. D. Urbanová.)

PLATE XI

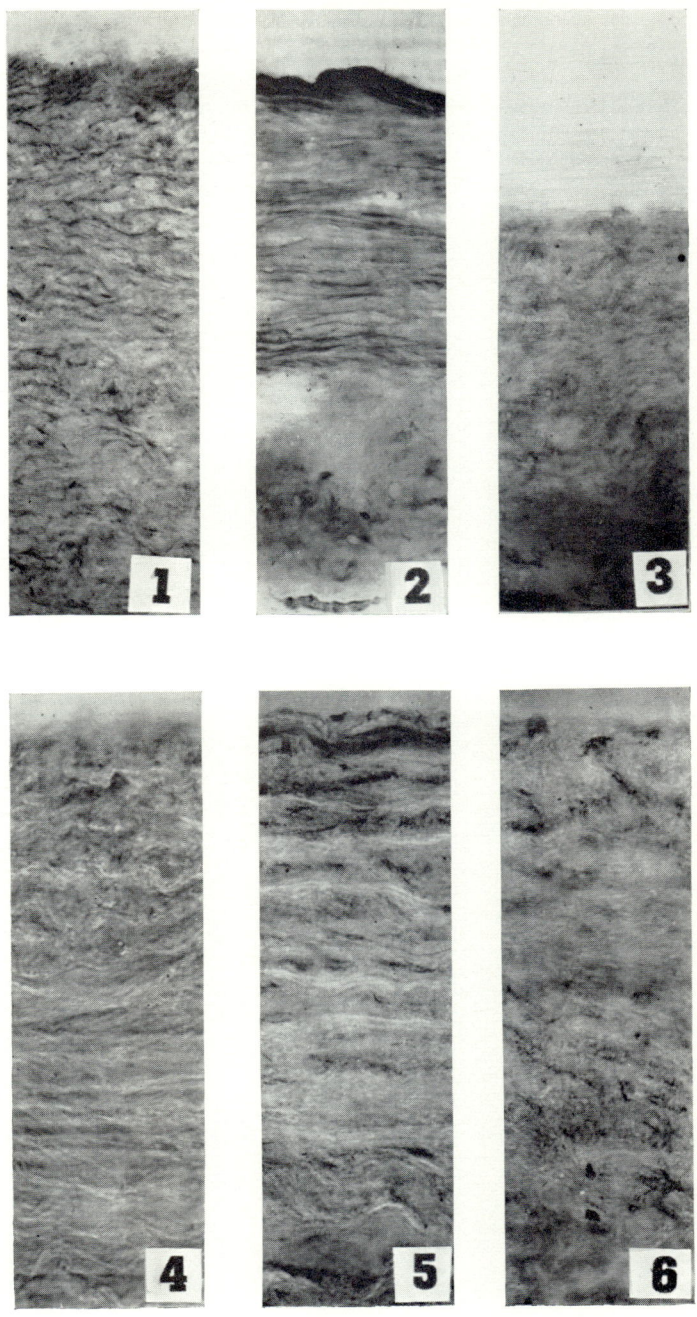

PLATE XII

1. 2. 3. Lactate dehydrogenase in the ascending aorta, abdominal aorta and pulmonary artery of a young rhesus macaque. Slightly stronger staining reaction in the abdominal than ascending aorta ($\times 250$).

4. 5. 6. Lactate dehydrogenase in the ascending aorta, abdominal aorta and pulmonary artery of a very young pig. Strongest staining reaction in the abdominal, weakest in the ascending aorta ($\times 120$). (Microphotographs by Dr. D. Urbanová.)

PLATE XII

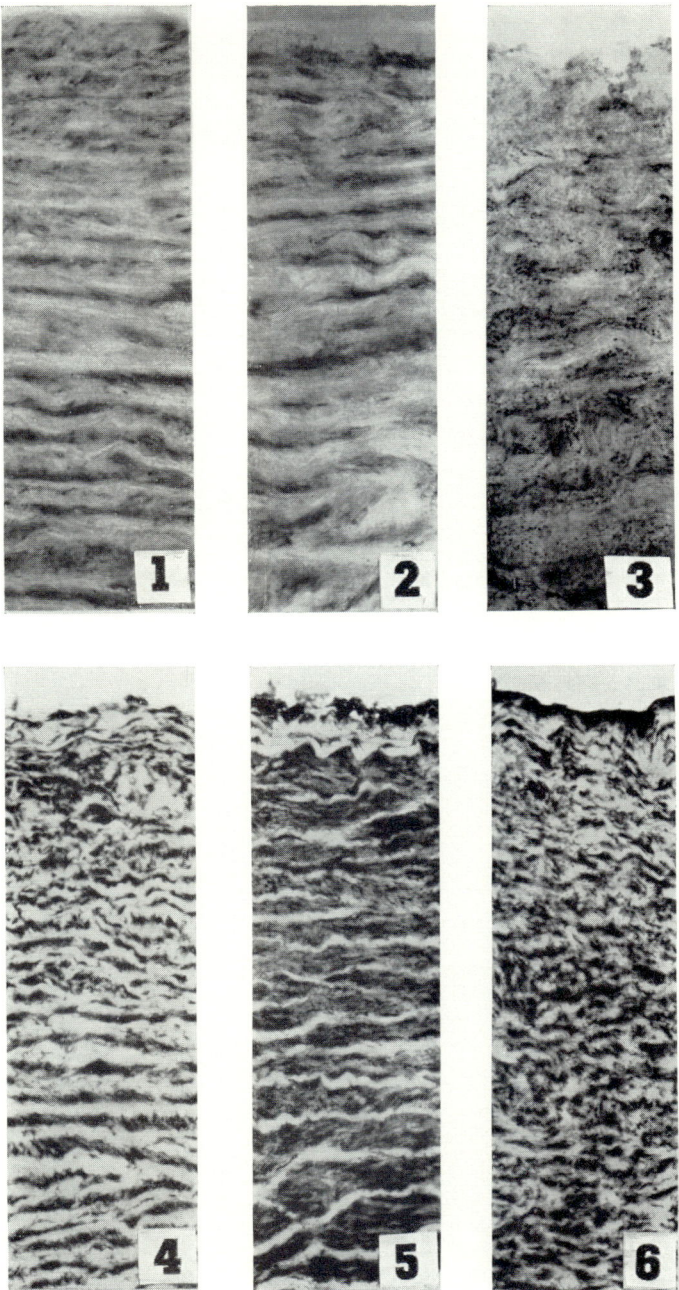

PLATE XIII

1. 2. 3. Acid phosphatase in the ascending aorta, abdominal aorta and pulmonary artery of a very young pig. Strongest staining reaction in the abdominal, weakest in the ascending aorta ($\times 120$).

4. 5. 6. Malate dehydrogenase in the same vessels as above. Strongest staining reaction in the abdominal, weakest in the ascending aorta ($\times 120$). (Microphotographs by Dr. D. Urbanová.)

PLATE XIII

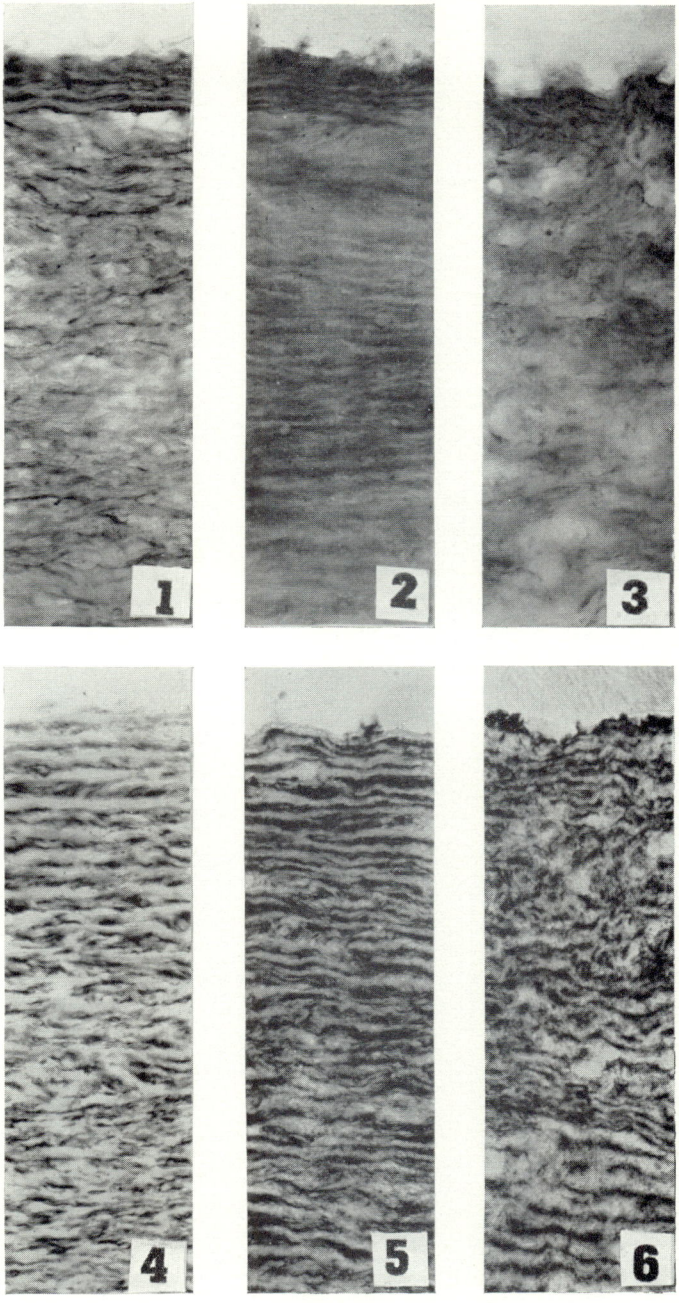

media and then declines in the outer parts. It is interesting to compare this gradient of changes with that of other enzymes. Figures 53 and 54 demonstrate that the gradient of ATPase and 5'-nucleotidase activities is exactly converse to that of LDH, the lowest activities being in the middle layers of the media. These biochemical findings are in excellent agreement with Adams and co-workers' histochemical results with ATPase and 5'-nucleotidase, and also with the quantitative data reported by these authors on LDH activity in the human aorta (see introduction to this chapter). A similar "radial gradient" of ATPase activity was histochemically observed in the aortas of 18-months-old purebred beagles (Higginbotham and Higginbotham, 1967).

As seen from Fig. 55, the activity of malate dehydrogenase constantly declines from inside to outside, without an expected elevation in the layers nearer to the adventitia. The activity of acid phosphatase (Fig. 56) is entirely "anomalous", since it remains practically unaltered in the different layers of the media.

Figure 57 shows an illustrative example of the electrophoretic separation of LDH isozymes in consecutive layers of a human aorta cut from the lumen to the adventitio-medial junction. The results of the quantitative determination of the ratio of the M type LDH isozymes to the total LDH activity of individual layers are set out in Fig. 58. In contrast to the changes in total LDH activity (compare Fig. 52), the average proportion of the M type does not change in the different layers of the media, being approximately 60 per cent of total LDH activity.

These results indicate that the augmented total LDH activity of the middle layers of the aortic media is due to an increase in both principal isozyme forms or, in terms of Kaplan and co-workers' theory, the increase is due to a side by side rise of both anaerobic and aerobic isozymes of LDH. This means that the presumed adaptation to hypoxic conditions in the middle zones of the human aorta seems to be brought about by increased activity of the M form of LDH isozyme, proportional to the increased total LDH activity (rather than by a preferential increase in the activity of the M isozyme types as expected on the basis of the above theory). Nevertheless, whatever the mechanism of the rise in M isozyme activity, it is compatible with the idea that the middle layers of the aorta, poorly supplied with oxygen, utilize glycolysis for their energy needs to a greater extent than do the other layers. The increased total LDH activity and especially the proportionally increased activity of the M types of isozymes, can reasonably be regarded as an adaptive response to tissue hypoxia (induction or derepression?—see Chapter I).

One of the main functions of lactate dehydrogenase is to regulate the intracellular ratio of NAD to $NADH_2$. Increased activity of the isozymic form that functions in the presence of excess pyruvate and lactate (i.e. the anaerobic M form) enables more NAD to be produced from $NADH_2$. Such "regenerated" NAD—in spite of oxygen lack—is again available for the triosephosphate dehydrogenase step of glycolysis and facilitates the subsequent important step, i.e. the "substrate-linked phosphorylation" of ADP to form ATP *via* the phosphoglycerate kinase reaction (see p. 21 and Fig. 2).

In addition to the *increased* activity of lactate dehydrogenase the *defect* in the activities of ATPase, 5'-nucleotidase and malate dehydrogenase also reflects the hypoxic damage to the middle layers of the aortic media, as suggested by

Adams. However, it is not yet clear why activity of malate dehydrogenase is also constantly decreased in the outer medial layers near the adventitia. More work is clearly needed to elucidate this question.

An interesting finding is the essentially unchanged activity of acid phosphatase in the consecutive layers of the media. In view of the decreased activity of the other enzymes investigated (except, of course, lactate dehydrogenase), it is tempting to speculate that another factor interferes, namely the tendency of acid phosphatase to increase as a result of lysosomal activation in damaged tissue (see Chapter XV). This interpretation is consistent with the presumed ischaemic damage to aortic tissue.

The increased rate of aortic glycolysis and the probable general decline of enzyme protein synthesis provides a wide field for speculation about atherogenesis. For example, as mentioned in Chapter I, the increased lactate production and resulting *acidosis* may be a pathogenic factor in atherosclerosis (Baló, 1963). As suggested by Adams (1964*a* and *b*) the enzymatic failure of the midmedial zone may lead both to *diminished lipid catabolism* and *impaired lipid transport* across the media and, thus, to accumulation of lipid within the inner layers of the arterial wall. In our own opinion, the decreased rate of *energy production*—as also occurs in other kinds of vascular injury—is probably the most important metabolic consequence of ischaemic damage: it must lead to further decline of protein (enzyme) synthesis and, thus, to deterioration of all the arterial wall's protective mechanisms (see Chapter XIX). In addition, it appears to be well established that local tissue hypoxia is a stimulus for the *activation of fibrogenic cell function*, especially for production of sulphate containing mucopolysaccharides and collagen (see Altschuler and Angevine, 1954; Krompecher, 1960; Kenny and Fink, 1966; Chvapil, 1967; Chvapil and Hurych, 1968 and many others). The distribution of ground substance within the human aortic wall seems to correspond to the presumed differences in oxygen tension (Bertelsen, 1963). This circumstance may also be of considerable importance in atherogenesis.

The work of Lojda and Frič (1966*a*, *b*), mentioned before, has revealed interesting details about vascular LDH isozymes. They showed variations in the aortic isozyme pattern among different mammalian species. The percentage proportion of the H type isozyme in total aortic LDH decreased in the following order: rabbit (76·7), guinea-pig (68·6), pig (67·6), rat (53·0) and man (27·0). Rats treated with methylthiouracil exhibited a shift towards the slower moving anaerobic aortic LDH fractions, whereas thyroid administration produced the converse effect. In male rats castration also caused a shift towards the anaerobic aortic fractions. In the child's aorta the activities of fast moving (aerobic) fractions were higher than in adult aortas. In human atherosclerotic plaques increased activity of the slower moving fractions was detected, whereas the activity of the fast-moving electrophoretic bands was almost negligible. In the light of the "aerobic-anaerobic" theory of LDH isozymes, the last findings seem to support the view that the rate of glycolysis is higher in the atherosclerotic plaque.

Recent data dealing with arterial "α-hydroxybutyrate dehydrogenase" activity also support this interpretation. It was mentioned in Chapter II (p. 23) that lactate dehydrogenase oxidizes other L-2-hydroxymonocarboxylic acids in addition to lactate. Rosalki and Wilkinson's (1960) work shows that LDH isozymes differ from each other in their relative affinities for α-oxobutyrate and pyruvate. The activity with α-oxobutyrate as substrate ("α-hydroxybutyrate dehydrogenase activity") has been found to provide a measure for the activity of fast-moving LDH isozymes (Wilkinson et al., 1961). For "α-hydroxybutyrate dehydrogenase" determination, Sanwald and Kirk (1966) used a method that is based on the reduction of α-oxobutyrate with $NADH_2$ and photometric estimation of the resulting decreased concentration of $NADH_2$ (see Optical Tests, Chapter XI). In supernatants of homogenates from human vascular tissue "α-hydroxybutyrate activity" was highest in the coronary artery, lowest in the aorta and intermediate in the pulmonary artery. Activity decreased in aortic tissue with advancing age, while a moderate reduction was observed when atherosclerotic segments were compared with normal parts of the same aorta. As there is good evidence of a relationship between LDH_1 and "α-hydroxybutyrate dehydrogenase" activity, Sandwald and Kirk's findings could be interpreted in the same way as Lojda and Frič's work.

It is concluded that the study of arterial LDH isozymes supports the idea that poorly nourished parts of the human arterial wall become hypoxic with advancing age and that the rate of glycolysis in these parts increases. The changes of the LDH isozyme pattern in atherosclerotic foci seem to reflect a similar metabolic situation.

Chapter XVII

COMPARISON OF ENZYME ACTIVITIES IN HUMAN VESSELS OR VASCULAR SEGMENTS DIFFERING IN SUSCEPTIBILITY TO ATHEROSCLEROSIS

It is a well-known clinical and morphological experience that different arteries and also different segments of the same artery show considerable differences in susceptibility to atherosclerosis.

For example, the extent and severity of atherosclerosis is higher in the aorta than the pulmonary artery and higher in the femoral artery than the brachial artery. It is generally assumed that differences in intravascular pressure are responsible for such disparities. Some authors also emphasize the primary importance of multiple small emboli as a prerequisite for the development of atherosclerosis in the pulmonary artery.

While an aetiological explanation can be advanced for the above differences, it is difficult to see why certain other vessels are spared, such as for example the internal mammary arteries (Duff and McMillan, 1951) and the diaphragmatic arteries (Wartman, 1933). Vascular haemodynamics are supposed partly to account for such disparities, as well as for the severe involvement of the coronaries in comparison with the renal arteries (Glagov et al., 1961). Likewise, the predilection of atherosclerosis for the abdominal aorta as compared with the thoracic part of the vessel (Dow, 1925; Sjövall and Wihman, 1934; Dock, 1950; Holman et al., 1958; Roberts et al., 1959; Glagov et al., 1961) could perhaps be explained by vascular haemodynamics and the erect posture, but the latter concept is weakened by similar findings in the aorta of many animals and birds (see Chapter XVIII). Dalith (1964) recently expressed the view that the different embryological development of the thoracic and abdominal aorta is the cause of their different susceptibility to atherosclerosis. He points out that, in contrast to the thoracic aorta, there are more loci of developmental structural deficiency in the abdominal segment, because a large number of mesonephric vessels atrophy and disappear from the region between the twelfth thoracic and the third lumbar segments of the distal aorta.

The concepts just mentioned seem to shed some light on the possible aetiology of the differences in the degree of atherosclerosis between various arteries or arterial segments, but they do not satisfactorily explain the pathogenic mechanism by which the susceptible vessels become atherosclerotic. We felt that such differences in susceptibility provide a unique opportunity not only for identification of the role played by some factors in the *localization* of atherosclerosis, but also provide an opportunity for the study of vascular factors in the *pathogenesis* of the disease.

First of all, it is apparent that the distribution pattern of the lesions in the artery depends on the type of plaque. As the relationship of fatty streaks to the more advanced plaques is a much discussed and controversial problem it is now necessary to discuss recent studies by the New Orleans investigators (the late Dr. Russell Holman's group) and by Mitchell and Schwartz in Oxford.

Three main types (stages) of lesion can be recognized on the luminal surface of the major arteries with the naked eye:

1. **Fatty streaks (flat sudanophilic lesions)**.—They show a wide variation in shape and size (dots, patches, streaks), are sharply demarcated from the intima and stain intensely with Sudan IV, which exposes those streaks that are not apparent in the unstained specimen. The lipid in the early fatty streak is contained mainly within macrophages in the subendothelial region (Adams, 1967a). A distinctive feature of these lesions is that transverse sections do not contain any pultaceous material (gruel) in their depths.

2. **Raised fibrous plaques** contain a variable amount of fat. According to Schwartz and Mitchell (1962a, b) these plaques macroscopically appear either as raised sudanophilic plaques or as raised white plaques. The former plaques are whitish yellow in the unstained artery and they stain variably with Sudan IV. In general they are larger than the fatty streaks and, when sectioned transversely, an area of pultaceous material is seen in their depths. The raised white "pearly" plaques have a firm surface that cannot be stained with Sudan IV. Nevertheless, transverse sections show that beneath the white surface-layer there is almost invariably a mass of pultaceous material.

3. **Complicated plaques** show calcification, ulceration, thrombosis or haemorrhage, and they complicate only raised plaques, never fatty streaks.*

As pointed out by Hudson (1965), it is easy to accept the progression of lesions from stage 2 to stage 3, but it is not absolutely certain whether stage 2 follows stage 1 or whether these are independent lesions.

The concept of the independence of the fatty streak and fibrous plaque has recently been very vigorously defended by the Oxford investigators (Mitchell and Schwartz, 1965). Their views are based first of all on the detailed study of the localization of the main types of lesions, especially in the aorta (Schwartz and Mitchell, 1962a, b). In an extensive necropsy study they observed that there was usually more fatty streaking in the thoracic than in the abdominal aorta, and occasionally these lesions were only present in the thoracic segment. Although the upper thoracic aorta was commonly the site of confluent areas of fatty streaking, it remained relatively free from raised lesions. In contrast, raised plaques were far more common in the abdominal aorta than in the thoracic segment. Another discrepancy in the distribution of these two types of plaque was observed at the ostia of the intercostal arteries. The fatty streaks exhibited a characteristic tendency to spare the aortic wall around and immediately distal to the ostia of the paired intercostal arteries, whereas these sites were a common site for raised fibrous plaques. A further difference in distribution was that fatty streaking showed a striking predilection for the mid-line posteriorly along the line of origin of the intercostal arteries, with a decided tendency to spare the lateral and anterior walls of the thoracic aorta in younger subjects, whereas,

* In the World Health Organization system (1958) lesions are classified into fatty streaks, fibrous plaques, fatty plaques (atheromas) and complicated lesions.

with advancing age and increasing surface involvement, it was in these latter areas of the aortic wall that considerable numbers of raised plaques developed.

In addition to these differences in distribution, Schwartz and Mitchell (1962a, b) emphasize that fatty streaks affect the intima alone, while in raised plaques all three arterial coats are involved, and the histological appearance is quite different.

Apart from these findings these authors pointed out that the area of fatty streaks had no relation to sex, age or diastolic blood pressure, but the area of raised plaques could be correlated with these factors.

In view of their evidence the Oxford investigators see little justification for considering that fatty streaks and fibrous plaques are part of one disease process.

However, neither differences in distribution nor differences in morphological appearance necessarily prove that two separate and independent processes are involved in the genesis of fatty streaks and fibrous plaques. The above observations could well be explained by assuming that not all lesions necessarily progress to the same stage and, in addition, by the fact that the initial lesions have a tendency to regress. In fact some of the epidemiological, histological and electron-microscopic results obtained by the New Orleans group seem to favour such possibilities; they are relevant to our theme.

In a much quoted paper Holman *et al.* (1958) analysed the findings in the aortas of a large number of necropsied individuals aged between 1 and 40 years. Intimal fatty streaks occurred in every case aged 3 years or older. In the age group 11 to 15 years the average percentage of surface involved in Negroes rose to 28·1 per cent, whilst the average for the Caucasian subjects rose only to 7·2 per cent. In subsequent years, although the difference became much smaller, the surface involvement among Negroes remained consistently higher. Fibrous plaques began to appear in an appreciable number in the third decade, particularly in Caucasians. In the last age group investigated (35 to 40 years) the extent of fibrous plaques in Caucasians was almost twice that in the Negro. The abdominal aorta was consistently more severely affected by fibrous plaques than the thoracic aorta, ring and arch.

Two important deductions were made from the findings:

(*a*) After adding together the average values of surface covered by fibrous plaques and those covered by fatty streaks, it was concluded that the "regression" of fatty streaks seen in the Caucasian population after 36 years of age appeared to be largely due to their conversion to fibrous plaques;

(*b*) As the greater extent of fatty streaks led to fewer fibrous plaques in Negroes, it was suggested that the factors responsible for the succeeding stages of atherosclerosis (viz. fibrous plaques, complicated lesions and clinically recognizable disease) might differ from those that initiate the fatty streak. Although some of the observations were consistent with the view that fatty streaks can regress (e.g. their decline during terminal illness), the evidence was not sufficiently strong for definite conclusions to be made.

Complicated plaques were rarely encountered, because only patients up to the age of 40 years were studied. However, the investigation was later extended to cases of all age groups (McGill *et al.*, 1963). As expected, in older age groups fibrous and complicated aortic plaques were again more extensive among Caucasians than Negroes.

In the coronary arteries fatty streaks were present in about 50 per cent of cases in the second decade, later practically all cases exhibited such coronary lesions (Strong and McGill, 1962). In Caucasian males the area of intimal surface occupied by fibrous plaques sharply increased in the third and fourth decades, while complicated plaques were first found at least a decade earlier in these coronaries than those of either Caucasian females or Negro males and females.

An electron microscopic study of *coronary* lesions (Geer et al., 1961) and a study of the fate of the fatty streak therein (Robertson et al., 1963) revealed many details of considerable importance. Stainable lipid in fatty streaks is found as a rule in those specimens with the thickest intima; it is most often in the elastic-hyperplastic or collagenous layers, and seldom is it in the musculo-elastic layer (see p. 190). The lipid is intracellular and the intimal cell involved is the smooth muscle cell. Thus, in contrast to statements by Mitchell and Schwartz (see above), even at this relatively early stage, lipids may be seen in smooth muscle cells. The lipid-containing cells in fatty streaks appear to be viable; this strongly suggested to Robertson et al. (1963) that the lipid changes are at this stage reversible, just as are similar changes in the liver or kidney.

However, in some more advanced fatty streaks, the cells become engorged with lipid, thus conforming to the description of foam cells. In these lesions the smooth muscle cells show many inclusions or become highly vacuolated. This type of foam cells is termed myogenic foam cells to distinguish them from the macrophage type (for details see French, 1966). In the most advanced stage it appears that certain lipid-containing cells rupture or die and release their lipid into the extracellular space. This extracellular lipid incites an inflammatory reaction that is regarded as a reaction to injury (Robertson et al., 1963; McGill et al., 1963). In the earliest stages of human atherosclerosis lipid accumulates at the intimomedial junction (Adams and Tuqan, 1961), the zone of potential mechanical weakness. The injury is followed by repair processes that lead to a progressive increase in connective tissue with formation of a superficial collagenous layer, the fibrous cap. Thus, the fatty streak is transformed into a fibrous plaque. (See also sclerogenic action of lipids, Chapters XIX and XX.) In Caucasian males aged 30 to 39 years inflammatory changes and collagenization occured in 47 and 83 per cent respectively, whereas in Negro males the corresponding figures were only 15 per cent and 38 per cent.

The histological and electron microscopic appearances of *aortic* fatty streaks were basically the same as in the coronary arteries, even though some additional morphological evidence was presented about intracellular lipid inclusions in aortic smooth muscle cells. The features of aortic fibrous plaques were identical to those in the coronary lesions (McGill et al., 1963). Although transitional aortic lesions were observed, it was technically more difficult histologically to study the transition from fatty streak to fibrous plaque (see corresponding data on pig arteries in Chapter XVIII).

In parenthesis it should be mentioned that intimal modified smooth muscle cells seem to be identical to one type of cell clone obtained by culture of human arteries (Lazzarini-Robertson, 1963). In diseased arteries, these "atherophils" were transformed into foam cells ("atherocytes"), whereas another type of cell, the "fibrophils", changed into fibrocytes. In cultures of intimal cells derived from atherosclerotic vessels, the atherocytes eventually died out and were replaced by fibrocytes (Page et al., 1966). It is difficult to extrapolate from *in vitro* to *in vivo* conditions but, according to the latter authors, this event resembles the "natural history of spontaneous atheroma, from fat infiltration to fibroplaque formation".

From a recent international survey in which the lesions in 23,000 coronary arteries and aortas were statistically analysed, McGill (1966) concluded that the extent of coronary fatty streaks in young Caucasian males is closely associated with the extent of coronary fibrous plaques in middle-aged persons from the same population. However, no such association could be found between fatty streaks and fibrous plaques in either Caucasian and Negro aortas or Negro coronary arteries.

In appears, therefore, that the behaviour and significance of fatty streaks depend on their anatomical site and are not identical in Caucasian and Negro populations.

Nevertheless, in view of the histological and electron-microscopic data the statement made by McGill *et al.* in 1963 seems to be still valid: "... there is no reason to believe that the pathogenesis of aortic and that of coronary lesions are significantly different except that regression of aortic fatty streaks may be more likely to occur, especially in Negroes". One could now add that fatty streaks in coronary arteries can probably regress particularly in Negroes.

The question now arises *what factors can either prevent the progression of fatty streaks to fibrous plaques or induce fatty streaks to regress?*

One probably important factor appears to be the metabolism of the vessel wall. In view of the disparities between the severity of atherosclerosis in dif-

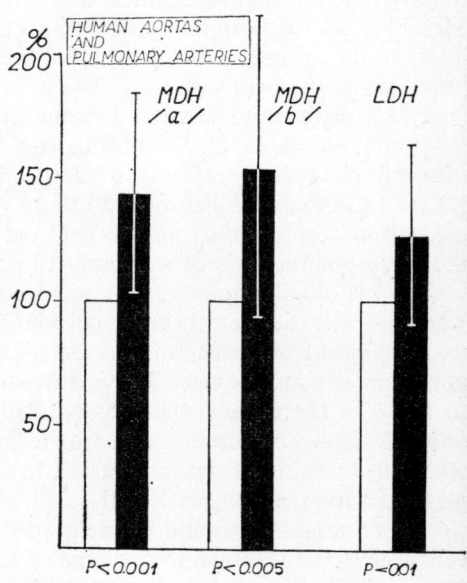

FIG. 59.—The activity of malate dehydrogenase and lactate dehydrogenase in 40 human pulmonary arteries (black columns) as compared with normal parts of the thoracic aortas from the same persons (=100 per cent). Upright line with bars ± S.D.* a = optical test and b = NT-PMS method in two different series of determinations (see Chapter XI). (Data from Zemplényi et al., 1965b.)

* In this and all subsequent similar experimental series the statistical significance of the difference between the calculated mean value and 100 per cent has been tested by Student's t-test.

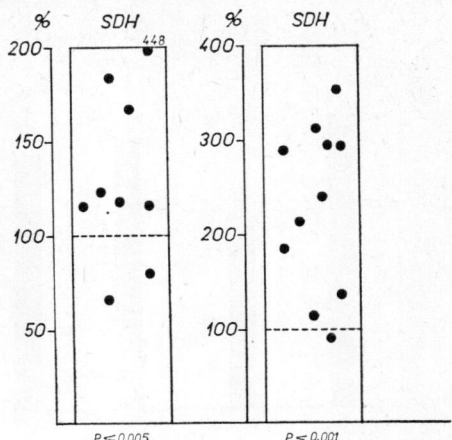

Fig. 60.—The activity of succinate dehydrogenase in a few human aortas and pulmonary arteries. Left column: The activity in the normal parts of the ascending aortas as compared with the normal parts of abdominal aortas (=100 per cent) from the same persons. Right column: The activity in the pulmonary arteries as compared with the thoracic aortas (=100 per cent) of the same persons. (Data from Zemplényi and Mrhová, 1964b; Zemplényi et al., 1965b.)

ferent arteries and arterial segments, we felt it opportune to investigate the role of local arterial metabolic differences in the pathogenesis of human atherosclerosis.

In the studies to be described, human arteries were obtained at necropsy 5 to 12 hours after death. Such material was provided by courtesy of several pathology departments in Prague. The vessels were immediately placed on ice and within 60 to 90 minutes were subjected to the further manipulations described in Chapter XI.

It must be pointed out that results obtained with necropsy material must be interpreted with reserve as far as enzymatic changes between different subjects are concerned. To avoid this pitfall we only compared arteries from the same subject.

As can be seen from Fig. 59, the *pulmonary artery* exhibits consistently higher malate dehydrogenase activity than do non-atherosclerotic segments of the *thoracic aorta*, the activity of which is arbitrarily set at 100 per cent. The same pattern is seen with lactate dehydrogenase and in the few experiments with succinate dehydrogenase (right column of Fig. 60). On the other hand, the activities of both acid phosphatase and 5'-nucleotidase are higher in the aorta than in the pulmonary artery (Fig. 61). However, the respective activities of ATPase and non-specific carboxylesterase do not differ in these two vessels (Zemplényi, 1964a, b).

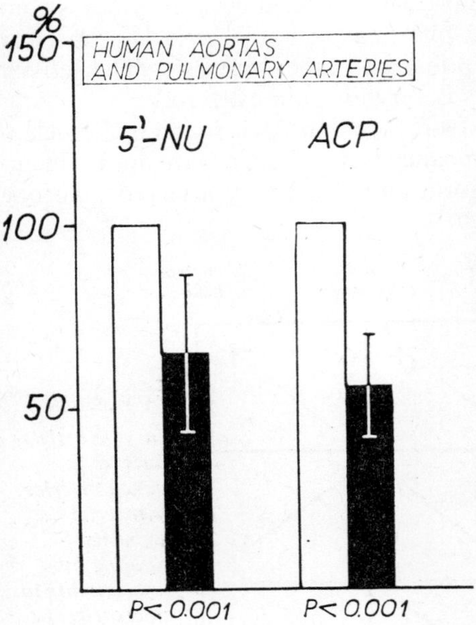

Fig. 61.—The activity of 5'-nucleotidase and acid phosphatase in the same human arteries as in Fig. 59.

Figure 62 shows that the activities of lactate and malate dehydrogenases

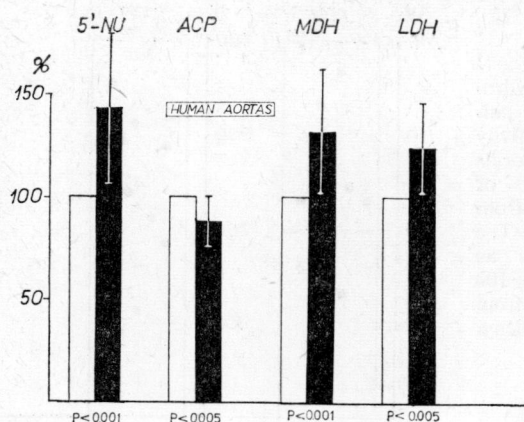

Fig. 62.—The activity of 5'-nucleotidase, acid phosphatase, malate dehydrogenase and lactate dehydrogenase in normal parts of 26 human ascending aortas (black columns) as compared with normal parts of the abdominal aortas from the same persons (=100 per cent). Upright line with bars ± S.D. (Data from Zemplényi and Mrhová, 1964b; Zemplényi et al., 1965b, 1966d.)

are higher in the *ascending* than in the *abdominal segments* of the same aortas; the same pattern obtains in the few experiments on succinate dehydrogenase (left column of Fig. 60). The activity of acid phosphatase is higher in the abdominal than ascending aorta. However, in contrast to the results of the comparative study of the pulmonary artery and aorta, the activity of 5'-nucleotidase is higher in the aortic segment with lower susceptibility to atherosclerosis (i.e. ascending > abdominal aorta) (Zemplényi and Mrhová, 1964b; Zemplényi et al., 1965b, 1966a, d).

The first two lines of Table IV summarize the more important findings of these investigations. The activity of the tricarboxylic acid cycle enzymes is consistently lower and the activity of acid phosphatase is higher in the vessel or vascular segment with the higher susceptibility to atherosclerosis. (The activity of alkaline phosphatase in these vessels is very low, almost zero.)

The only contradictory result in the two sets of studies is found in 5'-nucleotidase activity. The reason for this discrepancy is not clear, but we must remember that neither the structure of the aorta and pulmonary artery on the one

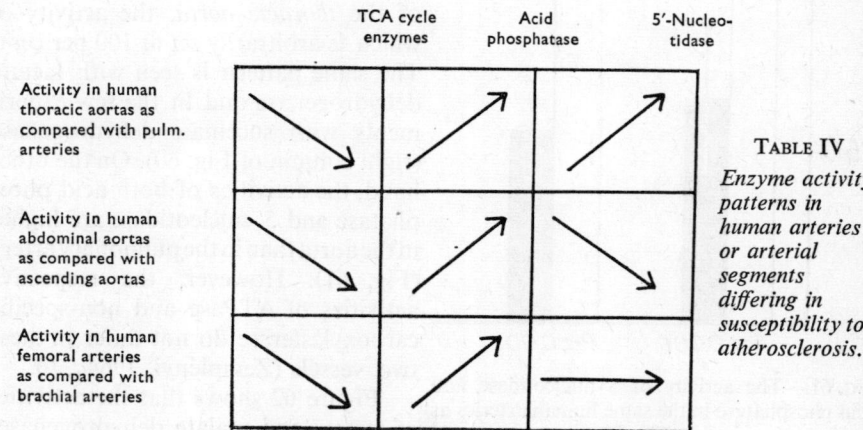

TABLE IV

Enzyme activity patterns in human arteries or arterial segments differing in susceptibility to atherosclerosis.

hand, nor the structure of the thoracic and abdominal aorta on the other hand are absolutely comparable. For example the medial elastin content is higher in the human aorta than pulmonary artery (Lansing et al., 1950), whereas the elastin content of canine and porcine aortas is lower in their abdominal than their thoracic parts (Harkness et al., 1957; Gillman, 1964). The elastin content of the human aorta also decreases in the distal direction and the collagen content relatively increases (see for details Chvapil, 1967 and Grant, 1967). Histological techniques show more smooth muscle and less elastic tissue in the human abdominal aorta than in its thoracic part (Gillman, 1964). Such structural

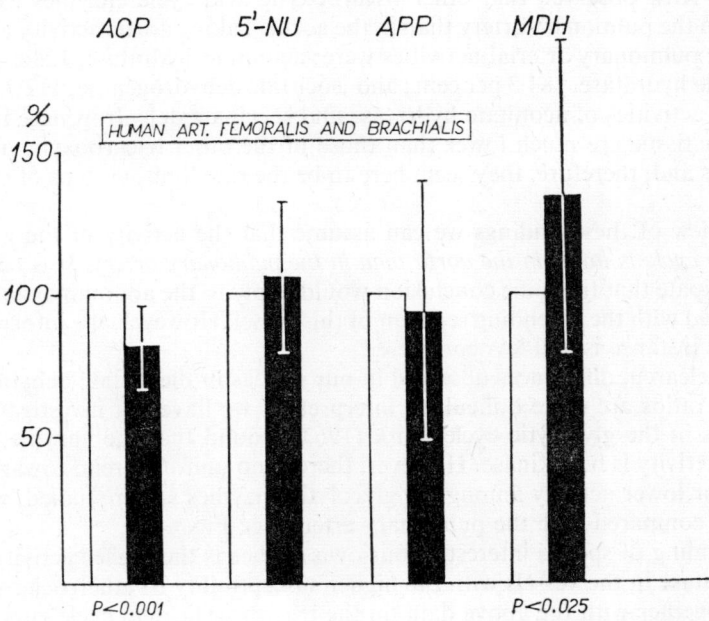

FIG. 63.—The activity of acid phosphatase, 5′-nucleotidase, ATPase and malate dehydrogenase in normal parts of 20 human brachial arteries (black columns) as compared with normal parts of the femoral arteries (=100 per cent) from the same persons. Upright line with bars ± S.D. (Data from Zemplényi and Mrhová, 1964b; Zemplényi et al., 1965b.)

differences could possibly be related to the "anomalous" 5′-nucleotidase results. As a corollary we can postulate that the differences in the other enzyme activities are not directly related to these differences in arterial structure, otherwise converse activity ratios would have been found in the two comparative studies. (See, however, enzymes in aortas of children and young subjects, p. 178).

Further experiments were carried out to compare enzymatic activity in vessels of a similar type, namely the *brachial* and *femoral* arteries.

Figure 63 summarizes the more important results indicating that the activity of malate dehydrogenase, a representative of the tricarboxylic acid cycle enzymes, is again higher in the artery less susceptible to atherosclerosis (i.e. the brachial artery), whereas the converse applies to acid phosphatase activity. No significant difference in the activities of ATPase and 5′-nucleotidase could be detected.

For comparison with the preceding series, the third line of Table IV schematically presents the results with the brachial and femoral arteries.

All three series of experiments show that the activities of tricarboxylic acid cycle enzymes (i.e. malate and succinate dehydrogenase) are definitely lower in vessels or vascular segments with higher susceptibility to atherosclerosis.

Our results on the differences between normal parts of the aorta and pulmonary artery are in good agreement with Kirk's findings (see Fig. 64), even though the latter were mostly obtained by comparison of different subjects and were usually calculated on a wet weight basis. In addition to malate dehydrogenase, Kirk observed that other tricarboxylic acid cycle enzymes were more active in the pulmonary artery than in the aorta. Taking aortic activity as 100 per cent the pulmonary arterial activities were: aconitate hydratase, 128·6 per cent; fumarate hydratase, 184·3 per cent; and isocitrate dehydrogenase, 112·7 per cent.

The activities of aconitate hydratase and succinate dehydrogenase in human vascular tissue are much lower than those of the other tricarboxylic acid cycle enzymes and, therefore, they seem here to be the rate limiting steps of the whole cycle.

In view of these findings we can assume that the activity of the *whole respiratory cycle is lower in the aorta than in the pulmonary artery*. It is reasonable to anticipate that the same conclusion would apply to the abdominal aorta when compared with the ascending segment of this vessel. However, the information in this last instance is still less complete.

The clearcut differences observed in our studies in the lactate dehydrogenase activity ratios are more difficult to interpret, as we have not investigated other enzymes in the glycolytic cycle. Kirk (1963a) found that the enzyme with the lowest activity is hexokinase. However, there is no uniform trend towards either higher or lower activity among the glycolytic enzymes so far studied, when the aorta is compared with the pulmonary artery (see Fig. 64).

A finding of special interest in our own studies is the higher activity of acid phosphatase in the vessels with the higher susceptibility to atherosclerosis. This point together with the above data on the tricarboxylic acid cycle enzymes will be considered in detail later.

It is important to stress the fact that our results were in most cases obtained from necropsied vessels of older persons. With such vessels it is often very difficult, if not impossible, to decide whether "normal" vascular segments are really free of discrete atherosclerotic lesions. In fact it is known that the lipid and especially cholesterol content increases very substantially in human arteries with advancing age. Lande and Sperry (1936) observed an increase of aortic lipids from an average of 152 mg./100 g. in the age group 11–29 years to an average of 1299 mg./100 g. in the age group 60–80 years. Likewise, Faber and Lund (1949) and Bürger (1954) reported a steady increase of cholesterol concentration in "normal" aortas from values of about 300 mg./100 g. in childhood to values over 2,000 mg./100 g. in advanced age.

Kirk (1962a, b) also found that the aortic cholesterol concentration is about six-fold higher in the age group 60–90 years than in the child. This is in line with our experience that aortic cholesterol concentration rises from about 0·12 per cent (on a wet weight basis) in the youngest age groups to about 1·15 per cent in subjects over 70 years.

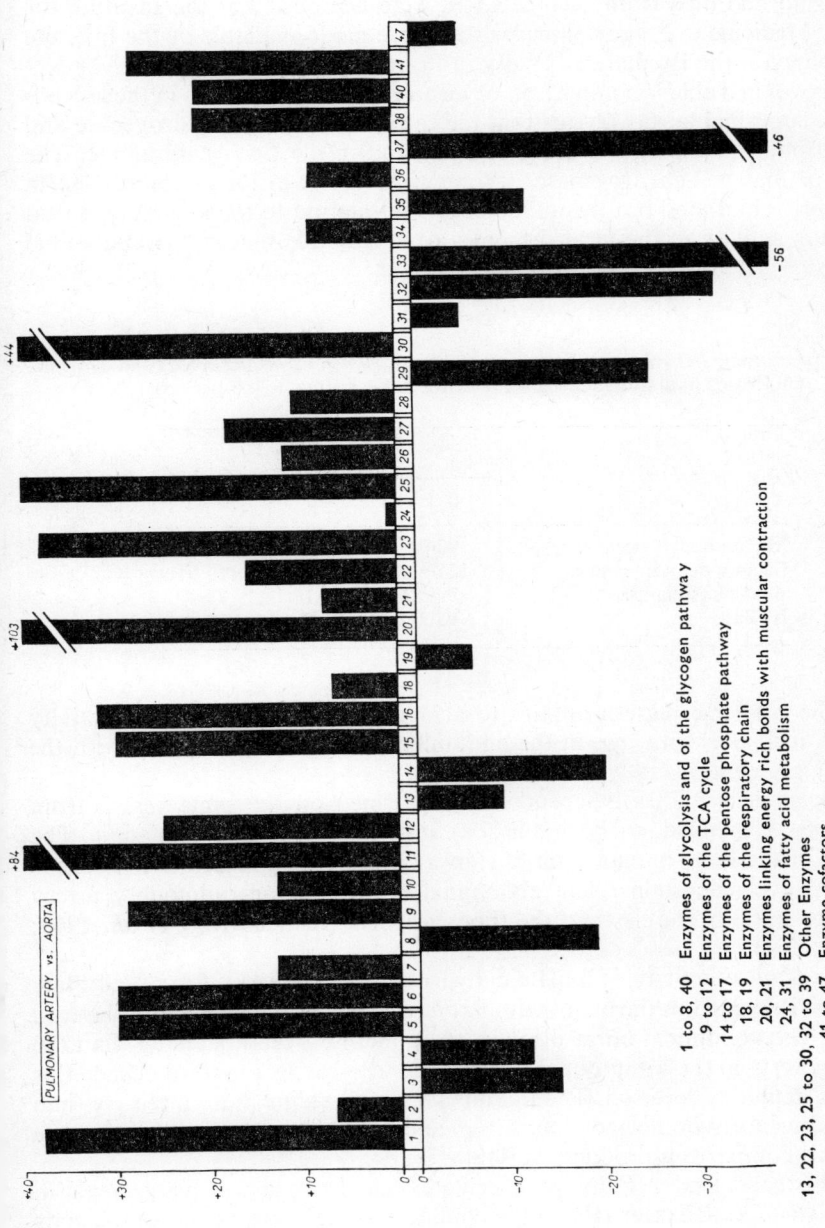

FIG. 64.—Differences in enzyme activities and cofactor levels between the human pulmonary artery and aorta. The enzyme activities or cofactor levels of aortas, calculated on a wet weight basis = 100 per cent. (Constructed according to data from Kirk, 1963a and some more recent publications by Kirk and co-workers.) For symbols see Fig. 9a and b.

It was therefore necessary to investigate the enzymatic pattern in arteries of young subjects and particularly in children's arteries. We were able to study by biochemical and histochemical techniques arteries from children and young people, who had died from accidents and were necropsied at the Institute for Forensic Medicine in Prague. Other vessels were made available by the Institute of Pathology in the Paediatric Faculty in Prague.

As shown in Table V comparison of mean enzymatic activities in the ascending and abdominal aortas reveals that the ratios for malate dehydrogenase and acid phosphatase undergo striking changes during the first years of life. The activity of malate dehydrogenase is somewhat higher in the abdominal aorta during early childhood but it soon decreases in relation to the ascending aorta, so that the activity in the latter becomes nearly double that in the abdominal

TABLE V

The relationship between some enzyme activities in human ascending aortas and age
(Activities in abdominal segments of the same aortas = 100 per cent)

Enzyme	Age groups		
	0–1	2–20	30–42
Malate dehydrogenase	90·6	125·5	191·7
Lactate dehydrogenase	123·9	120·7	119·6
Acid phosphatase	99·6	73·1	77·9
ATPase	95·2	96·8	73·2

aorta. The converse pattern applies to acid phosphatase, the relative activity of which increases with age in the abdominal aorta when compared with the ascending segment.

Histochemical studies independently performed on the same vessels from 47 subjects (Urbanová, to be published) shed much light on this topic. The structure of the large human arteries shows typical elastic structure in sections stained with elastic stains. The abdominal aorta and the pulmonary artery have more smooth muscle than the thoracic aorta (see also Rigg *et al.*, 1960; Gillman, 1964).

Histochemical activity of malate dehydrogenase (Plate IV, *4* to *6* and Plate VI, *4* to *6*) is located in the endothelium, muscle cells of the media and the *vasa vasorum*. The abdominal aorta displays more intense overall staining than the ascending aorta in the age group 2 to 10 years, whereas by the third decade this pattern is definitely reversed. The pulmonary artery stains more intensely than the ascending aorta in the age groups 2–10 and 11–20 years, but in the other age groups this comparison produces variable results.

The histochemical activity of succinate dehydrogenase is very weak in arterial tissue (see Chapter III) and is confined to the same tissue components as the other dehydrogenases. Unfortunately, it is impossible reliably to compare staining intensities of this dehydrogenase in the different arterial segments.

The activity of lactate dehydrogenase (Plate V) is located in the same sites as that of malate dehydrogenase, but the staining reaction for this glycolytic enzyme is more intense. Overall activity is higher in the abdominal than ascending aorta in the younger age groups (0–1 and 2–10 years), whereas in the age group 21–90 the reverse pattern obtains. In the age group 11–20 the comparative results are variable. Activity in the pulmonary artery is in all cases higher than in the ascending aorta.

Acid phosphatase activity (Plate IV, *1* to *3* and Plate VI, *1* to *3*) is found in all the arteries studied and is bound to the muscle cells of the tunica media. In endothelial cells the staining reaction is usually very weak. With the exception of the youngest age group (0–1 years), activity is uniformly higher in the abdominal than the ascending aorta. The ascending aorta in turn shows higher activity than the pulmonary artery in practically all cases.

Histochemical activity of ATPase is confined to endothelium and medial muscle cells; the staining reaction is stronger in the former than the latter. In all age groups the overall staining reaction for ATPase is stronger in the abdominal than ascending aorta. With the exception of the youngest age group (0–1 years), ATPase activity is more intense in the pulmonary artery than in the ascending aorta.

Activity of 5′-nucleotidase is histochemically localized as for the preceding enzyme, but the activity is also very marked in the *vasa vasorum*. Only aortas of the youngest age group (0–1 years) always show higher activity in the abdominal than ascending part, whereas in the other age groups the results are not uniform. However, in all age groups 5′-nucleotidase activity is greater in the ascending thoracic aorta than in the pulmonary artery (see the unexpected biochemical findings on p. 174).

The activity of non-specific carboxylesterase is confined to medial muscle cells and to a lesser degree to the intima. The adventitial *vasa vasorum* stain about as intensely as medial muscle cells. The differences in overall activity between the aortic segments are not uniform, whereas the staining reaction in the pulmonary artery is in practically all cases higher than in the ascending aorta.

The conclusions from comparing the biochemical and histochemical data are very relevant to the general theme of this chapter:

(*a*) The biochemical and histochemical findings are evaluated in very different ways. Thus, biochemical results are referred to the protein or deoxyribonucleic acid content of the tissue, whereas histochemical evaluation is achieved by inspection under the microscope. These approaches cannot be expected to yield identical details about overall enzyme activities. In spite of this reservation, the correlation in most instances is fairly good. The histochemical data confirm the differences between the pulmonary artery and aorta. In addition, they confirm and even extend the biochemical finding of an age-dependent relative decrease in some enzyme activities in the abdominal aorta (e.g. MDH and probably all dehydrogenases), whereas converse changes apply to other enzymes, such as acid phosphatase and perhaps 5′-nucleotidase. The only major discrepancy lies in ATPase activity, but it is not certain whether the histochemical technique detects exactly the same enzyme activity as does the biochemical assay (see Chapter VI).

(*b*) In the tunica media, staining reactions for all the enzymes investigated are confined to the smooth muscle cells; with the exception of ATPase and 5′-nucleotidase, they are much less intense in the intima. This is the chief reason for most of the enzymatic differences between the pulmonary artery and thoracic aorta (with the exception of 5′-nucleotidase—see p. 174), because the former contains more smooth muscle cells. For the same reason it is reasonable to assume that all the *age-linked changes* in aortic segmental enzyme patterns, as reflected in the biochemical results, are related to factors that affect the *vascular smooth muscle cell*. In view of the histological and electron microscopic evidence mentioned in this chapter (p. 171) and in Chapter XVIII (p. 190), this last conclusion is of special significance and we shall return to it later in Chapter XIX.

The age-linked changes in enzyme activity ratios in vascular segments that markedly differ in their susceptibility to atherosclerosis could well be a fundamental factor in determining the localization and perhaps also the progression of the disease.

In his earlier studies Kirk compared aortic enzyme activities with those in the pulmonary artery, but he did not compare activities in different segments of the same aorta. Recently, he and his co-workers reported higher activity of 3-hydroxyacyl-CoA dehydrogenase activity in the abdominal than thoracic aorta (Sanwald and Kirk, 1965a), and the same pattern was found in the biotin content of these segments (Kirk and Sanwald, 1966). Both the enzyme and cofactor play a role in lipid synthesis (see pp. 53 and 82) and, thus, these differences are of considerable interest with regard to arterial lipid metabolism.

The activity of the aminotransferase E.C. 2.6.1.16 (see Chapter VIII), which is an enzyme involved in the biosynthesis of mucopolysaccharides, was significantly higher in the abdominal than thoracic aorta (Haruki and Kirk, 1965), while the coenzyme A content of the latter segment was also higher (Sanwald and Kirk, 1965a).

These enzymes as well as both cofactors undergo significant changes with ageing (see Chapter X), but the activity ratios and their changes in various aortic segments cannot be calculated from the data so far published. Nevertheless, these results draw attention to still further differences between vascular segments that vary in their susceptibility to atherosclerosis, and it will be of great interest to identify the factors that could cause such differences.

Returning to our own results, the enzyme differences between vascular segments with different susceptibilities to atherosclerosis could be part of a general biological phenomenon. Therefore, we pursued the same line of investigation in several mammalian and avian species. The next chapter is concerned with these problems and then we shall try to answer the questions posed in the introduction to this chapter.

CHAPTER XVIII

PROBLEMS OF COMPARATIVE ATHEROSCLEROSIS AS RELATED TO THE DIFFERENT SUSCEPTIBILITY OF ARTERIES AND ARTERIAL SEGMENTS TO THE DISEASE

THE findings described in the preceding chapter have shown that in man the activities of several vascular enzymes differ according to the anatomical site; some of these differences seem to be correlated with local susceptibility to atherosclerosis.

One of the central questions in the process of atherogenesis—as pointed out earlier—is the relationship between the fatty streak and the raised fibrous plaque, as only the latter develops into the complicated lesion that is characterized by ulceration, calcification, haemorrhage or thrombosis. These last features are the main causes of clinically manifest disease.

In man it is well established that raised fibrous plaques are more frequent in the abdominal than in the thoracic aorta. However, opinions differ about the extent of fatty streaking in these two aortic segments. In order to elucidate this and other problems that arise from the study of human vessels it is important to consider related aspects of comparative atherosclerosis in animals.

Although birds are phylogenetically remote from man, we shall begin with the domestic chicken, because experimental and spontaneous lesions in this species can be clearly distinguished both by their nature and site. Related findings in ducks will also be presented.

ENZYMES OF THE VASCULAR WALL IN THE CHICKEN AND THE DUCK

From several thousand autopsies on birds, Fox (1933) described atherosclerosis in many sub-classes of the class *Aves*, in particular parrots, toucans, eagles, ducks, geese, ostriches and rheas. In the sub-class *Galliformes* he observed nine cases of atherosclerosis in the pheasant and turkey group; he quoted Yamagiwa and Adachi's (1914-1916) work in which were described atherosclerotic-like lesions in chickens. Nevertheless, Grünberg (1964) pointed out that the severe lesions often found in old parrots led this species in the past to be commonly accepted as the representative "atherosclerotic" among birds.

Extensive morphological studies of spontaneous chicken atherosclerosis were presented by Fahr (1935) and about nine years later in Dauber's (1944) classic paper. These findings, as well as those of Katz and Stamler (1953) and Weiss and Fisher (1959) revealed that in the chicken the abdominal aorta is by far the commonest site of spontaneous atherosclerosis. The lesions are characterized by raised, smooth, longitudinal, white or yellow, ridge-like thickenings, that are microscopically quite similar to human fibrous plaques (see below). According to Howard and Gresham (1964) spontaneous atherosclerotic lesions in the turkey's aorta are also chiefly confined to the abdominal segment and are predominantly fibrous rather than fatty.

In addition to fibrous abdominal lesions, the thoracic aortas of female chickens (Dauber, 1944) and some males (Siller, 1965) develop yellow sudanophilic flat intimal lesions, which seem to be similar to fatty streaks.

In contrast to *spontaneous* fibrous lesions, the *cholesterol-fed chicken* develops lesions predominantly in the ascending aorta and arch (Katz and Stamler, 1953).

Lindsay et al. (1955) carried out detailed studies of the genesis of the naturally occurring lesions and claimed that, in the abdominal aorta, the initial changes were fragmentation of the internal elastic membrane, proliferation of intimal fibroblasts and deposition of mucoid substance, resulting in the formation of intimal plaques. Later lipids were deposited in the deep parts of these plaques, but they apparently played no part in the pathogenesis of the earliest abdominal lesions. In contrast, the thoracic lesions result from a primary accumulation of lipids, which first appear in the intima and later in the media and which lead to the development of plaques at this site. If chickens are made hypercholesterolaemic, the primary fibrous abdominal lesions are modified by extensive secondary deposits of lipid and cholesterol (Lindsay and Chaikoff, 1963). Birds thus treated also show primary lipid lesions that consist of foam cell accumulations in all portions of the vascular system. Weiss (1959) also concluded that the distribution of spontaneous lesions along the aorta is the reverse of that observed in cholesterol-fed chickens.

However, the nature and localization of these spontaneous and experimental lesions may be related to the considerable structural differences between these aortic segments. Thus, the thoracic aorta is a typical elastic artery composed of connective tissue and alternating bands of smooth muscle cells that are separated from each other by thick elastic laminae. The abdominal aorta is a true muscular artery, in which the media mainly consists of smooth muscle. Although such structural differences between aortic segments are particularly conspicuous in birds, they seem to constitute a general phenomenon that is common to most mammals including man (Rigg et al., 1960).

In addition to the above-mentioned publications, the detailed histological picture of spontaneous chicken atherosclerosis has recently been summarized by Grollman et al. (1963) and Siller (1965). The latter author described the "natural history" of the spontaneous lesions and demonstrated that the mature fibrous abdominal plaque contains deposits of lipids, including anisotropic cholesterol crystals that mainly accumulate at its base just inside the internal elastic membrane. Although some authors regard these abdominal lesions as arteriosclerotic (not atherosclerotic) and, thus, not comparable with human atherosclerotic disease (Katz and Pick, 1963; Wissler, 1965 and others), there seems to be no valid reason to deny their similarity with the corresponding fibrous plaque in man. As pointed out by Gresham and Howard (1963) "... it is unreasonable to expect the human prototype to be faithfully reproduced in species that differ widely in metabolism and vascular structure."

The morphological pattern of advanced experimental lesions in the chicken, as summarized by Pick and Katz (1965), reveals many features common not only to the corresponding lesions in rabbits but also to those of man, including foam cell plaques, necrosis, atheromatous abscesses, fibrosis and hyalinization, and even ulceration (Katz and Pick, 1963). (See also Reiniš et al., 1961a, b.)

As we used ducks as well as chickens in our studies, a brief note is needed on atherosclerosis in the sub-class *Anseriformes*. Among 516 specimens of this sub-class (ducks, geese and swans), Fox (1933) observed evidence of atherosclerosis in 26 animals (5 per cent). The lesions took the form of clearly outlined, raised, single or confluent yellow masses that were particularly sited in the aortic sinus above the heart, at the origins of the brachiocephalic arteries and, to a lesser extent, at the origin of the renal arteries. In this sub-class Grünberg (1964) found vascular lesions, mainly of the fibrous type, in 7 out of 42 birds in captivity but only in 1 out of 27 shot in their natural habitat. This is in agreement with Vastesaeger and Delcourt's (1962) observations concerning birds living in captivity or in freedom, and with findings by Wolffe et al. (1949) showing that spontaneous atherosclerosis is much less frequent in wild ducks than in domesticated ducks and geese. The latter authors found that force-feeding the goose induces experimental lesions similar to those in man.

Naturally occurring lesions in this sub-class were characterized histologically by Fox (1933) as an intimal process, ranging from simple fibrillary proliferation and elastic reduplication to complete hyaline and fatty replacement, with calcification, cartilage formation, ossification and even formation of bone marrow. In advanced lesions the internal elastic membrane disintegrates, while the media is progressively thinned. Clearly the progress of the disease in these animals illustrates all the morphological features of the full range of atherosclerotic processes.

In the experiments to be described, we compared enzyme activities in the ascending and abdominal aortas of 30-day old chickens. In addition, we carried out similar comparisons in the aortas of 50–60 day old ducks. Enzyme activities in the pulmonary arteries were also studied by histochemical means.

Figure 65 summarizes the results obtained in 20 chickens' aortas. The activity of malate dehydrogenase is very significantly lower in the ascending than abdominal aorta. In contrast, the activities of alkaline and acid phosphatase, carboxylesterase and lactate dehydrogenase are higher in the ascending aorta, whereas no difference is observed in the activity of ATPase in these segments.

Figure 66 shows that in 36 specimens of ducks' aortas the activity of malate dehydrogenase was again significantly lower in the ascending than abdominal aortic segments, whereas carboxylesterase and lactate dehydrogenase reveal a reverse trend in their activity ratios. In contrast to the findings in the chicken's aorta, there is no difference in the activity of acid phosphatase between the two aortic segments, while the ATPase activity is lower in the ascending than abdominal aorta.

As demonstrated histologically by Dr. Urbanová, the abdominal segment of both the chicken's and duck's aorta (Plate VII) is unequivocally of muscular type, while the ascending aorta is predominantly of elastic structure. In the chicken's aorta the transition between elastic and muscular segments is less abrupt than in the duck's aorta.

The histochemical enzyme reactions are essentially confined to the same

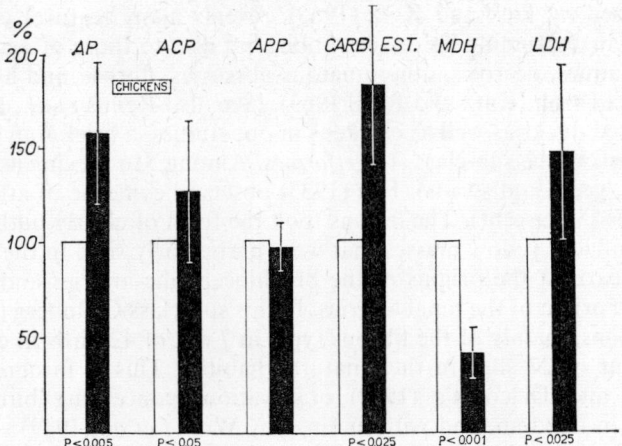

Fig. 65.—The activity of alkaline phosphatase, acid phosphatase, ATPase, non-specific carboxylesterase, malate dehydrogenase and lactate dehydrogenase in the ascending aortas of 20 chickens (black columns) as compared with the abdominal aortas (=100 per cent) of the same chickens. Upright line with bars ± S.D. (Data from Zemplényi et al., 1965c.)

vascular constituents as in the human aorta, but the staining intensities are different, usually in agreement with biochemically observed species differences (see also Lojda, 1965).

The activity of succinate dehydrogenase in the duck's aorta is clearly localized in muscular elements in the media. In agreement with the varying anatomical structure, the highest overall activity is exhibited by the abdominal and the lowest by the ascending aorta, the activity in the pulmonary artery being inter-

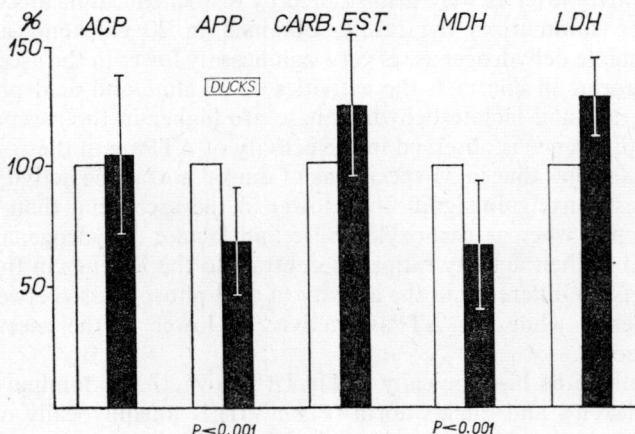

Fig. 66.—The activity of acid phosphatase, ATPase, non-specific carboxylesterase, malate dehydrogenase and lactate dehydrogenase in the ascending aortas of 18 ducks (black columns) as compared with the abdominal aortas (=100 per cent) of the same ducks. Upright line with bars ± S.D. (Data from Zemplényi et al., 1965c.)

mediate between the two. With other enzymes the same essential histochemical pattern was observed in various vascular segments in both the chicken and duck. These patterns are illustrated for malate dehydrogenase (Plate VIII, *4* to *6* and Plate X, *4* to *6*), lactate dehydrogenase (Plate IX, *1* to *6*) and acid phosphatase (Plate VIII, *1* to *3* and Plate X, *1* to *3*).

These histochemical results accord with the general conclusion about human vessels (see Chapter XVII), namely that differences in overall enzyme activities depend on the *density and, by inference, on the intact state of arterial smooth muscle cells*. The subsequent sections of this chapter will provide evidence that the same conclusion also applies to young mammals.

ENZYMES OF THE VASCULAR WALL IN THE RHESUS MACAQUE

As can be seen even from a cursory survey of the literature, data about the "natural history" of atherosclerosis in subhuman primates are extremely contradictory. In fact, until recently only very sketchy information was available about atherosclerosis in these animals—taxonomically the nearest relatives to man.

Most authors have observed lesions in primates that have the features of fatty streaks. Among 796 subhuman primates autopsied in the Philadelphia Zoo, Fox (1933) found such lesions in only 7 Old World monkeys (*Cercopithecidae*) and in 1 animal of the typical South American monkey family (*Cebidae*). The lesions took the form of superficial atheromatous patches in the sinuses of Valsalva, arch and thoracic aorta, but rarely extended down to the abdominal part.

However, in general, fatty streaking is reported more frequently in primates, especially after gross staining with Sudan dyes. This manoeuvre reveals intimal lesions that are otherwise not visible to the naked eye. For example in a similar report from the London Zoo, Finlayson and Symons (1933) found fatty streaking in 35 (44 per cent) of 79 primates, whereas true atheroma was only detected in 3 such animals. It is interesting that fatty streaking is more common in primates than in other animals: the average frequency of this lesion in captive mammals being 23 per cent and in birds 24 per cent. However, the criteria for differentiating fatty streaks from more advanced lesions are not firmly established in subhuman primates and this might to some extent explain the controversial views on the "natural history" of the lesions in these animals.

McGill *et al.* (1960) observed raised fibrous plaques in only 7 out of 163 baboons shot in Kenya, but microscopically these lesions differed from the corresponding human plaques. On the other hand sudanophilic intimal lesions were present in about 20 per cent of young animals and approximately 75 per cent of animals with permanent dentition had such lesions to some degree. On the basis of similar findings, Gresham and Howard (1963) concluded that "failure to find anything more than fatty streaks even in old primates might support the view that the fatty streak is not the precursor of the atheromatous lipid-laden plaque. Alternatively, it may indicate the absence of the stimulus that in man provokes the transformation of streak to plaque".

Fibrous plaques are probably not so rare as suggested by these data. Among 23 baboons captured in Kenya, Howard *et al.* (1966) observed not only sudanophilic aortic plaques in 14 animals, but also aortic fibrous plaques in 8 animals probably aged more than 4–6 years. The fatty streaks were particularly common around the origins of branches leading off the aorta, whereas white fibrous plaques were common in the mid-aortic region and at the bifurcation of the abdominal aorta. Histologically the lesions bore a very close resemblance to the early human fatty streak and fibrous plaque.

Spontaneous raised fibrous plaques are also not uncommon in the squirrel monkey, which appears to be a very promising animal "model" for the study of atherosclerosis (Middleton et al., 1964). This small and comparatively inexpensive subhuman primate develops not only aortic fatty streaks in almost 100 per cent of cases, but also raised aortic plaques and lipid-containing coronary lesions.

In apes, especially in chimpanzees, the lesions so far observed resemble human fatty streaks (Vastesaeger and Delcourt, 1961; Stare, 1963; Strong et al., 1966). However Lindsay and Chaikoff (1966) described fibrous plaques in the aortas and coronary arteries of these animals; such lesions eventually exhibit hyalinization, abundant collagen fibres and cholesterol deposition or calcification in their deeper parts.

Lindsay and Chaikoff (1966) carefully studied the aortas and coronary arteries in 17 families of subhuman primates, comprising 59 animals in captivity and 8 wild baboons from Africa. According to their findings, the basic initial process consists of a degenerative intimal process that is characterized by elastic degeneration and regeneration, mucoid deposition, and fibroblastic and smooth muscle proliferation. The authors claim that this connective tissue proliferation is primary and seems to be neither associated with, nor caused by, lipid infiltration of the intima; they considered that the pathogenesis of the lesions is basically the same in both subhuman primates and other mammals (see Lindsay and Chaikoff, 1963). These conclusions accord with previous findings by Gillman and Gilbert (1957), who studied the aortas of 85 baboons fed a low cholesterol diet and reached the conclusion that atherosclerotic-like lesions were related to preceding fibrotic changes in the intima.

Fox's (1933) observations suggest that the rhesus macaque (*Macaca mulatta*) is resistant to spontaneous atherosclerosis. Likewise, Lapin and Yakovleva (1963) observed some degree of aortic atherosclerosis in only 7 rhesus monkeys out of 976 necropsied; only one animal had gross coronary atherosclerosis. However, it seems that in this Old World monkey too, the occurrence of atherosclerosis depends much on age. Thus, Lindsay and Chaikoff (1963, 1966) observed well-developed aortic lesions in 7 rhesus monkeys aged over 15 years. The degree of intimal thickening was definitely higher in the abdominal than thoracic aorta, but seemingly there was no difference in the extent of atherosclerosis in these two segments.

As the rhesus monkey is commonly used in laboratory work, it is not surprising that most attempts to induce experimental atherosclerosis in subhuman primates have been directed at this animal. Rinehart and Greenberg (1949) succeeded in producing such experimental atherosclerotic changes by feeding the animal a pyridoxine-deficient diet. Feeding 500–800 mg. of cholesterol per day did not produce atherosclerotic lesions in the rhesus macaque (Kawamura, 1927; Hueper, 1946), but Mann and Andrus (1956) and Taylor et al. (1962)—

using higher doses of dietary cholesterol—succeeded in producing extensive atherosclerotic lesions in this species. These workers also showed that arterial injury and repair had a potentiating effect upon the development of such lesions.

The findings on primate disease have recently been summarized by Taylor (1965), who suggested that the "pathogenesis of diet-induced atherosclerosis might be summed up briefly as a four-stage process:
(1) Diffuse interstitial deposition of lipid droplets in the intima.
(2) Proliferation of subendothelial multipotential mesenchymal cells, stimulated by the presence of irritating lipid deposits in the intima. Some cells become macrophages and phagocytize the lipid droplets; others become fibrocytes and form collagen and elastic tissue; and *still others become smooth muscle cells*.
(3) Additional reactive proliferation of subendothelial cells with formation of larger masses of foam cells and thicker layers of intimal fibrous scar tissue. The large aggregates of foam cells begin to undergo degeneration.
(4) Medial destruction, further deposition of lipid-laden macrophages in both the intima and media, and profound reaction to the lipid deposits which now consist partly of extracellular, irritating lipid material, often deeply buried in a thickened intima. These reactive phenomena consist of extreme, dense intimal scarring with luminal narrowing, arteritis, vascularization of the media and intima, and even calcification and thrombosis".

This concept of the natural history of primate atherosclerosis is—of course—precisely the converse of that suggested by many students of spontaneous atherosclerosis in this species (see above), and at present it is extremely difficult to decide which of these views is nearer the truth. However, it must be kept in mind—as pointed out by McMillan (1965)—that the most complete model of human atherosclerosis is that demonstrated in the rhesus monkey by Taylor.

In addition, an important feature of these studies is the fact that the distribution of atherosclerosis throughout this monkey's arterial system is strikingly similar to that observed in man. The sequence of development of the disease in different arterial segments also closely follows the human pattern.

These studies on subhuman primates stimulated us to study enzyme activities in the aortas of rhesus macaques.

The arteries of 12 rhesus macaques of approximately 10–12 months of age were obtained immediately after killing the animals; they were transported on ice, and processed as described in Chapter XI. These vessels were made available to use by courtesy of Dr. Trčka (Research Institute of Pharmacology and Biochemistry, Prague).

Figure 67 shows that the activity of 5'-nucleotidase, ATPase and carboxylesterase is higher in the rhesus monkey's ascending aorta than abdominal aorta; the same pattern is also seen with lactate dehydrogenase (Fig. 68). Conversely, the activity of malate dehydrogenase and β-glucuronidase is higher in the abdominal than ascending aortic segment, but no such differences were observed with acid phosphatase and succinate dehydrogenase.

The structure of the large arteries of *Macaca mulatta* is much the same as those of man. The histochemical enzyme reactions are also localized to the same structures as in human vessels (Urbanová, to be published).

The histochemical activity of malate dehydrogenase is strongest in the endo-

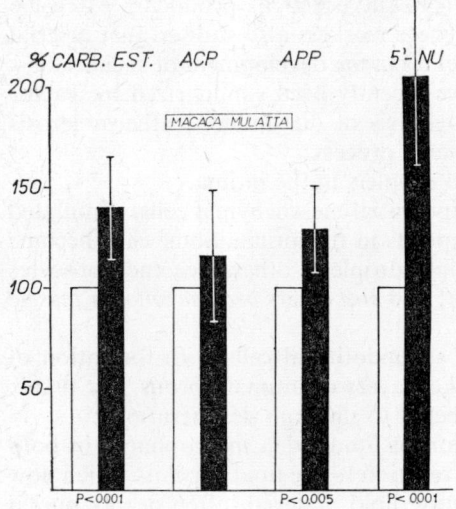

FIG. 67.—The activity of non-specific carboxylesterase, acid phosphatase, ATPase and 5′-nucleotidase in the ascending aortas of 12 rhesus macaques (black columns) as compared with the abdominal aortas (=100 per cent) of the same animals. Upright line with bars ± S.D. (Data from Zemplényi et al., 1965c.)

thelium and is also readily detectable in medial smooth muscle cells. The overall reaction is always stronger in the abdominal than ascending aorta, but the differences between the latter and the pulmonary artery are not consistent (Plate XI, 4 to 6).

The staining reaction for lactate dehydrogenase is very strong in both the muscle cells and endothelium. Overall activity is again higher in the abdominal than ascending aorta, whereas the differences between the ascending aorta and pulmonary artery are variable (Plate XII, 4 to 6).

Succinate dehydrogenase activity is very weak; it appears to be confined to endothelium and muscle cells and is highest in the abdominal aorta. A similar staining reaction was obtained with glycerolphosphate dehydrogenase.

Acid phosphatase activity is located in muscle cells (Plate XI, 1 to 3); differences in its overall activity between the aortic segments are not uniform.

A strong ATPase reaction is exhibited by muscle cells; activity is also found in the endothelium—if intact. Overall activity is always higher in the ascending

FIG. 68.—The activity of succinate dehydrogenase, malate dehydrogenase, lactate dehydrogenase and β-glucuronidase in the same macaque vessels as in Fig. 67.

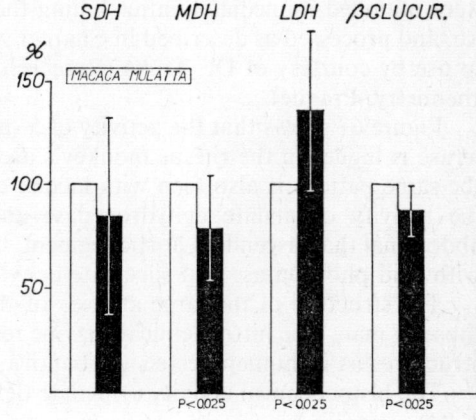

than in the abdominal aorta, but the differences between the ascending aorta and pulmonary artery are not constant.

The staining reaction for 5'-nucleotidase is located in the same tissue components as the preceding enzyme. Overall activity is higher in the ascending than in the abdominal aorta, but is consistently weaker in the pulmonary artery than in the ascending aortic segment.

Although we must remember the reservations mentioned on p. 179 when comparing enzyme patterns revealed by histochemical and biochemical techniques, it is nevertheless encouraging that the histochemical data—which were analysed absolutely independently—strongly supported the biochemical results. In general the same enzyme patterns were revealed in the biochemical and histochemical studies. Lactate dehydrogenase was the only exception; the cause of this discrepancy was not determined. The main value of the histochemical data is that, unlike the biochemical results, they enable enzymatic activity to be localized. Enzyme activity in the smooth muscle cells is clearly the decisive factor that determines overall activity in the arteries studied. As with human and avian arteries, we can anticipate that any change in overall enzyme activity, as determined quantitatively by biochemical assay, will mainly depend on the *quantity and condition of the muscle cells in the vascular wall.*

ENZYMES OF THE VASCULAR WALL OF THE PIG

In the last few years investigators have increasingly turned to the pig as an experimental model for the study of atherosclerosis. From the pioneer work of Gottlieb and Lalich (1954), Skold and Getty (1961), Jennings *et al.* (1961), French *et al.* (1963), Dahme (1963) and others it became evident that spontaneous, as well as the artificially aggravated arterial lesions that occur in the pig strongly resemble early human atherosclerotic lesions.

The "natural history" of atherosclerosis in the pig, as studied by Luginbühl and Jones (1965) and by Getty (1965), shows that fatty streaks appear in the major arteries even by the age of four months. With progression of these lesions, connective tissue fibres increase in the intima: concomitant degenerative changes are characterized by splitting and fragmentation of the internal elastic lamina, occasional fenestrations (? ruptures) in which allow smooth muscle cells to invade the intima. These changes are very similar to those observed in human vessels.

The abdominal aorta appears to be the arterial segment in which streaks and plaques can first be detected. In the older pig, streaks and plaques are evident in the thoracic aorta, but at no time is atherosclerosis as severe as in the abdominal aorta (Getty, 1965). Similar findings have previously been reported by Dahme (1964), Gillman (1964) and others. In the coronary arteries the lesions develop later in life, but they are less extensive than those in the aorta; the left coronary artery is more affected than the right. This distribution of changes together with the characteristics of the lesions in the cerebral arteries and many of the major blood vessels is very similar to that of man (Getty, 1965). However in contrast to man, no evidence has yet been found in the pig of the later complications of atherosclerosis, namely ulceration, thrombosis and haemorrhage into the plaque.

As in human atherosclerosis, an important role is ascribed to smooth muscle cells in the genesis of atherosclerotic lesions in the pig's arteries (French et al., 1963; Luginbühl and Jones, 1965). The Oxford investigators have clearly delineated the morphological changes that occur in the intima of large porcine arteries (French, 1964; French et al., 1963; French and Jennings, 1965). It is important to realize that, while in small laboratory animals the endothelium of the aorta and its main branches is separated only by a narrow space from the internal elastic lamina, in larger animals and in man such a simple structure can only be found in the foetus or at birth. Very early in life smooth muscle cells apparently penetrate the space between the endothelium and internal elastic lamina to form the so-called musculo-elastic layer of the intima.

Under the electron microscope the various stages of intimal thickening and early plaque formation can be conveniently studied in the pigs' arteries, whereas for technical reasons in man this has so far only been possible in the coronary arteries (Geer et al., 1961; Robertson et al., 1963; see p. 171). In view of the great similarity between human and porcine atherosclerosis, the sequence of events in the pigs' arteries as observed in the Oxford studies—is very relevant to the human disease.

As early as at six weeks of age, the proximal part of the descending branch of the left coronary artery exhibits protrusion of smooth muscle cells from the media into the intima through fenestrations in the internal elastic membrane, as well as an increase in the ground substance. The protruding muscle is capped by a layer of elastic tissue which may give the impression of a splitting of the internal elastic lamina. Further proliferation leads in older animals to the formation of intimal musculo-elastic thickenings that are visible to the naked eye. Under the electron microscope these thickenings are seen to consist of many layers of smooth muscle cells with spaces between the cells that contain collagen and elastic fibres. In the oldest animals the superficial layer reveals more compactly arranged collagen and elastin with relatively few cells among them. In these areas the cells cannot be identified with certainty, as smooth muscle cells and occasional round cells and cells resembling fibroblasts can all be detected.

In the thoracic aorta the sudanophilic streaking of the proximal parts and the thickenings in the mid-thoracic segments contain much fewer smooth muscle cells than do the corresponding lesions in the coronary arteries. However, the more extensive changes in the abdominal aorta, which is a muscular vessel, are more similar to those in the coronary arteries, but contain more collagen, fewer typical smooth muscle cells, and a less distinct separation into musculo-elastic and hyperplastic elastic zones. Cells resembling fibrocytes are commonly seen in the immediate subendothelial zone.

As pointed out by French and Jennings (1965), the intimal thickenings in the pig show approximately the same distribution as the lesions of human atherosclerosis, but it is difficult to draw any clear line between normality and disease. The important feature is, however, that lipid accumulation occurs at sites of the larger thickenings, initially either in the form of extracellular droplets or in the form of circular cytoplasmic inclusions in the endothelial cells. In the oldest pigs, fat is seen in cells in the elastic hyperplastic zone of the intima, while the largest plaques contain fat foam cells and extracellular lipid with cholesterol crystal clefts. According to French and Jennings (1965), electrons microscopy shows that not only do superficial macrophages or foam cells

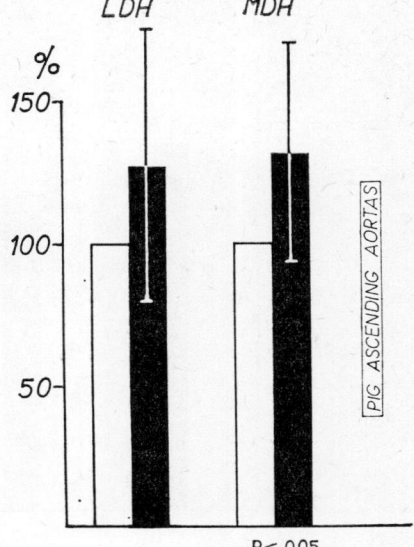

Fig. 69.—The activity of lactate dehydrogenase and malate dehydrogenase in 10 pig ascending aortas (black columns) as compared with the abdominal aortas (=100 per cent) of the same animals. Upright line with bars ± S.D. (Data from Zemplényi et al., 1965c.)

exhibit vacuoles corresponding to lipid droplets, but also many muscle cells in the deeper layers of the intima are vacuolated and probably contain lipid. Nevertheless, contrary to the views of Luginbühl and Jones (1965) and others (see p. 171), the Oxford investigators hesitate to regard all intimal foam cells as transformed smooth muscle cells, and they prefer to classify the vacuolated smooth muscle cell and the foam cell as two distinct cells (see discussion summary by Straus and Roberts, 1965).

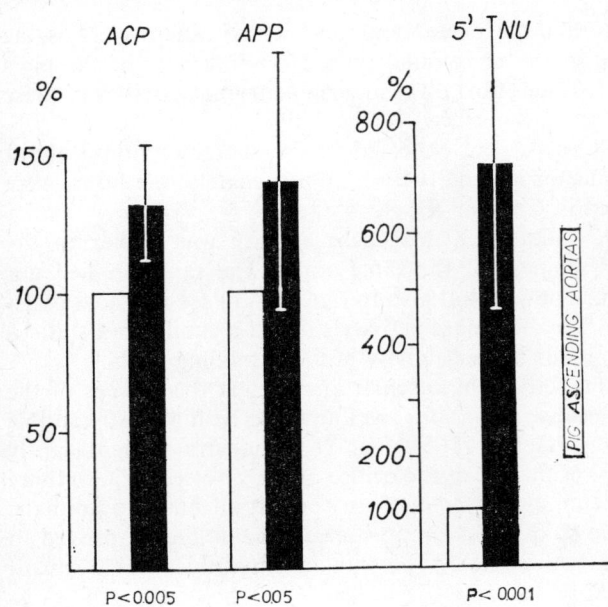

Fig. 70.—The activity of acid phosphatase, ATPase and 5′-nucleotidase in the same pig vessels as in Fig. 69.

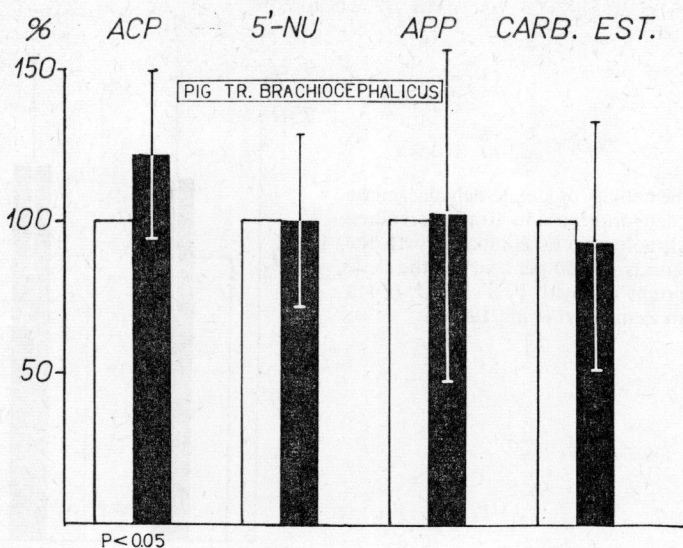

Fig. 71.—The activity of acid phosphatase, 5′-nucleotidase, ATPase and non-specific carboxylesterase in 10 pig aortas at *ostia* of the *truncus brachiocephalicus* (black columns), as compared with thoracic aortas (=100 per cent) of the same animals. Upright line with bars ± S.D. (Data from Zemplényi *et al.*, 1965c.)

As with human atherosclerosis (Chapter XVII), knowledge of the fine structure and histopathology of the pigs' vessels are of crucial importance for the proper interpretation of the vascular enzymatic changes in atherosclerotic lesions in this species.

As there is no doubt that the extent and severity of atherosclerosis, as discussed above, is higher in the abdominal than ascending part of the pig's aorta, we again concentrated our effort on comparing enzyme activities in these aortic segments.

Vessels from hogs and sows were collected in the slaughter house about 15 minutes after killing, put on ice and, within approximately one hour, were then processed as described in Chapter XI.

As shown in Fig. 69, malate dehydrogenase activity was higher in the ascending than abdominal segment of the same aorta. The same applied not only to the activity of 5′-nucleotidase but also to that of ATPase and acid phosphatase (Fig. 70). Lactate dehydrogenase activity showed a similar but statistically insignificant trend towards higher activity in the ascending aorta.

We also investigated the activity of vascular enzymes at the orifices of the brachiocephalic arteries, because these sites are known to be highly susceptible to atherosclerosis in the pig (Getty, 1965). Fig. 71 demonstrates significantly increased activity of acid phosphatase at the orifice of these vessels. The activity of carboxylesterase was definitely lower at these sites in all but two animals, nevertheless—probably due to this last circumstance—the difference proved to be statistically insignificant. Unfortunately, no data on dehydrogenases are available from this experimental series.

TABLE VI

Average activities of some enzymes in vascular tissue of man, other mammals and birds*

Enzyme†	Human aorta abd.	asc.	Macaque abd. aorta	Calf abd. aorta	Pig abd. aorta	Chicken abd. aorta	Duck abd. aorta	Human femor. art.
Lactate dehydrogenase	62·63	76·53	84·8	87·4	50·3	250·3	200·1	253·7
Malate dehydrogenase	23·76	30·63	37·8	24·1	24·1	47·7	36·4	26·23
Succinate dehydrogenase	36·6	65·3	213·0					
Alkaline phosphatase			0·25			0·88		
Acid phosphatase	0·53	0·46	0·82	0·40	0·24	0·39	0·52	0·33
ATPase	0·57	0·55	1·63	0·24	0·49	0·53	0·53	2·64
5'-Nucleotidase	0·61	0·85	0·46	0·48	0·20	0·11		0·91
β-Glucuronidase	8·2	9·08	4·20	3·34		21·13	11·8	
Non-spec. carb. est.	1·66	1·51	3·75		5·33	6·93	8·58	1·78

* Values corresponding to 100 per cent in figures of Chapters XVII and XVIII.
†Units: LDH   = μg diformazan/100 μg extract protein/hour
   MDH   = Same as above
   SDH   = μg diformazan/100 μg DNA/hour
AP, ACP   = μM phenol/100 μg extract protein/hour
ATPase, 5'-NU   = μg phosphate/100 μg extract protein/hour
β-Glucur.   = μg phenolphth./100 μg extract prot./16 hrs.
Non-spec. carb. est. = μg naphthol/1 mg. extract protein/hour

The histochemical results demonstrated a similar localization of enzyme activities as in the vessels of man and of the other animals investigated. Plate XII and XIII show some staining reactions in the arteries of a young pig.

Before ending this chapter, it is pertinent to mention some work on atherosclerosis and vascular enzymes in another domestic animal, the cow.

ENZYMES OF THE VASCULAR WALL IN CATTLE

The intima of the cow's aorta does not develop a clearly defined musculoelastic layer before the second year (Wolkoff, 1924). With advancing age the entire intima may become thick and nodular; the process is accompanied by elastic and collagen proliferation and the appearance of fine lipid droplets. In elderly cows, quite extensive intimal plaques may be formed in the proximal aorta. However, such intimal lesions are not common in the abdominal aorta, where saucer-shaped intimal depressions frequently overlie areas of Mönckeberg-like medial calcification (Fox, 1933). According to Alibasoglu et al. (1962, quoted by Lindsay and Chaikoff) two-thirds of young cattle condemned for Johne's disease exhibited gross arteriosclerotic disease that appears in the form of scattered, irregular, patchy, shell-like plaques in the abdominal aorta. However, the high incidence of arterial disease was believed to be related to tuberculous infection. Lindsay and Chaikoff (1963) observed fatty streaks and fibrous plaques in the aortas of old domestic cows.

Although the details about atherosclerosis (or arteriosclerosis) in this species are only very sketchy, some of the aortic lesions seem to share certain features with atherosclerosis, but the picture is confused by the frequently simultaneous occurrence of calcification.

In our own experiments we investigated some enzyme activities in the vessels of 4-week-old calves. Fig. 72 shows that the activity of malate dehydrogenase was significantly lower in the ascending than abdominal aorta, whereas no significant differences between these segments could be detected with the other enzymes investigated.

The ascending aortas displayed higher activities of acid phosphatase and 5'-nucleotidase than the pulmonary arteries of the same animals. For technical reasons enzymes of the tricarboxylic acid cycle could not be investigated, and histochemical results are also not available for this series.

Comparison of the results obtained with human and animal vessels discloses some common enzymatic features of arteries or arterial segments that are resistant to atherosclerosis. In the next chapter an attempt will be made to evaluate the significance of these features.

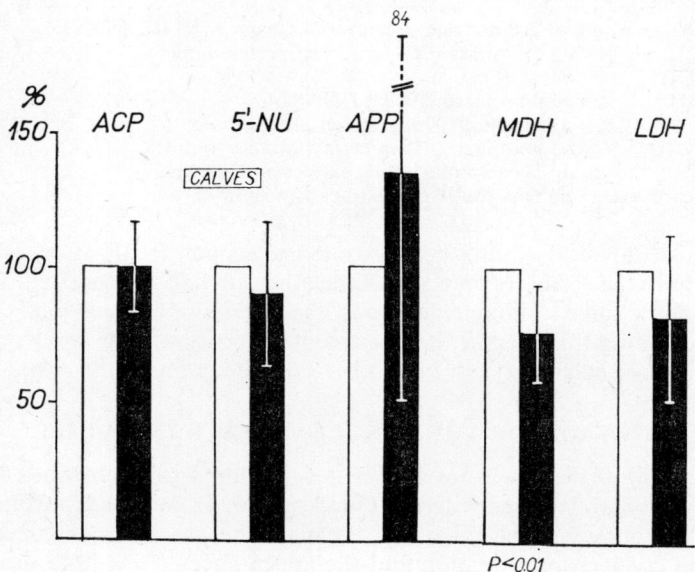

FIG. 72.—The activity of acid phosphatase, 5'-nucleotidase, ATPase, malate dehydrogenase and lactate dehydrogenase in 10 calf asc. aortas (black columns) as compared with the abdominal aortas (=100 per cent) of the same animals. Upright line with bars ± S.D. (Data from Zemplényi et al., 1965c.)

Chapter XIX

CONCLUDING REMARKS TO PART TWO: AN ATTEMPT TO EVALUATE THE SIGNIFICANCE OF LOCAL VASCULAR FACTORS IN ATHEROGENESIS

THE evidence assembled in the first part of this volume, as well as that presented in Chapters X and XII to XIV, conclusively shows that the arterial wall contains enzymes that catalyse the basic reactions in glycolysis, the tricarboxylic acid cycle, terminal oxidation and other metabolic "mainstreams".

The available data indicate that many arterial enzyme activities *change with age* (Figs. 9a and b). some of them exhibit clear-cut *sex-linked differences* and are probably regulated by hormones. The study of such regulatory factors remains a fruitful field for future research.

There is no doubt that the overall activities of most enzymes significantly alter in *atherosclerotic human vessels* (see Figs. 10a and b). These findings, which have mostly been obtained from advanced human lesions, are extremely difficult to interpret from the point of view of their role in the genesis of atherosclerosis. The study of *very early stages*, by the joint efforts of the biochemist and histochemist, might disclose interesting new information. In addition, the biochemical and histochemical study of vascular enzymes in the initial stages of spontaneous and experimental *animal atherosclerosis*, especially in those species that develop similar lesions to those in man (e.g. some subhuman primates and the pig), can reasonably be expected to add a great deal to our knowledge of the role played by vascular metabolic factors in the pathogenesis of the disease.

The striking differences in susceptibility to atherosclerosis shown by different arteries and arterial segments enabled us to delineate characteristic enzyme patterns in these vulnerable arterial sites; we feel that the results allow us to draw some tentative conclusions.

The results described in the preceding chapters disclose the fact that the tricarboxylic acid cycle enzymes exhibit a common pattern in the aortas of all the young mammals and birds studied, namely higher overall activity in the abdominal (more susceptible) segment than in the ascending (less susceptible) segment. This is summarized in the first four lines of Table VII. By contrast, in older human subjects, the arteries or arterial segments that are less susceptible to atherosclerosis (ascending aorta, brachial artery, pulmonary artery) exhibit higher activity of tricarboxylic acid cycle enzymes than the corresponding susceptible arteries (see last three lines of Table VII). In the susceptible human arteries the activity of acid phosphatase is higher, whereas in the above animals and birds the activity of this enzyme is either the same in both aortic segments or even is lower in the more susceptible abdominal segment.

Although interspecies differences must be taken into account, the probable explanation for this discrepancy is revealed by comparing enzyme activity patterns in the aortas of children and adults (see Chapter XVII). The biochemical and histochemical findings clearly show that the situation is the same in the very young human aorta as in animal aortas, but during subsequent years the pattern changes towards that seen in the older human artery. The pattern observed in arteries from pigs aged 6–8 months (Figs. 69 and 70) displays intermediate characteristics. In view of the early onset of lesions in pig arteries (see p. 189) it seems that the change from the "young" to "older" pattern in these animals develops quite early.

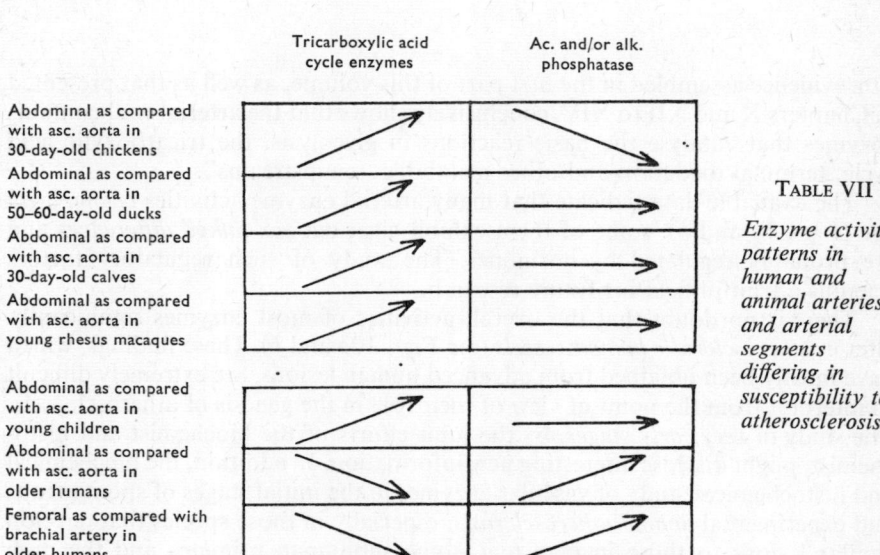

TABLE VII

Enzyme activity patterns in human and animal arteries and arterial segments differing in susceptibility to atherosclerosis.

The experiments described in Chapter XV disclose that *injury* to the vascular wall *is accompanied by increased acid phosphatase activity and reduced activities of tricarboxylic acid cycle enzymes and non-specific carboxylesterase. Likewise, with advancing age arterial segments that are more susceptible to atherosclerosis also display increased activity of acid phosphatase and decreased activity of tricarboxylic acid cycle enzymes when compared with more resistant segments.* Non-specific carboxylesterase activity cannot be reliably evaluated in this respect, because a decline in such activity would be masked by the increased lipolytic and esterolytic activities in response to lipid accumulation (see part three of this book).

In the light of these consistent changes in enzyme activity patterns, it seems justifiable to infer that the *vascular segments that are more susceptible to atherosclerosis are also those that are more exposed to damaging agents.*

Before considering more closely the question of what kinds of vascular injury may be responsible, the reader must be reminded of the probable morphological sequence in the development of atherosclerotic lesions (see McGill et al., 1963):

Fatty streaks → fibrous plaques → complicated plaques → clinically manifest disease.

This evolutionary pathway of progression is evidently accepted by most pathologists today (see Wissler, 1965), in spite of some occasional objections cited in preceding chapters.

There is a good deal of indirect but very reasonable evidence that the fatty streak is a reversible lesion. On the other hand, the transformation of the fatty streak to the fibrous plaque appears to be the critical step in the pathogenesis of atherosclerosis (McGill et al., 1963).

The results assembled in the preceding chapters and summarized above suggest that the changes in enzyme activity ratios (patterns) are probably indicative of some sort of damage to those vascular segments that are more susceptible to atherosclerosis. If so, this factor obviously must have a decisive role in the transformation of the "innocuous" fatty streak into the more advanced lesion.

The large number of factors that *may* injure the vessel wall were listed on p. 145 and the question now arises which of them actually *does*. The *damaging factor* (or interplay of factors) *must*, of course, *preferentially attack the susceptible arteries or arterial segments*.

There can be little doubt that vascular *haemodynamic stress*, including such variables as pulsatile flow, lateral pressure, suction pressure, wall tension, shearing strain, viscous drag and turbulence, plays an important part in this respect. However, this cannot be the whole story, as it has been repeatedly demonstrated that thoracic aorta homografts implanted into the abdominal aorta develop minimal atheromatous changes and, in contrast, abdominal homografts retain their higher susceptibility to atherosclerosis when implanted into the abdominal aorta (Haimovici et al., 1958; Haimovici and Maier, 1964, 1966). Similarly, in Woyda et al.'s (1960) experiments exchange autografts of pulmonary artery and abdominal aorta retained their original susceptibility to atherosclerosis. Such findings indicate that haemodynamic factors cannot be the only ones that determine susceptibility to atherosclerosis and the fate of the fatty streak.

The histochemical findings mentioned in connection with differences in susceptibility to atherosclerosis in man and animals clearly show that changes in overall activities of vascular enzymes reflect corresponding *changes in smooth muscle cells*, which are—while intact—the metabolically most active components of the arterial wall. However, they are also the vascular constituents that are most susceptible to injurious agents. This is not a new idea as other evidence indicates that *vascular muscle cells are highly vulnerable* (see Gillman, 1964; Constantinides, 1965). Therefore their metabolic activity gradually declines as a result of continuous vascular damage. Consequently, one can further infer that *those arterial segments that contain abundant muscle cells* and are thus particularly vulnerable, *are also eminently susceptible to atherosclerosis*.

Hence, dissimilarity in medial muscle-cell density presumably accounts for the different susceptibility of aortic segments to atherosclerosis in the chicken and pig as well as in other species, including man, although such divergences in segmental structure are less striking in the human aorta.

However, *intimal thickening* may also be significant in this regard. Advanced atherosclerotic lesions appear at sites characterized by accentuated *musculo-elastic intimal thickening* in the coronaries, aorta and arterial branchings (Gross *et al.*, 1934; Sappington and Cook, 1936; Dock 1946; Wilens, 1951; Movat *et al.*, 1958; Wagenvoort, 1954; Stehbens, 1960; Buck, 1963 and others). There is even some evidence that animal species that do not display such intimal thickening are more resistant to spontaneous atherosclerosis. Two further points are also relevant. *First*, there is good reason to believe that the main stimulus that causes the intima to thicken is in some way associated with fluid dynamics and vascular haemodynamic stress (Texon, 1957, 1963; Rodbard, 1959, 1962 and others*). *Secondly*, the musculo-elastic layer—by definition—is particularly rich in smooth muscle cells, which some authors consider invade the subendothelium from the media through discontinuities in the internal elastic lamina (Gross *et al.*, 1934; Buck, 1963; French, 1964, 1966 and others).

Consequently, it is justifiable to infer that haemodynamic factors of any magnitude have at least two important effects; they render certain parts of the vascular wall more *vulnerable* by causing intimal thickening and they *damage* the vascular wall if their action is protracted.

In this connection it must be emphasized that in agreement with Movat *et al.* (1958), Adams (1967), French (1966) and many others we regard progressive intimal thickening a normal age-related phenomenon that constitutes a prerequisite for the development of atherosclerosis but is not part of the picture of the disease. It may be added that progressive intimal thickening probably also alters the arterial wall by acting as a mechanical barrier that increasingly impairs the diffusion of oxygen into the inner and middle parts of the tunica media. This causes hypoxic or anoxic metabolic damage of this region and impaired lipid transport across the artery. Furthermore, local tissue hypoxia can be expected to stimulate increased production of mucopolysaccharides and collagen (see Chapter XVI).

As pointed out above, smooth muscle cells—while uninjured—are the metabolically most active components of the vascular wall. It follows that the protective mechanisms of the arterial wall, including protection against lipid accumulation and against the sclerogenic action of certain lipids (see later) are also mainly functions of these cells.

The above reasoning, although simplified, seems to be compatible with most observations on the relationship between the fatty streak and fibrous plaque, including the alleged differences in localization as described in Chapter XVII. For example, one can presume that the more abundant muscular components in the media and subendothelium of the abdominal aorta—in contrast to the ascending aorta—protect the former in young subjects against expansion of fatty spots and streaks. On the other hand, in older individuals, when the smooth muscle cells become damaged at sites of continuous haemodynamic stress, the tissue at these sites more readily undergoes fibrotic replacement by the action of sclerogenic lipids than do areas of relatively unthickened intima. Under certain

---

* Another factor that may cause or aggravate initimal thickening is incorporation of surface deposits—organization of *encrusted mural thrombi*. It is of great interest that fluid dynamics and haemodynamic vascular stress also appear to be important factors in thrombus formation and localization (Mustard *et al.*, 1964).

circumstances, of course, especially in some animal species, it is possible that haemodynamic vascular stress may alone cause the formation of pure fibrous lesions with morphological characteristics that differ from the usual human fibrous plaques.

Returning now to the problem of vascular enzymes, their study would be rather meaningless if it only revealed vascular injury. However, such is not the case, because the very nature of some of these enzymatic disturbances enables certain metabolic consequences to be predicted and allows the sequence of events in the atherosclerotic process to be partly explained.

The findings in those arteries that are susceptible to atherosclerosis reveal that the activity of tricarboxylic acid cycle enzymes decreases therein. There is good reason to believe that vascular injury also impairs respiration and oxidative phosphorylation in such vessels. A fundamental mechanism protecting the arterial wall against atherosclerosis must evidently be the *unimpaired production of available energy* within this tissue. As long as respiration and oxidative phosphorylation of vascular tissue is normal, the production of energy-rich ATP bonds is adequate for all energy-requiring processes, such as protein and enzyme synthesis. In contrast, *decreased ATP production necessarily results in lower synthetic activity* with reduced output of *enzyme proteins* as well as *phospholipids* and *reduced removal of sclerogenic lipids* (see part three). A vicious circle develops that finally induces the failure of all defence mechanisms of the artery against lipid accumulation and damaging agents of whatever origin.

In this connection it must be mentioned that we are far from being able to delineate all the defence mechanisms of the arterial wall. One of them indeed appears to be the local *synthesis of phospholipids* (Shore et al., 1955; Zilversmit and McCandless, 1959; Zilversmit et al., 1961a; Christensen 1961, 1962; Böttcher et al., 1960; Böttcher, 1964; Abdulla and Adams, 1965 and others). Phospholipids may exert a protective effect by facilitating the solubilization of hydrophobic high-surface tension lipids such as cholesterol and triglycerides (Friedman et al., 1957; Dixon, 1958; Felt et al., 1958; Day, 1962, 1964; Adams et al., 1963a, b; Dunnigan, 1964 and others). "Micellar fat", stabilized by phospholipid, appears to be the form in which fat is transported from the cell (Dixon, 1958). Lecithin may be also involved in cholesterol esterification (see Chapter XXIV).

Another defence mechanism of the arterial wall is its equipment of enzymes catalysing the *hydrolytic cleavage of lipids*. This is discussed in part three of this book.

The data also indicate that as a result of hypoxic damage the vessel wall presumably meets its energy needs by increased glycolysis with a concomitant augmented production of lactic acid and local acidosis. As mentioned before, according to Baló (1963) "acidification" of the artery may induce destruction of the elastic membranes and may thus be a factor in the pathogenesis of atherosclerosis.

In conclusion, it appears reasonable to believe that the problem of the pathogenesis of atherosclerosis is neither the question of the deranged plasmatic transport of lipids and other blood constituents, nor purely the question of local factors. The decisive factor is the *improper balance between the above blood constituents and the protective mechanisms of the vessel wall*. With normal plasma lipids and metabolically fully active arterial tissue, atherosclerosis will not develop. If for some reason the lipid (lipoprotein) spectrum of plasma becomes abnormal in quality or quantity, no atherosclerosis would ensue as long as the arterial wall is able to defend itself against protracted lipid accumulation. On the other hand, atherosclerosis will develop even with normal plasma lipids if the protective mechanisms of the vessel wall are sufficiently disturbed.

PART THREE

THE LIPOLYTIC ACTIVITY OF THE VASCULAR WALL

Chapter XX

INTRODUCTION

IT is evident from the preceding chapters that injury to the vessel wall is an important factor in the pathogenesis of atherosclerosis. The cause of vascular damage need not necessarily be of a mechanical character. Thus, lipids have a tendency to accumulate in the vascular wall as fatty streaks, and may under certain circumstances damage the vascular wall so that fatty streaks are converted into the more advanced fibrous plaques. Duff (1935) suggested that cholesterol damages the arterial wall and it was later shown that cholesterol provokes severe connective tissue sclerosis, both after subcutaneous implantation (Spain and Aristizabal, 1962; Adams et al., 1963b) and direct injection into the arterial wall (Christianson, 1939). Shimamoto (1963) claimed that, within 2 hours after administration of cholesterol to rabbits, a marked oedematous reaction develops in their aortas accompanied by fragmentation of elastic and collagen fibrils. It is, therefore, very likely that prolonged exposure to excess lipids, in particular cholesterol, damages the arterial wall and provokes fibroblastic activity and the development of fibrosis.

However, it was shown that cholesterol is rapidly mobilized and more readily absorbed when phospholipid is added to the subcutaneous implant (Adams et al., 1963b). As suggested by Adams (1967a) this effect may be the result of either dispersion of cholesterol by the surface-active phospholipids or by cholesterol esterification with polyunsaturated fatty acids in a reaction catalysed by lecithin: cholesterol fatty acid transferase (see Chapter XXIV). In this connection it is of great interest that in subcutaneous implants esters of cholesterol with polyunsaturated fatty acids (linoleate, linolenate and in particular arachidonate) display much less sclerogenic activity than free cholesterol or saturated cholesterol esters (Abdulla, et al., 1967, quoted by Adams, 1967a). The sclerogenic activity of triglycerides appears in such implants to be low. However, other evidence supports the view that the last class of lipids is not so harmless. Electron optic investigations by Moskowitz (1967) clearly show that tripalmitin and tristearin—in contrast to triglycerides containing predominantly oleic, linoleic or linolenic acids—induce irreversible degenerative changes in monolayer cell cultures. It may be deduced that probably the fatty acid composition determines the tissue injuring characteristics of both cholesterol esters and triglycerides.

It is reasonable to expect that the vascular wall is endowed with mechanisms for ridding this organ of excess lipids. While many excellent studies have been concerned with the biosynthesis of various lipids in the vascular wall, relatively little is at present known about lipid breakdown in this tissue. Of those enzymes that catalyse special reactions in arterial fatty acid anabolism and catabolism, only 3-hydroxyacyl-CoA dehydrogenase (see p. 52) has so far been investigated in detail. As the tricarboxylic acid cycle is the common final step in glucose, amino acid and fatty acid catabolism, diminished activity of this cycle would also be expected to impair lipid degradation in the vascular wall.

Tissue fatty acids are mostly present as esters and, before their oxidation, the esters must be hydrolysed. Such reactions are catalysed by certain hydrolases, the distribution and activity of which may be of considerable local importance in atherogenesis.

It must be emphasized, however, that in comparison with many other enzymes, the substrate-specificity of most fatty acid ester hydrolases ("fatty acid esterases") is low and such overlapping specificities have caused much uncertainty and confusion.

Enzymes with carboxylic ester hydrolysing activity are usually divided into three groups according to Aldridge (1953a, b) and Augustinsson (1958).

(1) A-Esterases (arom-esterases; arylesterases) catalyse the hydrolysis of carboxylic esters of aromatic alcohols, e.g. p-nitrophenyl acetate, phenyl acetate β-naphthol acetate, phenyl butyrate. Unlike aliesterases, arom-esterases exhibit little activity against tributyrin and are not inhibited by organophosphorus compounds such as di-isopropyl phosphorofluoridate (DFP) or paraoxon. They are also insensitive to physostigmine (eserine), but are inhibited by chelating agents and activated by calcium ions.

(2) B-Esterases or aliesterases (aliphatic esterases) catalyse the hydrolytic cleavage of both aliphatic and aromatic esters, and also tributyrin and methyl-butyrate, but not choline esters (see Skořepa, 1965). They are inhibited by organophosphorus compounds, but not by physostigmine.

(3) C-Esterases are even less specific than the preceding group, they attack not only choline esters, but also aliphatic and aromatic esters. They are not inhibited by DFP and are activated rather than inhibited by low concentrations of well-known sulphydryl inhibitors, such as p-mercuribenzoate (see Myers' review, 1960).

In contrast to the preceding poorly defined groups of esterases, some of the related hydrolases (such as acetylcholinesterase, pancreatic lipase, cholesterol esterase, phospholipase A, etc.) have been more closely defined and are usually placed in separate categories, although they share a common feature with the aliesterases in their susceptibility to inhibition by organophosphorus compounds of the paraoxon type. There is, however, general lack of agreement on the exact meaning of the term lipase (see Wills, 1965). According to one definition, enzymes splitting fatty acids from glycerol are lipases. According to another, enzymes splitting long-chain fatty acids from esters of several different alcohols are lipases. However, as pointed out by Wills (1965), an enzyme that hydrolyses tributyrin is a lipase according to the first definition but not according to the second one, while the reverse holds for an enzyme hydrolysing benzyl stearate. The most satisfactory definition has been advanced by Sarda and Desnuelle (1958). They consider that the characteristic property of true lipases is the ability to accumulate at the phase boundary of oil-water systems. Consequently, according to these authors, true lipases act only in a heterogeneous system—on emulsified substrates—and do not act, or act very slowly, on substrates in solution. The latter definition would include such enzymes as cholesterol esterase, which do not attack triglycerides; so Hofstee (1960) recommends that lipases should be delineated as fat splitting esterases that predominantly act on undissolved substrates.

INTRODUCTION

Lipases are very unspecific enzymes and their substrate spectrum overlaps with that of the equally unspecific esterases (Wills, 1965). This reservation also applies to lipoprotein lipase (see below). In the studies to be described it is sometimes difficult to decide which enzyme is responsible for the effect observed, so the less demanding terms "lipolytic" and "esterolytic" activity will be used in conformity with the basic reactions of lipolysis and esterolysis. We are fully aware of the incompleteness of such a terminology; attempts to differentiate such activities in the arterial wall into lipases, cholesterol esterase and phospholipase A will later be described in more detail.

The interest in the relationship between atherosclerosis and hydrolytic cleavage of fats originates from the fortuitous discovery of the heparin-clearing reaction by Hahn in 1943, who observed that intravenous injection of heparin to a highly lipaemic dog led to rapid clearing of the plasma. It is not intended to give a detailed account of the enormous literature upon this subject, because its mechanism, physiological importance, and relationship to problems of fat transport and fat utilization have been comprehensively reviewed (Nikkilä, 1953; Levy, 1958; Robinson, 1960, 1963; Engelberg, 1963 and others).

Extensive investigations have shown that heparin releases a lipase from the tissues that hydrolyses triglycerides contained in chylomicrons and low density lipoproteins. This lipase is now termed lipoprotein lipase or clearing factor enzyme (lipase) instead of the former name "clearing factor". Unlike other lipases, this enzyme does not convincingly hydrolyse triglyceride emulsions unless the triglyceride is part of a lipoprotein complex.

According to Korn (1958) heparin is an integral component of the system and enables the enzyme to be bound to the lipoprotein substrate. Heparin probably also increases the stability of the enzyme. *In vitro*, low concentrations of heparin activate but higher concentrations inhibit the clearing factor enzyme. Protamine sulphate and 1 M NaCl are specific inhibitors.

The free fatty acids released by the "clearing reaction" become bound to plasma albumin and the resulting complex is an essential source of energy for the organism. The reaction causes chylomicrons to disappear and, due to the solubility of the above fatty-acid-albumin complex, the plasma is cleared of its turbidity. However, not all the physicochemical events taking part in the clearing reaction are known. It is of some interest that the binding of free fatty acids to albumin can be followed both polarographically (Kubie *et al.*, 1956) and by the altered binding capacity of albumin for certain dyes (May, 1955; Zemplényi, 1956, 1958a, b). Nevertheless, the most reliable estimate of lipolysis is the determination of the amount of free fatty acid or glycerol liberated during the reaction.

Lipoprotein lipase is contained in several organs and its activity differs from tissue to tissue. Havel (1958) and Robinson and Harris (1959) assumed that such activity is principally confined to the capillary wall; this localization would be consistent with one function ascribed to the enzyme, namely facilitation of the passage of triglyceride fatty acids across the capillary wall (Robinson, 1963). In those tissues where an excess of fat is usually synthesized (e.g. adipose tissue and mammary gland), lipoprotein lipase seems to be involved in the fatty acid esterification process. However, this function is not necessarily apparent in those tissues that utilize fatty acids for their energy supply. In fasting animals enzyme activity substantially *decreases in adipose tissue* (Hollenberg, 1959; Páv and Wenkeová, 1960; Robinson, 1960), whereas activity in the *myocardium* substantially *increases* (Zemplényi and Grafnetter, 1959b, d; Hollenberg, 1960; Szabó et al., 1962). In view of Björntorp and Furman's (1962) experiments it seems that another enzyme that is more sensitive to fasting is also involved in the myocardium.

In the following sections only those aspects of the "clearing reaction" and "lipoprotein lipase" will be discussed that are relevant to the lipolytic and esterolytic activities of the vessel wall. Other related topics have been previously reviewed (Zemplényi, 1964a).

Chapter XXI

MAST CELLS AND ATHEROSCLEROSIS

It is assumed by many authors that heparin is synthesized and stored in the mast cells, which are regarded as "unicellular endocrines" that produce not only heparin, but also hyaluronic acid, serotonin, histamine and other pharmacologically-active substances. It is fair to add that the mast cell theory of heparin production is far from being unanimously accepted; the conflicting views are reviewed by Fulton *et al.* (1957) and Jaques (1961). Details of the morphology and functions of mast cells have been reviewed by Riley (1959), Rudzit (1959), Corbascio (1960), Phillips *et al.* (1960), Demopoulos *et al.* (1961), Smith (1963), Parish (1964), Bloom (1965) and others.

In acute experiments in rats fed olive oil after pretreatment with protamine, Fodor and Lojda (1956) observed degranulation of mesenteric mast cells. Similar findings were reported after intravenous injection of an olive-oil emulsion (Hill, 1957). Long-term fat feeding is said to cause a significant fall in the mast-cell count in the rat's mesentery (Fodor *et al.*, 1958a, 1960), in the mesentery and myocardium of some fish (Stolk, 1959a, b), and in the liver (Ahlqvist, 1960). On the other hand, the similar experiments of Grunbaum *et al.* (1957), Jennings *et al.* (1960) and Watson (1961) showed no clear-cut changes in either mast cell count or morphology.

With the positive findings, especially in acute experiments, it is possible that factors were involved that were not specifically related to the fat overload.

Asboe-Hansen and Wegelius (1956) demonstrated that living mast cells in the hamster's cheek pouch react by degranulation and sometimes disruption to administration of histamine-liberator compounds, peptone and histamine itself. Postnov (1959) reported similar findings in the rat's subcutaneous mast cells after histamine administration. Uvnäs (1958) suggested that a lecithinase in the mast cell membrane is normally prevented by an inhibitor from attacking the cell envelope. It may be imagined that the prerequisite for mast cell degranulation or disruption is the removal of this inhibitor with subsequent destruction of the lecithin in the cell membrane.

The inter-relationships between mast cells, histamine, heparin and, perhaps, clearing factor seem therefore to be of a complex nature. Although there is good evidence that heparin may be involved in the transport of fat between the blood and tissues, there is as yet no concrete evidence of a connection between fat transport and mast cells.

In view of the probable secretion of heparin by mast cells, morphological reports on changes in these cells in atherosclerosis are of some interest, even though such results are very conflicting.

In 48 senile atherosclerotic hearts Cairns and Constantinides (1954) observed a significantly lower mast cell count than in 46 young adult and in 48 senile non-atherosclerotic hearts. Among non-atherosclerotic adults the male hearts had a significantly lower mast cell count than the females; no age dependence was noted. Pollak (1956, 1957) reported that mast cells are rarely detected in early and never found in advanced human atheromatous lesions.

On the other hand Paterson and Mills (1958) found no relationship between the number of myocardial mast cells and the presence or severity of atherosclerosis, but the mast cell count fell in conditions complicated by thrombosis or infarction. Likewise, Lempert et al. (1961) found no correlation between coronary atherosclerosis and the number of mast cells. The possible protective action of mast cells against atherosclerosis was tested by Watson (1958). However, depletion of mast cells by compound 48/80 in a group of rats fed cholesterol failed to produce atheroma in these animals.

A definite increase of peri-arterial mast cells in human coronaries, proportional to the severity of atheroma in the underlying segment of the vessel, was reported by Pomerance (1958) in accord with previous findings by Hjelman (1954) and Sundberg (1955). Cachectic patients tend to have lower counts and diabetic patients higher counts than other patients with equivalent degrees of coronary atheroma. Around zones of recent thrombosis the number of mast cells is also markedly increased. Quite recently Pouchlev et al. (1966) confirmed the marked increase in periarterial mast cells in human aortas and coronary arteries, dependent on the degree of atherosclerosis. On the other hand, the mast cell count in the intima of the diseased vessels significantly fell.

It is probable that the decreased number of intimal mast cells, as observed by Pollak and Pouchlev et al., is a secondary phenomenon due to degenerative changes in the inner parts of the atherosclerotic vessel. The higher periarterial mast cell count has been interpreted by Parish (1964) as a non-specific inflammatory response in the damaged area. Pouchlev et al. (1966) believe that the increase is associated with the formation of new connective tissue during the repair processes involved in atherogenesis.

In summary it can be stated that although there is evidence that heparin—a probable component of lipoprotein lipase—may be involved in the transport of fat between the blood and tissues, there is as yet no unequivocal proof of a connection between fat transport and mast cells. The changes in the number of periarterial (adventitial) and intimal mast cells must be regarded as non-specific secondary phenomena, and there is no conclusive evidence that mast cells are in any way responsible for the development of atherosclerotic lesions.

The failure to demonstrate any clear-cut relationship between arterial mast cells and atherosclerosis does not exclude the involvement of vascular lipolytic activity in the process of atherogenesis. On the basis of the above negative findings it is reasonable to expect that vascular lipolysis, if detectable and if related to atherogenesis, will prove to be relatively independent of heparin (or similar compounds). Indeed, we shall see that analysis of vascular lipolysis reveals, in addition to lipoprotein lipase, the presence of other related hydrolases in the arterial wall.

Chapter XXII

THE RELATIONSHIP BETWEEN ELASTASE, ATHEROSCLEROSIS AND LIPOPROTEIN LIPASE

It is assumed by some authors that the pancreatic enzyme elastase (pancreaticopeptidase E, E.C. 3.4.4.7.) has a role in the metabolism of intact elastic tissue (Baló, 1963). Gore and Larkey (1960) state that "... teleologically, there seems to be a need for elastase activity in tissues rich in elastic fibres, especially when, as in the aorta, their integrity must be maintained in the face of the constant pounding of arterial pulsation".

The latter authors reported that a crude mucopolysaccharide extract from human aortas displayed weak elastase activity. The enzyme was assayed with orcein-stained elastin powder as substrate and the release of orcein from combination with elastin was the "index" of enzyme activity. (The method used was that described by Sachar et al., 1955, and later a similar more precise method was designed by Banga, 1963).

To the author's knowledge there are no other reports in the literature about the presence of elastase in arterial tissue and the above findings await confirmation. Some findings, however, suggest a possible connection between elastase, lipoprotein lipase and atherosclerosis; this evidence will be briefly reviewed.

The presence of a specific elastolytic enzyme in extracts of fresh ox pancreas was described by Baló and Banga (1948, 1949, 1950) and they coined the name elastase for it. Using fresh aortas as substrate they found that the pancreatic enzyme dissolved the elastic fibres and left the collagen unchanged. They maintained that elastase did not liberate amino acids but that the insoluble elastin was changed into a soluble protein. Hence, the action of the enzyme was not considered as proteolytic but rather as depolymerizing; fibrous elastin being transformed into globular form by disruption of hydrogen bonds (Banga and Nowotny, 1951b). However, subsequent work (Partridge et al., 1955; Robert and Samuel, 1957; Gilfillan et al., 1960; Naughton and Sanger, 1961) showed that the enzyme exhibits proteolytic activity. The proteolytic component of elastase (see below) was found to hydrolyse peptides, especially at bonds adjacent to neutral aminoacid residues.

Soon after the discovery of pancreatic elastase, further analysis disclosed that elastase preparations contain at least two separate elements; the above discrepancies were, therefore, probably caused by the use of crude or partly purified preparations. Hall (1953, 1955) and Banga and Baló (1956) isolated two types of elastase, but the purification and complete separation of these fractions has only lately been achieved by means of column chromatography with DEAE-Sephadex (Loeven, 1963c).

The first component is a proteolytic enzyme (E_2 or elastoproteinase), while the other component (E_1 or elastomucase) exhibits only slight elastolytic activity but dissolves a mucoprotein out of substrates such as ligamentum nuchae or human aorta. Elastoproteinase (real elastase) is active by itself, but elastomucase seems to potentiate the action of the proteolytic enzyme against elastin.

In a series of elegant investigations Hall and Czerkawski (1961*a*, *b*) have shown that the last reaction may be divided into at least three separate phases which result in the conversion of solid elastin into degradation products of lower molecular weight: (1) An initial adsorptive phase during which the enzyme-substrate complex is formed and which is accompanied by a marked decrease in the rigidity of elastic fibres. (2) The solubilization reaction that produces soluble α-elastin as a result of the fission of a limited number of peptide bonds. (3) Further breakdown of α-elastin with considerable release of amino groups.

Substantial impetus to the study of elastase was given by Baló and Banga's (1953) findings that in human atherosclerosis the elastase content of the pancreas is markedly low at autopsy. It was suggested that the decrease of pancreatic elastase results in a disturbance of the metabolism of vascular elastic fibres. This was consistent with previous studies from the same laboratory attributing to arterial elastic damage a pathogenic role in atherosclerosis.

These results (and those to be discussed below) stimulated several groups of investigators to study the effect of elastase in the prevention and treatment of experimental and human atherosclerosis. In addition, the level of the specific serum elastase inhibitor was studied under various conditions connected with atherosclerosis. Unfortunately, the results were extremely conflicting (see Loeven's review, 1963*b*).

Another important impetus to the study of the enzymes of the elastase complex was given by findings which seemed to indicate a close relationship between *elastomucase* and the *clearing factor enzyme*. This hypothesis is essentially based on the following observations:

(1) Several authors (Lansing *et al.*, 1952; Sachs, 1954; Labella, 1957; Adams, 1959; Loomeijer, 1961 and others) have reported that elastin is in fact a lipoprotein that, according to Labella, contains plasmalogen and unsaturated lipid. However, Loomeijer maintained that the lipid is a saturated fatty acid with a chain length of about 12 carbons. On this basis it is logical to expect that some component of the elastase complex might possess lipolytic activity.

(2) Hall *et al.* (1952), Banga and Schuler (1953) and Baló *et al.* (1954) observed that acid mucopolysaccharides are released during the digestion of elastic fibres by elastase. As acid mucopolysaccharides apart from heparin could play a part in the clearing reaction, the above observations stimulated attempts to produce the clearing factor *in vitro*.

Saxl (1957) incubated human aortas and ox ligamentum nuchae with an elastase preparation rich in elastomucase (E_1). Globules of visible fat appeared in the medium after 20 hours incubation, and were stained pink with Nile Blue sulphate. However, after the addition of a serum $\alpha_1 + \alpha_2$-globulin fraction and reincubation, the fat globules disintegrated and they stained blue with Nile Blue sulphate. This was considered as proof that the fat liberated from the tissues had been hydrolysed. The reaction was also tested for esterase activity using α-naphthol acetate and naphtol AS acetate as substrates. The incubation medium, which was originally opalescent due to the presence of an emulsion, rapidly became less turbid. (Unfortunately, no quantitative data were provided). Electron micrographs were also interpreted as proving the presence of a tissue clearing reaction. On the basis of these findings the author concluded that ". . . elastomucase, its substrate (the mucopolysaccharide) and a factor inseparable from the $a_1 + a_2$-globulin fraction of serum, interact to produce a complex which is capable not only of causing a diminution in the opacity of lipaemic serum, but also of bringing about the degradation of neutral fat deposits in both normal connective tissue and in atheromatous tissue".

In the same laboratory Hall (1958) tested the *in vitro* production of plasma clearing activity by using lipaemic serum or synthetic emulsion as substrates. In addition to the uncharacterized mixture of polysaccharides liberated from elastic tissue by elastase (AES, see below), heparin and chondroitin sulphuric acid were also employed. The production of "clearing factor" required the simultaneous presence of E_1, an acid mucopolysaccharide and a serum component (a_1-globulin). Limited activity could be obtained with E_1 alone acting on human lipaemic serum, but not with the other substrates. This difference was explained by the adequate polysaccharide content of human lipaemic serum. It was concluded that the interaction of the mucolytic component of elastase, an acid polysaccharide and a_1-globulin results in the production of lipoprotein lipase activity. Unfortunately, the results were expressed only in terms of percentage reduction in optical density and no attempt was made further to characterize such activity.

(3) Subsequent work by Hall (1961*a, b*) provided evidence that elastomucase (E_1) itself exhibits esterolytic activity. In view of the lability of pure elastomucase, a slightly less pure preparation (AES, "acetate extract-soluble material") was employed and the enzymatic characteristics were studied using Tween 20 substrate dissolved in ammediol buffer (pH 7·8). In contrast to pancreatic lipase, which was also tested on the same substrate, the elastomucase preparation was activated by low concentrations of heparin but inhibited by higher concentrations. Increased ionic strength caused activity to fall, whereas preincubation with total serum proteins increased the activity of AES. Some evidence indicated that the effect of serum proteins involved β-lipoprotein (Sf $_{0-20}$ fraction).

On the basis of these and previous results mentioned above (Hall, 1958), it was suggested that the AES preparation, which contains elastomucase as its major active component, has ". . . lipase activity when tested against fatty acid esters in the presence of protein and polysaccharide, whereas in the presence of the proteolytic enzyme elastase it releases polysaccharide from elastic tissue and therefore appears to have mucase activity". Hall (1961*b*) concluded that E_1 represents the enzyme that is liberated by the injection of heparin into a normal animal. Consequently, he suggested the name "elastolipoproteinase" for the enzyme (Hall, 1964*a, b*).

(4) Banga (1961, 1962) and Banga and Baló (1962) reported that the human aorta contains a specific substrate for such a lipolytic enzyme, namely a mucolipoprotein containing neuraminic acid as a component of the acid mucoid.

(5) In the above experiments only a partly purified pancreatic enzyme preparation was used and the lipolytic activity was ascribed to its elastomucase

content. Loeven (1963a, 1964) found, however, that at least three enzymes with different elastomucase activity could be separated from crude elastase preparations by means of DEAE-Sephadex column chromatography, but none of these elastomucases displayed lipoprotein lipase activity. On the other hand, by changing the elution technique, a further fraction could be isolated that was considered to be real elastolipoproteinase. This means that "elastomucase" and "elastolipoproteinase" are not identical. Loeven (1963a, 1964) also isolated from human plasma an elastomucase fraction of γ-globulin character, and another fraction of β-globulin nature that exhibited properties ascribed to elastolipoproteinase; both fractions exerted a synergistic effect on the activity of elastoproteinase.

Loeven (1964, 1967) also tested the effect of purified elastase components on experimental cholesterol atherosclerosis. Elastomucase and elastoproteinase had no influence on lipid transport and regression of atherosclerosis, whereas the aortas of the animals treated with "elastolipoproteinase" exhibited only traces of atheromatous lesions and their livers were nearly fat-free. Loeven believes that the contradictory results of previous authors, who used elastase preparations to prevent or treat experimental atherosclerosis, were caused by the differing elastolipoproteinase content of the preparations used.

From this discussion it is clear that there is growing evidence to indicate a possible role of enzymes of the elastase complex in lipolysis. However, some of the findings on which this relationship to lipoprotein lipase are founded, are open to objection. For example, the experiments on *in vitro* "production" of clearing factor by the interaction of elastomucase, aortic tissue and $a_1 + a_2$-globulin took no account of the spontaneous lipolytic and esterolytic activity of vascular tissue. A simple alternative explanation could be that acid mucopolysaccharide liberated by the action of elastomucase activated the aortic lipoprotein lipase in a way similar to the action of heparin. Szabó and Cseh's (1962) findings appear to be compatible with such an explanation. Furthermore, caution should be exercised in accepting changed Nile Blue staining as an indicator of lipolysis (see above).

Much of the evidence for the hypothesis about the lipolytic action of elastase stems from the similar properties of the impure elastomucase preparation and lipoprotein lipase when Tween is used as substrate. However, Tweens are not suitable for detecting lipoprotein lipase activity (Grafnetter and Zemplényi, 1958, 1959; Páv *et al.*, 1961; Zemplényi, 1964a). This substrate only enables non-specific esterase ("Tween esterase") activity to be detected; all attempts to relate the activity of elastase fractions to lipoprotein lipase should be based on the use of lipoprotein-containing substrates and on the unequivocal demonstration of typical inhibition characteristics.

Nevertheless, it is possible that elastolipoproteinase and the clearing factor enzyme are identical, or at least closely related. Although later research might prove this relationship, many aspects of this hypothesis are at present obscure. Thus, the site of elastase production in the pancreas has not been definitely established. The evidence so far accumulated rather supports its origin from the exocrine acinar cells and this would appear to preclude a systemic function. It is difficult to accept Hall and Wilkinson's (1963) suggestion that reabsorption from the intestinal mucosa would enable the enzyme to act systemically. Nevertheless, Loeven's recent identification of elastomucase and elastolipoproteinase in the plasma (see above), suggests that these enzymes may originate in endocrine cells of the pancreas.

Finally, the postulated pathogenic significance of elastase in atherosclerosis is based on the assumption that the elastoproteinase fraction primarily damages arterial elastic tissue, whereas elastomucase (or elastolipoproteinase?) prevents secondary lipid infiltration. "The maintenance of the *status quo* in arterial tissue may be dependent on the continuation of a very finely balanced equilibrium between elastase, E_1, acidic polysaccharide, plasma proteins, lipid and calcium. . . ." (Hall, 1961b). The only evidence for this attractive theory is Gore and Larkey's (1960) finding mentioned in the introduction to this chapter. Unfortunately, in their paper, these authors only made the following brief statement about this problem: "slight elastase activity was displayed by crude aortic mucopolysaccharides but not by chondroitin sulphate B or the other specific moieties". It would be highly desirable to reinvestigate this important question with the exact modern methods now available.

It is the author's belief that the possibly important relationship between the enzymes of the elastase complex, lipoprotein lipase and atherosclerosis deserves further concentrated research. Reverting to the topic of vascular enzymes, there is at present *insufficient evidence to permit the conclusion that elastase or enzymes of the elastase complex are primarily involved in the lipolytic activity of vascular tissue.*

Chapter XXIII

THE LIPOLYTIC ACTIVITY OF THE VESSEL WALL

In the experiments to be described vascular lipolytic activity was estimated by incubating weighed amounts of minced tissue or homogenate in lipaemic human serum, diluted 1:1 with Sörensen phosphate buffer (pH 7·35). In some experiments the substrate was either "activated" Ediol* or 3 per cent rat chyle in bovine albumin with added glucose (120 mg./100 ml.). Incubation was carried out for 150 minutes (in later experiments only 90 minutes) in a water bath at 37° C. with constant agitation.

The amount of free fatty acids (FFA) liberated was estimated by Dole's (1956) method modified by using Nile Blue as an indicator. The extent of lipolysis was determined by the difference between the initial and final concentrations of FFA in the incubation mixture. Enzyme activity was expressed either as mEq. FFA/1./g. tissue or μmoles FFA/g. tissue (for further details see Zemplényi and Grafnetter, 1958a, b, 1959a, b).

With this method we first investigated the relationship between aortic lipolytic activity and the animal's *age*. We compared lipolytic activity in the aortas of male rats of different age groups. The average activity of 7-month-old rats (i.e. adult but not old rats) was taken as 100 per cent (Zemplényi and Grafnetter, 1959c). We found (Fig. 73) that aortic lipolytic activity in 24-month-old rats is significantly lower than that in 7-month-old animals ($P < 0.02$), but the decreased activity in 12-month-old animals is only of borderline significance. Aortic lipolytic activity in quite young rats is significantly lower than that in adult rats ($P < 0.001$). Estimations of the dry substance and the nitrogen content in the aortas did not reveal any differences between the various age groups.

Mallov (1964) and Adams *et al.* (1966) using similar methods confirmed in other species that aortic lipolytic activity falls with advancing age. In this connection Dury's (1961) findings are instructive. Using our technique (but with Ediol as substrate), he showed that heparin injection into young rats leads to increased aortic lipolytic activity. In old rats no such effect could be observed.

In other experimental series we investigated whether there is a difference in aortic lipolytic activity in *animal species* that are *susceptible to experimental atherosclerosis*, as compared with the rat which is relatively resistant (Zemplényi and Grafnetter, 1958a, b). Figure 74 presents the results of studies where the activity of the aortas of various animal species was compared with that of rat aortas. Each value was obtained from a different animal. The rat's aorta has clearly

---

* Ediol (SchenLabs) is a 50 per cent coconut oil emulsion and also contains glyceryl monostearate (1·5 per cent) and polyoxyethylene sorbitan monostearate (2 per cent). Activated Ediol is prepared by preincubation with a small amount of human serum for 30 minutes at 30° C.

THE LIPOLYTIC ACTIVITY OF THE VESSEL WALL 215

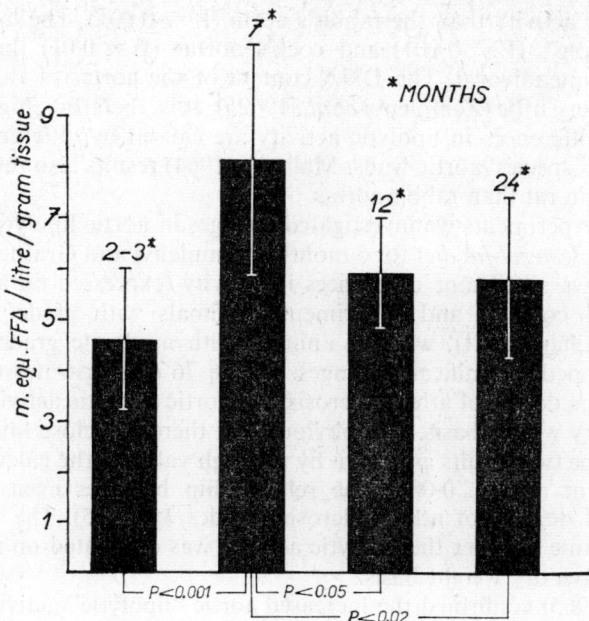

FIG. 73.—The effect of age on the lipolytic activity of rat aortas. Upright line with bars ± S.D. (Data from Zemplényi and Grafnetter, 1959c.)

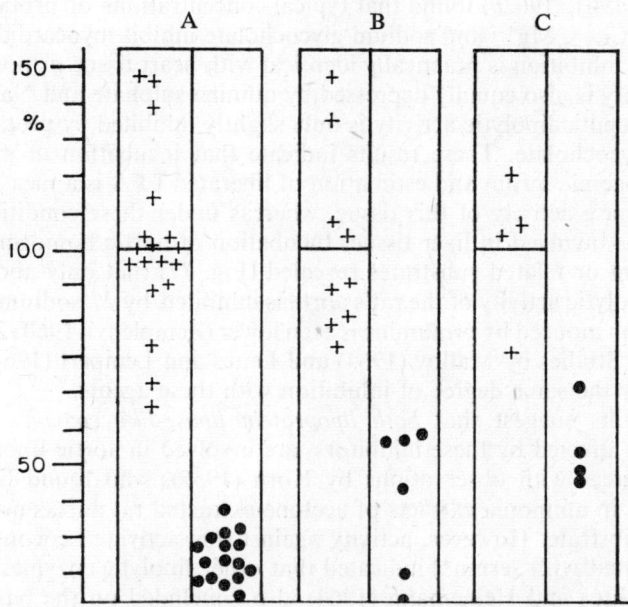

FIG. 74.—Interspecies differences in aortic lipolytic activity. A = rat-rabbit; B = rat-g. pig; C = rat-cock. The av. rat aorta activity = 100 per cent. (From Zemplényi et al., 1959a.)

higher lipolytic activity than the rabbit's aorta ($P < 0.001$). The lower activity in the guinea-pig's ($P < 0.001$) and cock's aortas ($P < 0.01$) than the rat's aortas is also unequivocal. The DNA content of the aortas of these different species varies very little (Zemplényi et al., 1962b). It is, therefore, highly improbable that the differences in lipolytic activity are caused by different cellularity of these various species' aortic walls. Mallov's (1964) results also revealed much higher activity in rat than rabbit aortas.

In further experiments we investigated changes in aortic lipolytic activity in *rabbits fed a cholesterol-fat diet* for 6 months (Zemplényi and Grafnetter, 1959a). Figure 75 shows significant differences in activity (expressed on a wet weight basis) between controls and experimental animals with grade 3–5 atherosclerosis (see Chapter XI), whereas animals with moderate grades of atherosclerosis developed insignificant changes. Figure 76 demonstrates the relationship between the degree of atherosclerosis and aortic lipolytic activity expressed on a fat-free dry weight basis. It is obvious that there is a close linear relationship between the two results as shown by the high value of the calculated correlation coefficient at $P < 0.001$. The relationship becomes even closer after excluding mild degrees of atherosclerosis (grades 1 to 2.5). The results were basically the same whether the lipolytic activity was calculated on a wet or dry weight or fat-free dry weight basis.

Lempert (1965) confirmed the increased aortic "lipolytic" activity in rabbits fed cholesterol for a period of 120 days. However, Tween-60 was used as substrate in these experiments. Experimental stress significantly altered lipolytic activity (see below).

Turning now to *inhibition-activation characteristics*, Grafnetter and Zemplényi (1959, 1961, 1962b) found that typical concentrations of protamine sulphate, NaCl, Ca^{++}, Mg^{++} and sodium glycocholate inhibit myocardial lipolytic activity. Such inhibition is practically identical with heart tissue or post-heparin plasma; activity is also equally depressed by quinine sulphate and NaF. On the other hand, hepatic lipolytic activity is only slightly inhibited by protamine sulphate and glycocholate. These results indicate that incubation of myocardial tissue with lipaemic serum and estimation of liberated FFA is a measure of the lipoprotein lipase activity of this tissue, whereas under these conditions a different lipase is involved in liver tissue. Incubation of aortic homogenates with lipaemic serum or related substrates revealed (Fig. 77) that only about 55 per cent of the lipolytic activity of the rat's aorta is inhibited by M sodium chloride, while inhibition induced by protamine is even lower (Zemplényi, 1960; Zemplényi et al., 1960b). Studies by Mallov (1964) and Leites and Lempert (1964) showed approximately the same degree of inhibition with these agents.

These results suggest that *both lipoprotein lipase-like enzymes and other esterases*, less affected by these inhibitors, are involved in aortic lipolysis. This conclusion agrees with observations by Korn (1955), who found lipoprotein lipase activity in ammonia extracts of acetone-extracted rat aortas using chylomicrons as substrate. However, activity against non-activated coconut oil (i.e. not preincubated with serum*) indicated that other lipolytic enzymes were also involved. Zsoldos and Heinemann (1964) also concluded on the basis of perfusion experiments that in rabbit aorta the enzyme shares many but not all of

* See footnote p. 214.

THE LIPOLYTIC ACTIVITY OF THE VESSEL WALL 217

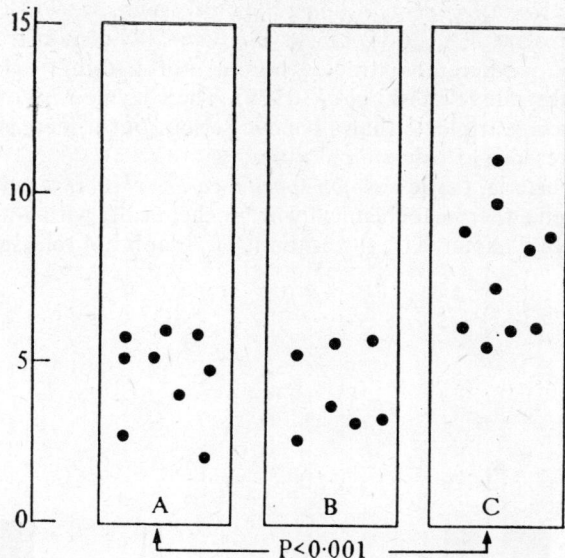

FIG. 75.—Lipolytic activity in aortas of cholesterol-fat fed rabbits (in μM of free fatty acids liberated per gram of wet tissue). A = aortas of control rabbits. B = aortas of grade 1–2·5 atherosclerosis. C = aortas of grade 3–5 atherosclerosis. (From Zemplényi and Grafnetter, 1959a.)

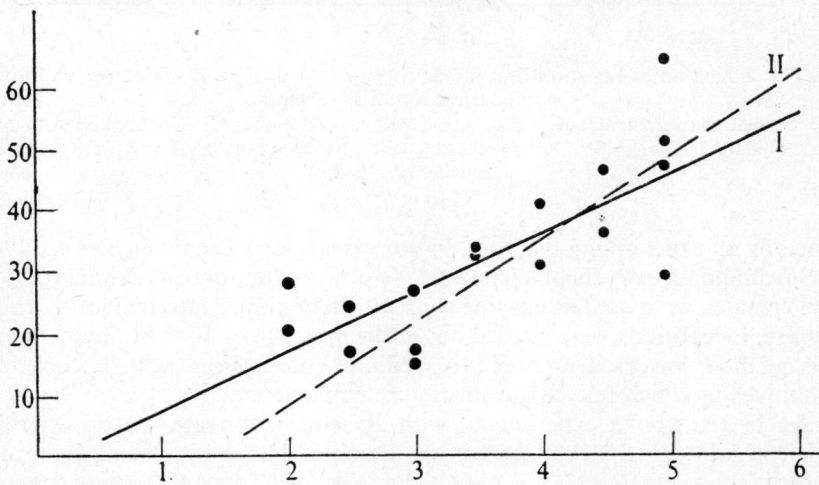

FIG. 76.—Correlation of aortic lipolytic activity (fat free dry weight basis) with degree of atherosclerotic lesions in cholesterol-fat fed rabbits. Regression line I (for degrees of atherosclerotic lesions 0–5): $Yi = -3.48 + 9.93 Xi$, $r = 0.7842$, $P < 0.001$. Regression line II (for degrees of atherosclerotic lesions 3–5): $Yi = -18.7 + 13.42 Xi$, $r = 0.7902$, $P < 0.01$. (From Zemplényi and Grafnetter, 1959a.)

the characteristics of lipoprotein lipase. The presence of "classical" lipase (glycerol-ester hydrolase, E.C. 3.1.1.3.) was unequivocally demonstrated in aqueous extracts of acetone-butanol-extracted porcine aorta, using a glycerol trioleate hydrosol as substrate (Patelski et al., 1966). The enzyme activity was distinctly lower with glycerol trioleate than lipaemic serum, but it increased to the same value in the presence of calcium chloride.

Moreover, arterial tissue has *non-specific carboxylesterase* activity which can be demonstrated either histochemically or biochemically with β-naphthol acetate as substrate (see Chapter XI); the amount of β-naphthol released can be deter-

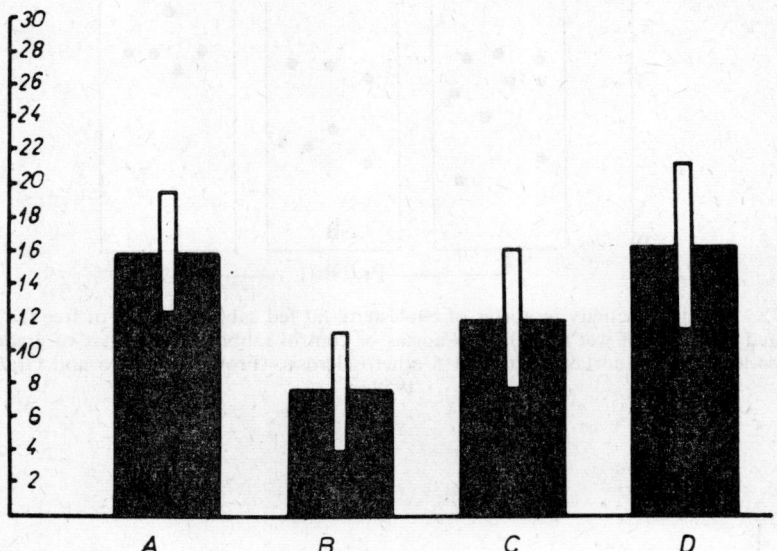

Fig. 77.—Inhibition of rat aorta lipolytic activity (in μM of FFA liberated per gram tissue) with lipaemic serum as substrate.
A = control measurements; B = incubation with 1·0 M NaCl; C = incubation with protamine sulphate 1·25 mg./ml.; D = incubation with physostigmine 0·125 mg./ml. (From Zemplényi et al., 1960b.)

mined by an azocoupling reaction. In our experiments (Zemplényi et al., 1962b, 1963b) ammonia extracts of aortic acetone-powders or aqueous extracts of aortic homogenates were used as enzyme sources. To minimize interference by cholinesterase, incubations were carried out in the presence of 10^{-6} M physostigmine. This method was used in order to facilitate comparisons with histochemical techniques in which related substrates are employed.

As in the above experiments with lipaemic substrate, aortic esterolytic activity against β-naphthol acetate was much higher in rat's aorta than in the rabbit's, chicken's (Fig. 78) or pig's aorta (Fig. 79). These differences cannot be attributed to variable cellularity in these species' aortic walls, because the deoxyribonucleic acid content of the aortas was practically uniform.

A similar correlation between lipolytic and esterolytic activity was also revealed in *experiments with calciferol* (see Chapter XV). Grafnetter and Zemplényi (1962b) observed that aortic lipolytic activity is unequivocally decreased in vitamin D-fed rats 2 to 3 weeks after the last dose of calciferol. Esterolytic activity in acetone-powder extracts from such aortas was also significantly reduced when activity was referred either to wet weight or DNA content.

It is well known that aortas of calciferol intoxicated rats are characterized by calcification and by changes in the connective tissues and mucopolysaccharides. In this respect it is interesting that aortic lipolytic activity is significantly inhibited by $0.01\ M\ Ca^{++}$ (Grafnetter and Zemplényi, 1961), whilst some mucopolysaccharide mixtures also inhibit the enzyme (Gerö et al., 1962a—see below).

In what follows further information is summarized about the lipolytic and esterolytic activities of the arterial wall.

Gore and Larkey (1960) described "clearing activity" by the aorta and, in addition, they observed elastase activity in this tissue (see Chapter XXII).

Sex differences in lipolytic activity were reported by Szendzikowski et al. (1961, 1962) from Pearse's laboratory. Lipolytic activity in the female rat aorta was shown to be approximately twice that in the male. The authors assumed that a high proportion of such activity was due to cholinesterase (see p. 225). Using α-naphthol butyrate as substrate Lacuara et al. (1962) found higher esterolytic activity in extracts of the male rat's aorta than in those of females; gonadectomy decreased such activity in males but increased it in females. In our own experiments with β-naphthol acetate (Zemplényi, 1964a) we were unable to detect significant sex differences in aortic esterolytic activity amongst rats of various ages. Mallov's (1964) findings were also negative. The reason for these contradictory results is not clear and no final conclusion can be made about the effect of sex on vascular lipolytic and esterolytic activities.

It is now appropriate to consider the *effect of stress* on vascular lipolytic activity. Leites and Chow-Su (1962a) modified Zemplényi and Grafnetter's method by using 2.5 per cent Tween-60 with 1 per cent albumin as substrate. Rats stressed by immobilization according to Selye's technique showed a significant fall in aortic esterolytic activity. (As Tween was used as substrate, we prefer in this case to speak of esterolytic rather than lipolytic activity). Adrenalectomy did not prevent the decreased esterolytic activity in stressed animals and, therefore, it was concluded that the phenomenon was not linked with increased secretion of adrenal hormones. This conclusion is supported by the observation that in acute experiments the administration of cortisone ($2.5\ \mu g./100$ g. body wt.) to normal or adrenalectomized rats did not affect aortic esterolytic activity. Injection of noradrenaline ($50\ \mu g./100$ g. body wt.) was followed by a fall in the esterolytic activity, while administration of neuriplegic and ganglion-blocking agents prevented the effect of stress on aortic esterolytic activity. The effect of stress on the aorta could be detected as early as 1 hour after starting the experiments (Leites and Chow-Su, 1962b). These authors' results suggest that the fall in aortic esterolytic activity accompanying this type of stress is mediated through the sympathetic nervous system and is independent of secretion from the adrenal cortex or medulla.

In the same laboratory Lempert and Leites (1963) studied aortic lipolytic activity against an artificial lipaemic substrate in cholesterol-fed rats which had been exposed to daily stress by immobilization. Cholesterol feeding, when

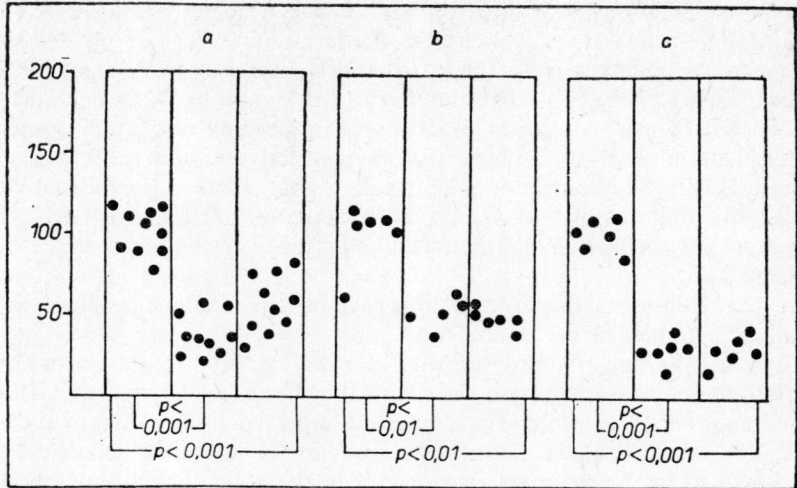

Fig. 78.—Non-specific carboxylesterase activity in rat's, rabbit's and cock's aortas. In experimental series *a* aortic acetone butanol powders were employed, in series *b* and *c* supernatants of aortic homogenates were employed. Results expressed as percentage differences; average rat's aorta activity = 100 per cent. (From Zemplényi et al., 1963b.)

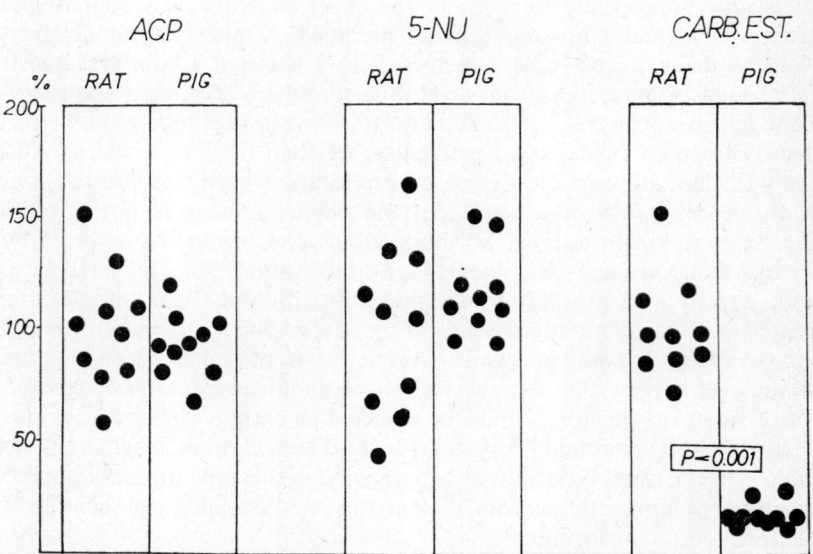

Fig. 79.—The activity of aortic acid phosphatase, 5'-nucleotidase and non-specific carboxylesterase in rat and pig aortas. Results expressed as percentage differences. (From Zemplényi et al., 1965b.)

combined with stress, significantly reduced lipolytic activity; a parallel histochemical study revealed a fall in non-specific esterase activity in the cellular elements of the aorta. The degree of lipid infiltration in the aorta correlated with the decrease in lipolytic activity.

Prolonged *treatment with ACTH* significantly reduced aortic lipolytic activity in $3\frac{1}{2}$-month-old male rats but not in repeatedly bred 10 to 11-month-old female rats (Patelski and Szendzikowski, 1962a). In some parallel histochemical investigations no definite change was seen in the activities of esterases, "$NADH_2$ and $NADPH_2$ diaphorases" and β-hydroxybutyrate dehydrogenase.

In rats with *alloxan diabetes* of 18 to 88 days duration, Leites and Chow-Su (1962a) observed decreased aortic esterolytic activity against Tween-60 as compared with control rats or rats with diabetes of shorter duration. Aortic esterolytic activity returned to normal in five out of eight animals, in whom the blood glucose spontaneously reverted to its usual level.

Coming now to other factors, Patelski *et al.* (1962) investigated the effect of *dietary lipids* on rat aortic lipolytic activity. Fat restriction or diets enriched with vegetable oils of different saturation (iodine numbers 111 and 68) produced no change. They then studied the effect of adding 1 per cent cholesterol–0·4 per cent cholic acid to these diets 18 weeks after the beginning of the experiment. Aortic lipolytic activity decreased in hyperlipaemic animals whose diet was low in unsaturated fats, whereas it increased with the more unsaturated diet. The authors suggested that, during adaptation to higher lipid (cholesterol) consumption, vascular lipolytic activity is inversely related to the saturation of the plasma fatty acids.

Gerö *et al.* (1962a) observed that a *mucopolysaccharide mixture* isolated from human atherosclerotic aortic intima inhibits rabbit aortic lipolytic activity against lipaemic serum. Under identical experimental conditions crystalline hyaluronic and chondroitin sulphuric acids inhibited such lipolytic activity less than did the aortic mucopolysaccharide extract. The authors suggested that chemical or physico-chemical changes in the mucopolysaccharide composition of the intimal ground substance lead to reduced vascular lipolytic activity, which in turn promotes atheroma formation.

In Chapter XV we presented evidence that both renal and DOCA *hypertension* cause reduced non-specific esterase activity in the rat's aorta. However, Mallov (1964) found a significant increase of aortic lipolysis against activated Ediol in hypertensive rats and rabbits. This is surprising, because in our experience lipolytic and esterolytic activities usually undergo parallel changes. It is not really clear whether this discrepancy is an exception to this rule, or whether the divergent results are to be attributed to differences in the experimental design, such as duration of hypertension and dietary fat content.

Another factor that some investigators consider to be atherogenic is prolonged exposure to *carbon disulphide*. Szendzikowski and Patelski (1962) reported that subcutaneous administration of this poison to rats significantly decreased aortic lipolytic activity against lipaemic serum. No change in non-specific esterase activity was detected by histochemical means.

Up to now emphasis has been laid on results obtained by biochemical methods. As pointed out in part two of this book, "classical" biochemical techniques do not reveal the cellular localization of enzymatic activities in the vascular wall; until recently no reliable method was available for the histochemical detection of lipolytic activity (see below). However, there are reliable histochemical reactions for non-specific esterases, and these have been used to study vascular esterolytic activity.

Gomori (1946) could not in general detect esterase ("lipase") activity in blood vessels from various animal species, using Tween-40 or -60 as substrate. Only the intima and adventitia of medium-sized and large arteries in pregnant rats exhibited an occasional positive reaction. Richterich (1952a, b) observed an occasional positive reaction with Tween-60 in the adventitia and less frequently in the media of mouse arteries. In atherosclerotic rabbits McMillan et al. (1954) found increased activity in the aortic endothelium, especially in lipophages, with Gomori's Tween method. However, Narpozzi (1957) used the same method but was unable to detect activity in either normal or atherosclerotic rabbit aortas.

The recorded results with indoxylacetate (Barrnett, 1952), α-naphthol acetate and naphthol-AS-acetate (Nachlas and Seligman, 1949a, b; Chessik, 1953) are not uniform. These authors reported a greater or lesser degree of such esterase activity in the arterial media and intima, or found no activity at all.

In co-operation with Lojda we showed that the histochemical reactions for non-specific esterase, with the use of α-naphthol acetate and AS-acetate as substrates, revealed similar species differences as did the biochemical determinations of lipolytic activity (Lojda and Zemplényi, 1958a, b, 1960; Zemplényi et al., 1958, 1959a). Even naked-eye inspection of the microscopic preparations showed that the rat's aorta reacts much more intensely than the rabbit's aorta, while the guinea-pig's and hamster's aortas react only very weakly.

The muscle cells in the tunica media of the rat's aorta stain strongly with the esterase and AS-esterase methods, but the intima only occasionally reacts. There is practically no intimal staining in the rabbits' aorta, but the medial muscle cells between the elastic membranes show a positive reaction that is more marked in the external layers. The intensity of staining is, however, much weaker than in the rat's aorta. The staining reaction in the media of the golden hamster's and guinea-pig's aortas is also very weak, but in this case the endothelial cells in the tunica intima react more intensely. In all species the adventitial fibrocytes show a strong reaction.

Histochemical observations on arterial non-specific esterase activity in experimental rabbit atherosclerosis, in calciferol-treated rats and in various normal segments of mammalian and avian vessels were mentioned in Chapters XIII, XV, XVII and XVIII.

In other experiments with α-naphthol acetate, naphthol AS-acetate, indoxyl acetate and 5-bromoindoxyl acetate substrates, it was shown that non-specific esterase activity is chiefly localized in the muscle cells of the aortic media, but preparations incubated in toto also displayed a pronounced reaction in the endothelium. In rat aortas such activity was completely inhibited by E600 ($10^{-6}\,M$) and partly by eserine ($10^{-6}M$ and $10^{-3}M$). Heparin, protamine sulphate, sodium chloride and taurocholate did not inhibit, so the enzyme activity demonstrated with these histochemical techniques is clearly not lipoprotein lipase.

With the aid of acetyl- and butyrylthiocholine iodide histochemical substrates, Lojda (1962) went on to show that cholinesterase is present in medial muscle fibres (see also p. 60). Similar results were obtained in rat aortas by Hess and Stäubli (1963) and in rat and rabbit aortas and pulmonary arteries by Navaratnam and Palkama (1965). Such "cholinesterase" activity shows species differences that are analagous to those found with non-specific esterase (Lojda, 1962). These results, together with inhibition studies, suggest that the histochemical reactions are partly due to cholinesterase as well as non-specific carboxylesterase. Analogous species differences were reported by Szendzikowki and Patelski (1960). With acetyl- and butyrylthiocholine substrates, they found much stronger activity in the rat's aorta than in the rabbit's and guinea-pig's aortas, while the chicken's aorta was inactive. The authors maintained that the rat's aorta contains mainly cholinesterase and only a small amount of non-specific esterase. We shall return to this problem later (see p. 225).

Leites reported interesting results about the induction of non-specific esterase activity in granulomas produced by intramuscular injection of cholesterol in rats. In young animals, the macrophages and giant cells of the granuloma exhibited increased esterolytic activity within 6 days after the injection of cholesterol. In old, stressed, diabetic or hypothyroid animals this period was prolonged to about 18 days, but in hyperthyroid adult animals increased esterolytic activity appeared as early as $1\frac{1}{2}$ days (Leites, 1963b, c). Further studies revealed that ageing, immobilization stress and hypothyroidism were associated with markedly reduced esterase staining reactions in rat aortas (Leites, 1963d). In alloxan diabetic rats Leites (1965a) found an early increase in aortic esterase activity, but it was followed by an unequivocal decrease in the later stages of the experiment.

Simple fat feeding also appears to affect aortic non-specific esterase activity (Leites, 1963f). After feeding sunflower-seed oil or butter to rats for 2 months, aortic esterase activity increased; the increment was higher in the aortas of animals fed the vegetable oil.

As in previous histochemical studies on rats (Leites, 1963a), Leites and Fedoseev (1964) investigated non-specific esterase (a-naphthol acetate esterase) activities in dogs that had been fed for 45–60 days on cholesterol, methylthiouracil or both cholesterol and methylthiouracil. Aortic enzyme activity, as compared with that in the normal dog, was significantly elevated in cholesterol-fed animals, even though atherosclerosis had not developed. Aortic enzyme activity was unequivocally decreased in methylthiouracil treated animals. The aortas of the animals that had been fed both cholesterol and methylthiouracil exhibited atherosclerotic-like lesions; their enzyme activity was intermediate between the two extremes and appeared to be related to the extent of disease.

Although ageing and those conditions that predispose to atherosclerosis—such as arterial injury (see Chapter XV), hypothyroidism etc.—cause arterial non-specific carboxylesterase activity to decline, the situation changes with the development of atherosclerotic lesions. In *experimental rabbit atherosclerosis*, the early plaque constituents (endothelium, foam cells, transitional cells) exhibit increased non-specific esterase staining with all histochemical methods for this enzyme (see Chapter XIII; McMillan et al., 1954; Lojda and Zemplényi, 1961; Fouquet, 1961; Zemplényi et al., 1963c; Cavallero, et al., 1963; Gonzales, 1963 and others). Similar histochemical results have been obtained in fatty lesions in chickens (Lojda, 1962), rats (Hess and Stäubli, 1963) and dogs (Gonzales and

Furman, 1965). These observations accord well with biochemical estimates of both lipolytic activity in rabbit aorta (see p. 216) and carboxylesterase activity (β-naphthol laurate esterase) in atherosclerotic rabbit and dog aortas (Maier and Haimovici, 1965b).

In human atherosclerotic lesions increased non-specific esterase staining has been observed by many authors (Gomori, 1946; Tischendorf and Curri, 1959; Müller and Neumann, 1959; Deribas et al., 1960; Levonen et al., 1960; Turolla, 1962; Gonzales, 1963; Leites, 1963e, and others). Kirk's (1965b) biochemical data confirm these histochemical findings. Using β-naphthol acetate and β-naphthol laurate as substrates he found that such activity is 48 per cent higher in fatty lesions of human aortas and coronary arteries than in "normal" parts of these vessels. As could be expected from corresponding histochemical evidence fibrous lesions showed no change in activity. It should be added that full-thickness "normal" segments of the abdominal aorta displayed lower mean activity than "normal" thoracic segments. Neither our own investigations (see Chapter XVII) nor those of Maier and Haimovici (1965a) revealed such clear-cut differences in carboxylesterase activity between these aortic segments. This discrepancy may be due to the use of different criteria for establishing the "normality" of arterial tissue from older human subjects.

Histochemical discussion has so far centred on non-specific esterase activity. In our experiments with Lojda we were unable to find a reliable histochemical method for detecting lipoprotein lipase. Adams (personal communication) also failed to get reliable results. However, two independent groups of investigators have recently modified Gomori's method by using triglyceride-containing substrates. It is claimed that prolonged 24 to 48 hours incubation at 37° C. with calcium chloride at pH 8·6 favours the hydrolytic activity of lipoprotein lipase and the enzymatically released fatty acids are captured by the available Ca^{++} and deposited as insoluble calcium soaps at the sites of enzyme activity. With such techniques Moskowitz and Moskowitz (1965) and Moskowitz (1967) investigated "lipoprotein lipase" and "hormone sensitive lipase" activities in adipose tissue and in monolayer cell cultures. Leites and her co-workers studied "lipoprotein lipase" activity in many tissues, including the vascular wall. Using substrates of lipaemic serum (Leites and Lempert, 1964) or an activated fat emulsion (Leites and Lempert, 1965; Leites and Golosovskaya, 1966), they reported that lipoprotein lipase activity was highest in myocardium, adipose tissue and pancreas. No activity was detected in the liver. Venous tissue exhibited higher activity than arterial tissue. Activity appeared to increase with increasing calibre of the artery. In human arteries activity steadily decreased with age. In human atherosclerotic plaques activity was confined to the endothelium covering the lesions and the lipophages, but activity was very low in necrotic lesions. In cholesterol-fed rabbits the activity pattern differed in "susceptible" and "resistant" animals (Leites, 1965c). Activity was *focally* increased in the atherosclerotic plaques of the susceptible animals, while in the resistant animals *diffusely* increased activity was observed in the unaffected aorta.

These findings are of considerable interest, but it is desirable to obtain convincing evidence (especially inhibition-activation characteristics) that lipoprotein lipase is being detected by those histochemical techniques. Moreover, tissue calcium may interfere in the method employed.

Returning to the well-established evidence about increased lipolytic and esterolytic activities in atherosclerotic vessels, Leites and Fuks's (1966) recent findings provide information about the mechanisms involved. Mice were injected subcutaneously with a suspension of cholesterol in sunflower-seed oil. The activity of esterases in the connective tissue cells was considerably reduced in animals treated with Actinomycin D in comparison with animals not receiving the antibiotic. As Actinomycin D blocks the synthesis of RNA-nucleotidyl-transferase (see Chapter I), it is reasonable to assume that the injection of lipid was followed by a substrate-linked induction of lipolytic enzymes (see Chapter I).

One has to be very cautious in drawing conclusions from experimental data of such very different kinds. Nevertheless, all the findings support the general hypothesis that the *increased esterolytic and, perhaps, lipolytic activity* of lipid-rich atherosclerotic lesions *is an adaptation* (enzyme induction—see Chapter I) *to increased penetration of lipids into the vessel wall*. This phenomenon may be interpreted as a manifestation of the arterial wall's protective mechanism against accumulation of excess lipids (see p. 199).

Before proceeding further with this discussion of lipid-cleaving enzymes in the vessel wall, the question of cholinesterase activity must be examined.

Thompson and Tickner's (1953) and Rosenberg and Dettbarn's (1965) observations on arterial cholinesterase activity were discussed on p. 60. There is no doubt that this enzyme activity can be detected in the vessel wall and, in view of the well-known overlapping specificities of the carboxylic ester hydrolyzing enzymes, it is reasonable to expect that cholinesterase activity would interfere in estimations of vascular lipolytic activity. For example, the above-mentioned authors recorded that $10^{-7}M$ eserine reduces tributyrin hydrolysis by 25 per cent in the rabbit aorta and by 5 per cent in the human carotid artery.

In our own experiments with rat aortic homogenates, eserine inhibited the hydrolysis of β-naphthol acetate by 36–54 per cent. In order to minimize interference by cholinesterase, esterolytic activity was determined in the presence of $10^{-6}M$ eserine (Zemplényi et al., 1963b). Much histochemical evidence (e.g. see p. 223) has also shown that histochemical methods for non-specific esterases cross-react with cholinesterase. However, practically no eserine inhibition is encountered in biochemical assays when lipaemic serum or activated Ediol are used as substrates (Zemplényi et al., 1960b, 1962a). Thus it is reasonable to assume that in these biochemical assays practically no interference with cholinesterase activity takes place. (See also Fig. 77.)

A contrary view has been expressed by Szendzikowski et al., (1961–1962), who suggested on the basis of inhibition studies with eserine and tetra-iso-propylpyrophosphoramide that at least 50 per cent of rat aortic lipolytic activity against lipaemic serum is due to cholinesterase. In our experience, freezing aortas with liquid nitrogen causes a pronounced fall in many enzyme activities. In this connection Szenzikowski et al. used liquid air and their reported enzyme activities are unequivocally lower than those in our experiments. Therefore, the high percentage of cholinesterase activity observed by these authors could be due to the lower vulnerability of this enzyme by deep freezing.

Chapter XXIV

THE CHOLESTEROL ESTERASE AND PHOSPHOLIPASE A ACTIVITY OF ARTERIAL TISSUE

RECENT evidence indicates that arterial tissue manifests clear-cut cholesterol esterase and phospholipase A activity. The investigation of these vascular enzymes is closely related to the problems discussed in the preceding chapter.

CHOLESTEROL ESTERASE, E.C. 3.1.1.13.
(Sterol ester hydrolase)

This enzyme catalyses the hydrolytic cleavage of cholesterol esters into cholesterol and the corresponding acid. In addition to cholesterol, esters of cholestanol, dihydrocholesterol and some other sterols (e.g. dehydroandrosterone) are hydrolysed by the enzyme. The enzyme from pancreas and dog serum displays both hydrolysing and esterifying activity. In contrast, the human serum evinces only esterifying activity (Sperry and Stoyanoff, 1938). Cholesterol esterase is confined to the microsomes. The most thoroughly investigated enzyme is that from the pancreas, but it has also been studied in the liver, duodenal juice, intestinal mucosa, adipose tissue, kidney, adrenals, ovary, testes and other tissues. The hydrolytic activity of the pancreatic enzyme is highest with the cholesterol esters of short-chain fatty acids, while long-chain acids are more readily esterified (Swell and Treadwell, 1955 and others). Unsaturated fatty acids seem to be more readily esterified than saturated fatty acids (Hernandez and Chaikoff, 1957).

The pancreatic enzyme is activated by bile acids, in particular cholic acid, whereas the enzyme from other tissues is somewhat inhibited by bile constituents. In some tissues, e.g. the liver, hydrolysing activity has been reported to be enhanced by lecithin and esterifying activity by phosphate ions (Nieft, 1949). Organophosphorus compounds inhibit cholesterol esterase.

It must be pointed out that recent studies on cholesterol ester metabolism have revealed that several different mechanisms are involved in *cholesterol esterification* in different tissues. For example cholesterol esterification in liver and adrenals is brought about by the reaction of free cholesterol with an activated fatty acid, i.e. with a fatty acyl coenzyme A thiol ester and is ATP dependent (Mukherjee *et al.*, 1958; Dailey *et al.*, 1962 and others).

Plasma cholesterol esterifying activity in most animal species including man appears to be due to another enzyme system. LeBreton and Pantaléon (1947) and Etienne and Polonovski (1960) reported that in incubated plasma the level of phospholipid decreases as cholesterol esterification increases. Subsequently, Glomset *et al.* (1962) and Glomset (1962) put forward the hypothesis that the esterification which occurs in plasma is catalysed by a lecithin:cholesterol fatty acid transferase. The work of other investigators (Rowen and Martin, 1963; Rowen, 1964; Shah et al., 1964; Aftergood and Alfin-Slater, 1967 and others) has confirmed the existence of such an enzyme which appears to act specifically to transfer highly unsaturated fatty acid from the 2-position of lecithin to the hydroxyl group of free cholesterol. Perhaps this reaction accounts for the high proportion of polyunsaturated fatty acids in plasma cholesterol esters. The reaction is accompanied by an apparent conversion of lecithin to lysolecithin and appears to be regulated by catecholamines (Gherondache, 1963), sex hormones (Aftergood and Alfin-Slater, 1967) and probably other hormones.

For obvious reasons the investigation of cholesterol ester hydrolysis and cholesterol esterification in the vascular wall is of special interest.

Patelski (1964) studied cholesterol esterase (hydrolytic) activity in glycerol-water extracts from acetone-butanol powder of porcine aorta. Activity was assayed by continuous titration of the fatty acids using colloidal emulsions of cholesterol esters as substrates. Calcium, magnesium, sodium and potassium ions inhibited such activity. Low concentrations of sodium taurocholate and glycocholate enhanced activity, but higher concentrations were inhibitory. Deoxycholate elicited no activating effect and in higher concentrations definitely inhibited. The pH optimum was 8·6. Thermal inactivation was achieved at 60° C.

Tetra-isopropyl pyrophosphoramide and p-chloromercuribenzoate both inhibited the enzyme, but in the latter case it could be reactivated with glutathione. These findings indicate the importance of enzyme-OH and -SH groups for catalytic activity.

Patelski et al. (1966) later investigated the effect of several "atherogenic" diets on aortic cholesterol ester hydrolysing activity in rats and rabbits. The diet containing 40 per cent peanut oil and, to a lesser degree, that containing butter caused a decline in rat aortic cholesterol esterase activity. This effect was substantially accentuated when thiouracil was added to the diet. Likewise, in rabbits a semisynthetic "atherogenic" diet (Howard et al., 1965) depressed enzyme activity. This change could be counteracted in the earlier stages by simultaneous administration of "Lipostabil" (polyenoic phosphatidyl choline plus pyridoxine). Unlike lipase or non-specific esterase activities (see Chapter XXIII), aortic cholesterol ester hydrolysing activity never increased and seemed to be negatively correlated with lipase activity.

Day and Gould-Hurst (1966) investigated cholesterol esterase activity in a few normal and experimentally induced atherosclerotic rabbit aortas. Pieces of aorta were incubated either with free $[4-{}^{14}C]$-cholesterol or $[4-{}^{14}C]$-cholesterol oleate. After 4 hours' incubation both the aortic tissue and the medium were extracted with chloroform-methanol and, after separation on alumina columns, free and ester cholesterol were determined in each fraction. Both normal and atherosclerotic aortas *hydrolysed* cholesterol oleate, but no *esterifying* activity could be demonstrated. In contrast to Patelski's findings no difference could be found in the hydrolytic activity of atherosclerotic and control aortas. This was a somewhat unexpected finding in view of the abundance of macrophages in such atherosclerotic lesions and the known cholesterol esterase activity in such cells (Day et al., 1963). Possibly the discrepancy can be attributed to the use of whole tissue in these studies instead of homogenates or extracts used by Patelski and co-workers.

On a much larger scale Shyamala *et al.* (1966) investigated the hydrolysis of cholesterol-7α-H^3-oleate by homogenates of chicken aortas. Definite cholesterol esterase activity was found and the percentage of hydrolysed labelled cholesterol oleate increased with advancing age. In the abdominal aortas of older birds activity was about double that in thoracic segments; female aortas showed much higher activity than the male aortas. It is interesting that the increase in aortic cholesterol esterase activity with advancing age was associated with an increased cholesterol ester content in the aorta and a decreased cholesterol ester level in the plasma.

The findings outlined above indicate that so far only hydrolytic cholesterol esterase activity has been demonstrated in the arterial wall. In fact Stein and Stein (1962) showed that slices of rabbit's, dog's, rat's and baboon's aortas incubated with $[1-{}^{14}C]$-linoleic acid almost entirely failed to acylate free cholesterol. Perfusion of rabbit aortas with labelled fatty acids revealed negligible synthesis of cholesterol esters (Bowyer *et al.*, 1966). On the other hand, in the latter authors' perfusion experiments considerable differences were found in the hydrolysis rate of cholesterol $[1-{}^{14}C]$-palmitate (12 per cent), stearate (9 per cent), oleate (4 per cent), as compared with linoleate (33 per cent). Such studies, as well as those with tissue culture cells (Rothblat *et al.*, 1967) corroborate the view that differences in tissue cholesterol ester content may be partly a reflection of the rate at which these esters are hydrolysed by cells.

Selective hydrolysis may also explain the results by Elspeth Smith (1965) showing that in human fatty lesions—as compared with serum—cholesterol oleate increases from 29 to 45 per cent, linoleate drops to half and eicosatrienoate (a minor component in serum) very distinctly increases.

Nevertheless, in spite of the above negative findings, it is possible that cholesterol esterification plays an important role in cholesterol metabolism of vascular tissue. Some indirect evidence supports this view. Lofland *et al.* (1965) observed that in susceptible pigeons known to develop spontaneous atherosclerosis (White Carneau) the cholesterol ester is the fraction showing maximal percentage change as the aorta progresses from the undiseased to the diseased state. In incubation experiments the percentage of total radioactivity found in aortic phospholipid reached a maximal value rather rapidly and then declined, whereas the activity of cholesterol fatty acid continued to increase. This was interpreted as evidence that the aorta of this species has the enzyme system necessary to esterify cholesterol and that Glomset's fatty acid transferase (see p. 226) may be involved.

Recently fatty acid transferase has been studied in human arterial tissue homogenates but until now only in the "de-esterifying" direction; homogenates incubated at pH 5 with ^{14}C-labelled cholesterol esters transferred the labelled fatty acid to lysolecithin to form radioactive lecithin (Abdulla, 1967, cited by Adams, 1967a). It will be of great interest to elucidate whether this system also works in the esterifying direction. As polyunsaturated fatty acids predominate at the 2-position of lecithin, the reaction catalysed by such an arterial fatty acid transferase may render "innocuous" the highly sclerogenic free cholesterol in this tissue (Adams, 1967b) (see Chapter XX).

Phospholipase A, E.C. 3.1.1.4.
(Phosphatide acyl-hydrolase)

This enzyme catalyses the hydrolytic cleavage of phosphatidylcholine (lecithin) by removing the fatty acid attached to the 2-position to produce lysolecithin:

$$\text{lecithin} + H_2O = \text{lysolecithin} + \text{fatty acid ion}$$

However, conflicting views have been expressed about the specificity of the enzyme for the 2-position and the required degree of unsaturation in the fatty acid chain to be hydrolysed. In addition, the enzyme acts on phosphatidylethanolamine, choline plasmalogen and phosphatidates by removing the fatty acid attached to the 2-position.

The richest sources of phospholipase A are snake, bee and wasp venoms. It is also present in many mammalian tissues; high levels are found in the pancreas and small intestine; moderately high levels in the testes, spleen, liver and lung; and low levels in brain, kidney, heart and skeletal muscle (Gallei-Hatchard and Thompson, 1965). Phospholipase A from snake venom and pancreas is characterized by its resistance to heat at pH 4, but it is rapidly destroyed when boiled at a more alkaline pH. However, phospholipase A from intestine, spleen and testes shows a marked sensitivity to heat and pH, which indicates that enzymes from different tissues have divergent properties (Gallei-Hatchard and Thompson, 1965).

The lyso-derivatives produced in tissues by the action of phospholipase A can be acylated by specific acylating systems and, in this way, new molecular patterns of phospholipid can be synthesized. Such acylating systems have been detected in the rat's liver, lung, intestine and kidney, and in human and rat brain (Lands, 1960; Webster, 1962; Webster and Alpern, 1964; Stein and Stein, 1966). The possible role of disordered acylation in multiple sclerosis has been considered by Webster and Thompson (see Thompson, 1963).

Fluoride, citrate and some metal ions inhibit the enzyme, whereas calcium ions and albumin activate it. Working with the pancreatic or cobra-venom enzyme, Ibrahim *et al.* (1964) found that deoxycholate accelerates hydrolysis of lecithin but strongly inhibits that of phosphatidylethanolamine. The aortic enzyme behaves differently (see below).

There are several methods for determining phospholipase A activity. Some of them are based on the measurement of liberated fatty acid by the manometric Warburg technique or by titration. In another method, enzymatic hydrolysis is assessed by thin-layer chromatography, which also enables the lyso-derivatives produced in the reaction to be measured.

Interest in aortic phospholipase A activity has been stimulated by the important studies of Stein and Stein (1962) and Stein *et al.* (1963) on fatty acid incorporation in the dog's, rabbit's, rat's and baboon's aortas. In studies concerned with fatty acid incorporation into lipids of aortic slices, these authors observed that $[1 - {}^{14}C]$-linoleic acid was incorporated more into phospholipids (especially lecithin) than neutral lipids. When glucose was omitted from the incubation medium, incorporation into the neutral lipid fraction declined, but the decreased incorporation into phospholipids was not so constant and in the dog aorta no decrease could be detected at all. As lysolecithin was always found in the aortic wall (also see Zilversmit *et al.* 1961*a*), it was suspected that endogenous lysolecithin might serve as an immediate precursor (acylation substrate) in phospholipid synthesis.

Further experiments were designed to compare the stimulating effects of glycerol 3-phosphate and lysolecithin on the incorporation of $[1-{}^{14}C]$-linoleic acid into aortic lipids. These experiments unequivocally confirmed the presence in the aortic wall of the above pathway for phospholipid synthesis. Glycerol 3-phosphate was indispensable for triglyceride but not for phospholipid synthesis. On the other hand, much more linoleic acid was incorporated into phospholipids in the presence of lysolecithin than glycerol 3-phosphate. Both triglyceride synthesis and the acylation of lysolecithin were ATP- and CoA-dependent (i.e. required energy). The finding that addition of purified lecithin to the incubation mixture resulted in massive incorporation of the labelled fatty acid into phospholipid strongly suggests that phospholipase A is present in aortic tissue.

Waligóra (1966) and Patelski *et al.* (1966) investigated the activity of phospholipase A in extracts from acetone-butanol powder of porcine aorta; they estimated such activity by continuous titration of the fatty acids released from purified egg yolk lecithin and from polyenoic phosphatidylcholine. Highest activities were found in glycerol-water extracts, which contained two factors that could be distinguished by their differential resistance to heat. They studied the properties of the thermostable enzyme; its pH optimum was 8·0. The highest specific activity was attained with a lecithin concentration of $2 \times 10^{-3} M$. Activity was inhibited by calcium, sodium and potassium ions and by di-isopropyl phosphorofluoridate. Sodium tauro- and glycocholate were found to be strong inhibitors, while phlorhidzin totally inhibited enzyme activity. On the other hand, n-butanol, ethylenediaminetetraacetate and triethylamine activated the enzyme. A 300-fold purification was achieved by precipitation at pH 5·0 and salting out with ammonium sulphate.

Activity was found to be much higher in the rat's than rabbit's aorta. "Atherogenic" diets containing thiouracil and either butter or peanut oil depressed aortic phospholipase A activity in rats, whereas feeding only peanut-oil increased enzyme activity.

In view of the probable protective role played by locally synthesized phospholipids in the pathogenesis of atherosclerosis (see p. 199), the demonstration of lecithin cleavage by aortic phospholipase A, together with the demonstration of the lysolecithin acylation reaction, is of special interest and deserves more study.

Chapter XXV

CONCLUDING REMARKS TO PART THREE

In Chapter XIX an attempt was made to evaluate the significance of local metabolic vascular factors in atherogenesis. It was mentioned that, in addition to other protective mechanisms, the arterial wall's equipment of enzymes that catalyse the hydrolytic cleavage of lipids can be regarded as a further defence barrier against the accumulation of lipids and against the transformation of the fatty streak into the fibrous plaque.

The evidence assembled in Chapter XXIII—in particular with regard to species differences, effect of ageing, stress, hypothyroidism, alloxan diabetes, adaptive changes to fat feeding and lipid accumulation in the vascular wall—support such a hypothesis. It would be very unwise, of course, to simplify the complicated problem of atherogenesis and to expect that deranged vascular lipolytic activity would be the only pathogenic factor in the disease. It is not surprising to find that certain conditions closely related to atherosclerosis do not evoke definite changes in lipolytic activity (for example see sex differences, p. 219).

The main function of the raised lipolytic activity apparently consists in *protecting the arterial wall* against accumulation of potentially injurious triglycerides (see Chapter XX).

In addition, triglyceride cleavage may serve the purpose of *providing fatty acids* for acylation of lysolecithin and cholesterol in a tentative reaction sequence as follows:

1. Lipolysis of arterial or plasmatic triglycerides and release of free fatty acids.
2. Coenzyme A- and ATP-dependent acylation of lysolecithin with the polyunsaturated fraction of free fatty acids from reaction 1, and production of lecithin.

[3. Esterification of free cholesterol catalysed by lecithin:cholesterol fatty acid transferase and release of lysolecithin for reaction 2.]

4. The lysolecithin involved in the above carrier function between reactions 2 and 3 may also be supplemented by the catalytic action of phospholipase A on arterial or plasmatic phospholipids.

Unfortunately, reaction 3 has had to be put into parentheses, because it has not at present been unequivocally demonstrated in vascular tissue. Therefore, this cholesterol "neutralizing" arterial defence mechanism can only be regarded as an attractive working hypothesis that awaits verification.

REFERENCES

ABDULLA, Y. H. (1967). *Biochim. biophys. Acta (Amst).* In press.
ABDULLA, Y. H., and ADAMS, C. W. M. (1965). *J. Atheroscler. Res.,* **5,** 504.
ABELL, L. L., LEVY, B. B., BRODIE, B. B., and KENDALL, F. E. (1952). *J. biol. Chem.,* **195,** 357.
ACHOR, R. W. P., BERGE, K., BERKER, N. O., and MCKENZIE, B. F. (1958). *Circulation,* **17,** 479.
ACKERMAN, R. F., DRY, T. J., and EDWARDS, I. E. (1950). *Circulation,* **1,** 1345.
ADAMS, C. W. M. (1959). *Lancet,* **1,** 1075.
ADAMS, C. W. M. (1964a). *Biol. Rev.,* **39,** 372.
ADAMS, C. W. M. (1964b). In *Biological Aspects of Occlusive Vascular Disease,* eds. D. G. Chalmers and G. A. Gresham. London: Cambridge Univ. Press.
ADAMS, C. W. M. (1967a). Vascular Histochemistry: In Relation to the Chemical and Structural Pathology of Cardiovascular Disease. London: Lloyd-Luke.
ADAMS, C. W. M. (1967b). *J. Atheroscler. Res.,* **7,** 117.
ADAMS, C. W. M., ABDULLA, Y. H., BAYLISS, O. B., MAHLER, R. F., and ROOT, M. A. (1966). Paper read at the Internat. Symp. on Recent Advances in Atherosclerosis, Athens, May 30–June 2.
ADAMS, C. W. M., BAYLISS, O. B., and IBRAHIM, M. Z. M. (1962), *Lancet,* **1,** 890.
ADAMS, C. W. M., BAYLISS, O. B., and IBRAHIM, M. Z. M. (1963a). *J. Path. Bact.,* **86,** 421.
ADAMS, C. W. M., BAYLISS, O. B., IBRAHIM, M. Z. M., and WEBSTER, M. W., Jr. (1963b). *J. Path. Bact.,* **86,** 431.
ADAMS, C. W. M., and TUQAN, N. A. (1961). *J. Path. Bact.,* **82,** 131.
AFTERGOOD, L. and ALFIN-SLATER, R. B. (1967). *J. Lipid Res.,* **8,** 126.
AHLQVIST, T. (1960). *Acta Pathol. et Microbiol. Scand.,* **50,** Suppl. 142.
AHMED, Z., and REIS, T. L. (1958). *Biochem. J.,* **69,** 386.
ALBRINK, M. T., MAN, E. B., and PETERS, J. F. (1955). *J. clin. Invest.,* **34,** 147.
ALDRIDGE, W. N. (1953a). *Biochem. J.,* **53,** 110.
ALDRIDGE, W. N. (1953b). *Biochem. J.,* **54,** 442.
ALEKSEEVA, A. S., and USHKALOV, A. F. (1963). *Enzymol. Biol. Clin.,* **2,** 108.
ALEKSEEVA, A. S. (1964). In *Ateroskleróz i Trombóz,* p. 10, by A. L. Mjasnikov. Moskva: Izd. "Medicina". (Russian).
ALEKSEEVA, A. S., and NEKRASOVA, A. A. (1963). *Cor et Vasa (Praha),* **5,** 190.
ALIBASOGLU, M., DUNNE, H. W., and GUSS, S. B. (1962). *Amer. J. vet. Res.,* **23,** 49.
ALTSCHUL, R. (1950). *Selected Studies on Arteriosclerosis.* Springfield, Ill.: C. C. Thomas.
ALTSCHULER, G. H., and ANGEVINE, D. M. (1954). In *Connective Tissue in Health and Disease,* p. 178, ed. G. Asboe-Hansen. Copenhagen: Munksgaard.
AMES, B. N., and MARTIN, R. G. (1964). *Ann. Rev. Biochem.,* **33,** 235.
ANITSCHKOW, N. (1913). *Beitr. Path. Anat.,* **56,** 379.
ANITSCHKOW, N. (1914). *Beitr. Path. Anat.,* **57,** 201.
ANITSCHKOW, N. N. (1912). *Ber. Ges. Russ. Ärzte in Petersburg,* **80,** 1.
ANITSCHKOW, N. N. (1921). *Verhandl. Virchow-Tagung russ. Pathologen,* p. 46. Leningrad.
ANITSCHKOW, N. N. (1933). In *Arteriosclerosis,* p. 271, ed. E. V. Cowdry. New York: Macmillan Co.
ANITSCHKOW, N. N., and CINZELRING, V. D. (1955). In *Ateroskleróz,* p. 7, ed. N. N. Anitschkow. Moskva: Medgiz. (Russian).
ANSON, M. L. (1938). *J. gen. Physiol.,* **22,** 79.
ANTONINI, F. H., and WEBER, G. (1951). *Arch. "De Vecchi",* **16,** 985.
APPELLA, E., and MARKERT, C. L. (1961). *Biochem. biophys. Res. Commun.* **6,** 171.
ARDLIE, N. G., KINLOUGH, R. L., GLEW, G., and SCHWARTZ, C. J. (1966). *Aust. J. exp. Biol. med. Sci.,* **44,** 105.
ASBOE-HANSEN, G., and WEGELIUS, O. (1956). *Acta physiol. scand.,* **37,** 350.
ASCHOFF, L. (1924). *Lectures on Pathology.* New York: Hoeber.
ASK-UPMARK, E. (1962). *Acta med. scand.,* **172,** 129.
AUGUSTINSSON, K. B. (1958). *Nature (Lond.).* **181,** 1786.

REFERENCES

AUGUSTINSSON, K. B. (1960). In *The Enzymes*, vol. IV, p. 521, eds. P. D. Boyer, H. Lardy, and K. Myrbäck. New York: Academic Press.
AXELROD, J. (1959). *Pharmacol. Rev.* **11**, 402.
BAILIE, M. J., and CHRISTIE, G. S. (1959). *Biochem. J.*, **72**, 473.
BAKER, H., and SOBOTKA, H. (1962). *Advanc. clin. Chem.*, **5**, 189
BALÓ, J. (1963). *Internat. Rev. Connect. Tissue Res.*, **1**, 241.
BALÓ, J., and BANGA, I. (1948). *Orv. Hetil.* (In Hungarian), **89**, 465.
BALÓ, J., and BANGA, I. (1949). *Schweiz. Z. Path.*, **12**, 350.
BALÓ, J., and BANGA, I. (1950). *Biochem. J.*, **46**, 384.
BALÓ, J., and BANGA, I. (1953). *Acta physiol. Acad. Sci. hung.*, **4**, 187.
BALÓ, J., BANGA, I., JOSEPOVITS, Gy. (1948–49). *Z. Vitamin- Hormon-u. Fermentforsch.*, **2**, 1.
BALÓ, J., BANGA, I., and SCHULER, D. (1954). *Acta morph. Acad. Sci. hung.*, **4**, 141.
BANGA, I. (1961). Paper read at the Vth Internat. Congr. Biochem., Moscow.
BANGA, I. (1962). *Biochem. J.*, **84**, 116 P.
BANGA, I. (1963). *Acta physiol. Acad. Sci hung.*, **24**, (1), 1.
BANGA, I., and BALÓ, J. (1956). *Nature (Lond.)*, 178, 310.
BANGA, I., and BALÓ, I. (1962). *Acta physiol. Acad. Sci. hung.*, **21**, 301.
BANGA, I., and NOWOTNY, A. (1951a). *Acta physiol. Acad. Sci. hung.*, **2**, 317.
BANGA, I., and NOWOTNY, A. (1951b). *Z. Vitamin- Hormon-u. Fermentforsch.*, **4**, 54.
BANGA, I., and SCHULER, D. (1953). *Acta physiol. Acad. Sci. hung.*, **4**, 13.
BANGHAM, A. D. (1964). In *Biological Aspects of Occlusive Vascular Disease*, p. 237, eds. D. G. Chalmers and G. A. Gresham. London: Cambridge Univ. Press.
BARANOVSKI, T., ILLINGWORTH, B., BROWN, D. H., and CORI C. F. (1957). *Biochim. biophys. Acta (Amst.)*, **25**, 16.
BARR, D. P. (1953). *Circulation*, **8**, 641.
BARRNETT, R. J. (1952). *Anat. Rec.*, **114**, 577.
BARROWS, Ch. H. Jr., and CHOW, B. F. (1959). In *The Arterial Wall*, p. 192, ed. A. I. Lansing. Baltimore: Williams and Wilkins Co.
BARTALOS, M., and GYORKY, F. (1963). *J. Amer. Geriat. Soc.*, **11**, 21.
BASSI, M. (1960). *Exp. Cell. Res.*, **20**, 313.
BEACONSFIELD, P. (1962). In *Metabolismus Parietis Vasorum*, p. 174, eds. B. Prusík, Z. Reiniš and O. Riedl. Prague: State Medical Publ. House.
BEHAL, F. J., KANAVAGE, C. B. and HAMILTON, R. D. Proc. Soc. exp. Biol. (N.Y.), 109, 411.
BEINERT, H. (1960). In *The Enzymes*, vol. II, p. 340, eds. P. D. Boyer, L. Hardy and K. Myrbäck. New York: Academic Press.
BEISENHERZ, G., BOLTZE, H. J., BÜCHER, Th., CZOK, R., GARBADE, K. H., MEYER-ARENDT, E., and PFLEIDERER, G. (1953). *Z. Naturforsch.*, **8b**, 555.
BELICER, V. A. (1939). *Enzymologia*, **6**, 1.
BELL, E. T., and CLAWSON, B. J. (1928). *Arch. Path.*, **5**, 939.
BENDITT, E. P., WONG, R. L., ARASE, M., and ROEPER, E. (1955). *Proc. Soc. exp. Biol. (N.Y.)*, **90**, 303.
BENEŠ, K., LOJDA, Z., and HOŘAVKA, B. (1961). *Histochemie*, **2**, 313.
BENNETT, J. C., and DREYER, W. J. (1964). *Ann. Rev. Biochem.*, **33**, 205.
BERGMANN, M. (1942). *Advanc. Enzymol.*, **12**, 191.
BERGMEYER, H. U. (1963). *Methods in Enzymatic Analysis*. New York: Academic Press.
BERGMEYER, H. U., and BERNT, E. (1963a). In *Methods in Enzymatic Analysis* p. 757, ed. H. U. Bergmeyer. New York: Academic Press.
BERGMEYER, H. U., and BERNT, E. (1963b). In *Methods of Enzymatic Analysis* p. 837, ed. H. U. Bergmeyer. New York: Academic Press.
BERGMEYER, H. U., and BERNT, E. (1963c). In *Methods of Enzymatic Analysis*, p. 846, ed. H. U. Bergmeyer. New York: Academic Press.
BERGMEYER, H. U., BERNT, E., and HESS, H. (1963). In *Methods of Enzymatic Analysis*, p. 736, ed. H. U. Bergmeyer. New York: Academic Press.
BERNHARD, E., and ROTHENBERG, A. (1951). *Proc. Soc. exp. Biol. (N.Y.)*, **78**, 533.
BERNICK, S., and PATEK, P. R. (1961). *Arch. Path.*, **72**, 321.
BERNICK, S., PATEK, P. R., ERSHOFF, B. H., and WELLS, A. (1962). *Amer. J. Path.*, **41**, 661.
BERTELSEN, S. (1963). In *Atherosclerosis and Its Origin*, p. 119, eds. M. Sandler and G. H. Bourne, New York: Academic Press.
BIHARI-VARGA, M., GERGELY, J., and GERÖ, S. (1964). *J. Atheroscler. Res.*, **4**, 106.

BJÖRNTORP, P., and FURMAN, R. H. (1962). *Amer. J. Physiol.*, **203**, 323.
BLASCHKO, H. (1963). In *The Enzymes*, p. 337, eds. P. D. Boyer, H. Lardy and K. Myrbäck. New York: Academic Press.
BLOOM, G. D. (1965). In *The Inflammatory Process*, p. 355, eds. B. W. Zweifach, L. Grant and R. T. McCluskey. New York: Academic Press.
BLUMENTHAL, H. T., HOROWITZ, S. T., HEMERLINE, A., and ROSEMAN, S. (1955). *Bact. Proc.*, 137.
BLUMENTHAL, H. T., LANSING, A. I., and GRAY, S. H. (1950). *Amer. J. Path.*, **26**, 989.
BLUMENTHAL, H. T., LANSING, A. I., and WHEELER, P. A. (1944). *Amer. J. Path.*, **20**, 665.
BORN, G. V. R. (1956*a*). *Biochem. J.*, **62**, 33P.
BORN, G. V. R. (1956*b*). *J. Physiol. (Lond.)*, **133**, 61P.
BORN, G. V. R. (1962). *Nature (Lond.)*, **194**, 927.
BORN, G. V. R. (1965). *Ann. roy. Coll. Surg. Engl.*, **36**, 200.
BORN, G. V. R., and CROSS, M. J. (1963). *Nature (Lond.)*, **197**, 974.
BÖTTCHER, C. J. F. (1964). In *Evolution of the Atherosclerotic Plaque*, p. 109. Urbana: Univ. Chicago Press.
BÖTTCHER, C. J. F., BOELSMA-VAN HOUTE, E., TER HAAR ROMENY-WACHTER, C. Ch., WOODFORD, F. P., and VAN GENT, C. M. (1960). *Lancet*, **2**, 1162.
BOUCEK, R. J. (1964). In *Biological Aspects of Occlusive Vascular Disease*, p. 136, eds. D. G. Chalmers and G. A. Gresham. London: Cambridge Univ. Press.
BOUNAMEAUX, Y. (1961). *Thrombos. Diathes. haemorrh. (Stuttg.)*, **6**, 504.
BOWYER, D. E., HOWARD, A. N., GRESHAM, G. A., BATES, D., and PALMER, B. V. (1966). Paper read at the Internat. Symp. on Recent Advances in Atherosclerosis, Athens, May 30–June 2. (In press.)
BRANDSTRUP, N., KIRK, J. E., and BRUNI, C. (1957). *J. Geront.*, **12**, 166.
BRANWOOD, A. W., and CARR, A. J. (1960). *Lancet*, **2**, 1254.
BRAUNSTEIN, A. E., and KRITSMANN, M. G. (1937). *Enzymologia*, **2**, 129.
BRIGGS, F. N., CHERNICK, S., and CHAIKOFF, I. L. (1949). *J. biol. Chem.*, **179**, 103.
BRINKMAN, R. (1934). *J. Physiol. (Lond.)*, **80**, 171.
BRONK, J. R. (1963). *Biochem. biophys. Acta (Amst.)*, **69**, 375.
BRUNS, F. H., and BERGMEYER, H. V. (1963). In *Methods of Enzymatic Analysis*, p. 724, ed. H. V. Bergmeyer. New York: Academic Press.
BRUNS, F. H., NOLTMANN, E., and WILLEMSEN, A. (1958). *Biochem. Z.*, **330**, 411.
BÜCHER, Th., and KLINKENBERG, M. (1958). *Angew. Chemie*, **70**, 552.
BUCK, R. C. (1963). In *Atherosclerosis and Its Origin*, p. 1, eds. M. Sandler and G. H. Bourne. New York: Academic Press.
BUCK, R. C., and ROSSITER, R. J. (1951). *Arch. Path.*, **51**, 224.
BURCH, H. B. (1957). In *Methods in Enzymology*, Vol. 3, p. 960, eds. S. P. Colowick and N. O. Kaplan. New York: Academic Press.
BÜRGER, M. (1954). *Altern und Krankheit*, 2nd edit. Leipzig: G. Thieme.
BURN, J. H., PHILPOT, F. J., and TRENDLENBURG, U. (1954). *Brit. J. Pharmacol.*, **9**, 423.
BURSTONE, M. (1959). *J. Histochem. Cytochem.*, **7**, 112.
CAHN, R. D., KAPLAN, N. O., LEVINE, L., and ZWILLING, E. (1962). *Science*, **136**, 962.
CAIRNS, A., and CONSTANTINIDES, P. (1954). *Science*, **120**, 31.
CARPENTER, K. J., and KODICEK, E. (1950). *Biochem. J.*, **46**, 421.
CARR, C. J., BELL, F. K., and KRANTZ, J. C., Jr. (1952). *Proc. Soc. exp. Biol. (N.Y.)*, **80**, 323.
CARR, C. J., BELL, F. K., HURST, J. N., and KRANTZ, J. C., Jr. (1954). *Circulat. Res.*, **2**, 516.
CARR, C. J., BELL, F. K., REHAK, M. J., and KRANTZ, J. C., Jr. (1955). *Proc. Soc. exp. Biol. (N.Y.)*, **89**, 184.
CASTELLANI, A. A., and ZAMBOTTI, V. (1956). *Nature (Lond.)*, **178**, 313.
CAVALLERO, C., and TUROLLA, E. (1960). *Giorn. Gerontol.*, Suppl. 22, 25.
CAVALLERO, C., TUROLLA, E., and SOLCIA, E. (1963). *Rivista di Istochimica*, **9**, 5.
CERIOTTI, C. (1952). *J. biol. Chem.*, **198**, 297
CHANCE, B., and WILLIAMS, G. R. (1956). *Advanc. Enzymol.*, **17**, 65.
CHANG, Y. O., LAURSEN, T. J. S., and KIRK, J. E. (1955). *J. Geront.*, **10**, 165.
CHARVÁT, J. (1952). In *Steroid Hormons*. Prague: State Medical Publ. House. (In Czech.)
CHARVÁT, J., and ŠILINK, K. (1949). *Čas. Lék. čes.*, **88**, 219, 255, 286.
CHATTOPADHYAY, D. P. (1961). *Nature (Lond.)*, **192**, 660.
CHESSIK, R. D. (1953). *J. Histochem. Cytochem.*, **1**, 471.

CHRISTENSEN, S. (1961). *J. Atheroscler. Res.*, **1**, 140.
CHRISTENSEN, S. (1962). *J. Atheroscler. Res.*, **2**, 131.
CHRISTIANSON, O. O. (1939). *Arch. Path.*, **27**, 1011.
CHRISTIE, R. W., and DAHL, L. K. (1957). *J. exp. Med.*, **106**, 357.
CHVAPIL, M. (1967). *Physiology of Connective Tissue.* London: Butterworths; and Prague: Czechoslovak Medical Press.
CHVAPIL, M., and HURYCH, J. (1968). *Internat. Rev. Connect. Tissue Res.* (In press.)
CIOTTI, M. M., and KAPLAN, N. O. (1957). In *Methods in Enzymology*, Vol. 3, p. 890, eds. S. P. Colowick and N. O. Kaplan. New York: Academic Press.
CLAGOV, S., ROWLEY, D. A., and KOHUT, R. I. (1961). *Arch. Path.*, **72**, 558.
CLARCSON, T. B. (1963). In *Advances in Lipid Research*, Vol. I, p. 211, eds. R. Paoletti and D. Kritechevsky. New York: Academic Press.
COLEMAN, R., and HÜBSCHER, G. (1961). *Biochem. J.*, **80**, 11.
CONCHIE, J., and LEVVY, G. A. (1963). In *Methods of Separation of Subcellular Structural Components* (Biochem. Soc. Symposium), **23**, 86.
CONNOR, W. E., HOAK, J. C., and WARNER, E. D. (1965). In *Genetics and the Interaction of Blood Clotting Factors*, p. 89, eds. F. Koller and F. Streuli. Stuttgart: F. K. Schattauer.
CONSTANTINIDES, P. (1953). *Science*, **117**, 505.
CONSTANTINIDES, P. (1961). *J. Atheroscler. Res.*, **1**, 374.
CONSTANTINIDES, P. (1963). *Circulation*, **28**, 653.
CONSTANTINIDES, P. (1965). *Experimental Atherosclerosis.* Amsterdam: Elsevier Publ. Co.
CONSTANTINIDES, P., BOOTH, T., and CARLSON, G. (1960). *Arch. Path.*, **70**, 712.
COOKSON, F. B., ALTSCHUL, R., and FEDOROFF, S. (1967). *J. Atheroscler. Res.*, **7**, 69.
CORBASCIO, A. N. (1960). *Circulat. Res.*, **8**, 390.
COURTICE, F. C., and SCHMIDT-DIEDRICHS, A. (1963). *Brit. J. exp. Path.*, **44**, 339.
COX, G. E., TRUEHEART, R. E., KAPLAN, J., and TAYLOR, C. B. (1963). *Arch. Path.*, 76, 166.
CRANE, R. K. (1962). In *The Enzymes*, Vol. 6, p. 47, eds. P. D. Boyer, H. Lardy, and K. Myrbäck. New York: Academic Press.
CRANE, R. K., and SOLS, A. (1953). *J. biol. Chem.*, **203**, 273.
CROOK, E. M., and LAW, K. (1952). *Biochem. J.*, **65**, 1.
DAHME, E. (1963). *Verh. dtsch. Ges. Path.*, **47**, 382.
DAHME, E. (1964). *Bull. Soc. Roy. Zool.*, **34**, 55.
DAILEY, R. E., SWELL, L., and TREADWELL, C. R. (1962). *Proc. Soc. exp. Biol. (N.Y.)*, **110**, 571.
DAKSHINAMURTI, K., and MISTRY, S. P. (1963). *J. biol. Chem.*, **238**, 294.
DALITH, F. (1964). *J. Atheroscler Res.*, **4**, 239.
DAUBER, D. V. (1944). *Arch. Path.*, **38**, 46.
DAVIDSON, E., HOWARD, A. N., and GRESHAM, G. A. (1961). *Brit. J. exp. Path.*, **42**, 195.
DAVIDSON, E., HOWARD, A. N., and GRESHAM, G. A. (1962). *Brit. J. exp. Path.*, **43**, 166.
DAVIS, D., and KLAINER, M. J. (1940). *Amer. Heart J.*, **19**, 185.
DAVIS, R. P. (1958). *J. Amer. chem. Soc.*, **80**, 5209.
DAVIS, R. P. (1961). In *The Enzymes*, Vol. 5, p. 545, eds. P. D. Boyer, H. Lardy, and K. Myrbäck. New York: Academic Press.
DAWSON, D. M., GOODFRIEND, T. L., and KAPLAN, N. O. (1964). *Science*, **143**, 929.
DAY, A. J. (1962). *J. Atheroscler. Res.*, **2**, 350.
DAY, A. J. (1964). *J. Atheroscler. Res.*, **4**, 117.
DAY, A. J., and GOULD-HURST, P. R. S. (1966). *Biochim. biophys. Acta (Amst.)*, **116**, 169.
DAY, A. J., FIDGE, N. H., GOULD-HURST, P. R. S., and RISELEY, D. (1963). *Quart. J. exp. Physiol.*, **48**, 298.
DEMOPOULOS, H., KALEY, G., and ZWEIFACH, B. W. (1961). *Circulat. Res.*, **9**, 845.
DERIBAS, V. I., FUKS, B. B., and SCHISCHKIN, G. S. (1960). *Dokl. Aked. Nauk SSSR (Otd. Biokh.)*, **134**, 443.
DERRIEN, M. (1911). *Bull. Soc. Chim.*, **9**, 110.
DIANZANI, M. U. (1954). *Biochem. biophys. Acta. (Amst.)*, **14**, 514.
DIANZANI, M. U. (1957). *Biochem. J.*, **65**, 116.
DIANZANI, M. U., and SCURO, S. (1956). *Biochem. J.*, **62**, 205.
DICKENS, F., and SALMONY, D. (1956). *Biochem. J.*, **64**, 645.
DIXON, K. C. (1958). *Quart. J. exp. Physiol.*, **43**, 139.
DIXON, K. C. (1961). *Amer. J. Path.*, **39**, 65.
DIXON, T. F., and PURDOM, M. (1954). *J. clin. Path.*, **7**, 341.

DIXON, M., and WEBB, E. C. (1964). *Enzymes*, 2nd edit. London: Longmans.
DOCK, W. (1946). *J. Amer. med. Ass.*, **131**, 875.
DOCK, W. (1950). *Bull. N.Y. Acad. Med.*, **26**, 182.
DOLE, V. B. (1956). *J. clin. Invest.*, **35**, 150.
DONNER, L. (1962). *J. Atheroscler. Res.*, **2**, 88.
DONNER, L., and HEYROVSKÝ, A. (1964a). *Čas. Lék. čes.*, **103**, 617.
DONNER, L., and HEYROVSKÝ, A. (1964b). *Thrombos. diathes. haemorrh. (Stuttg.)*, **9**, 476.
DORFMAN, R. I. (1952). *Vitam. and Horm.*, **10**, 331.
DORFMAN, R. I. (1964). In *Actions of Hormones on Molecular Processes*, p. 470, eds. G. Litwack and D. Kritchevsky. New York: J. Wiley & Sons.
DOW, D. R. (1925). *Brit. med. J.*, **2**, 162.
DUFF, G. L. (1965). *Arch. Path.*, **20**, 81, 257, 1935.
DUFF, G. L., and MCMILLAN, G. C. (1951). *Amer. J. Med.*, **11**, 92.
DUGUID, J. B. (1930). *J. Path. Bact.*, **33**, 697.
DUGUID, J. B. (1946). *J. Path. Bact.*, **58**, 207.
DUGUID, J. B. (1948). *J. Path. Bact.*, **60**, 57.
DUNNIGAN, M. G. (1964). *J. Atheroscler. Res.*, **4** 144.
DURY, A. (1961). *J. Geront.*, **16**, 114.
DURY, A., LEIGHLY, F., and DURY, M. (1957). *Science Studies, St. Bonaventura Univ., St. Bonaventura, N.Y.*, **19**, 37.
DUVE, C. DE (1959). In *Subcellular Particles*, p. 128, ed. T. Hayashi, New York: Ronald Press.
DUVE, C. DE (1963). Discussion of paper by Conchie and Levvy. In *Methods of Separation of Subcellular Structural Components*, (Biochem. Soc. Symposium), **23**, 107.
DUVE, C. DE (1964). In *Ciba Foundation Symposium on Cellular Injury*, p. 369, eds. A. V. S. de Reuck and J. Knight. London: Churchill.
DYRBYE, M., and KIRK, J. E. (1956). *J. Geront.*, **11**, 33.
EDER, H. A. (1958). In *Hormones and Atheroslcerosis*, p, 335, ed. G. Pincus. New York: Academic Press.
ELSON, L. E., and MORGAN, W. T. J. (1933). *Biochem. J.*, **27**, 1824.
EMMELOT, P., and BENEDETTI, E. L. (1960). *J. biophys. biochem. Cytol.*, **7**, 393.
ENGELBERG, H. (1963). *Heparin, Metabolism, Physiology and Clinical Application.* Springfield, Ill.: C. C. Thomas.
ENGELHARDT, U. A., and LJUBIMOWA, M. N. (1939). *Nature (Lond.)*, **144**, 668.
EPINOSA, E., and INSUNZA, I. (1962). *Proc. Soc. exp. Biol. (N.Y.)*, **8**, 174.
ESTERLEY, J. A., and GLAGOV, S. (1963). *Amer. J. Path.*, **43**, 619.
ETIENE, J., and POLONOVSKI, J. (1960). *Bull Soc. Chim. biol. (Paris)*, **42**, 857.
EVANS, S. M., IHRIG, H. K., MEANS, J. A., ZEIT, W., and HAUSHALTER, E. R. (1952). *Amer. J. clin. Path.*, **22**, 354.
FABER, M., and LUND, F. (1949). *Arch. Path.*, **48**, 351.
FAHR, H. O. (1935). *Die Arteriosklerose beim Haushuhn.* (Thesis, Giessen.)
FALCONER, G. F., and ADAMS, C. W. M. (1965). *Guy's Hosp. Rep.*, **114**, 130.
FARBER, E. (1962). *J. Histochem. Cytochem.*, **10**, 657.
FELT, V., REICHL, D., ROEHLING, S., and VOHNOUT, S. (1958). *Gerontologia (Basel)*, **2**, 247.
FILLIOS, L. C., ANDRUS, S. B., MANN, G. V., and STARE, F. J. (1956). *J. exp. Med.*, **104**, 539.
FILLIOS, L. C., KAPLAN, R., MARTIN, R. S., and STARE, F. J. (1958). *Amer. J. Physiol.*, **193**, 47.
FINE, I. H., KAPLAN, N. O., and KUFTINEC, D. (1963). *Biochemistry*, **2**, 116.
FINLAYSON, R., and SYMONS, C. (1964). In *Biological Aspects of Occlusive Vascular Disease*, p. 333, eds. D. G. Chalmers and G. A. Gresham. London: Cambridge Univ. Press.
FISCHER, E. R., and GELLER, J. H. (1960). *Circulat. Res.*, **8**, 820.
FISHMAN, W. F. (1951). *Vitam. and Horm.*, **9**, 213.
FISHMAN, W. H. (1947). *Science*, **105**, 646.
FISHMAN, W. H. (1951). *Ann. N.Y. Acad. Sci.*, **54**, 548.
FISHMAN, W. H., and BAKER, J. R. (1956). *J. Histochem. Cytochem.*, **4**, 570.
FLORKIN, M., and STOTZ, E. H. (1965). *Comprehensive Biochemistry*, Vol. 13, Enzyme Nomenclature, 2nd edit. Amsterdam: Elsevier Publ.
FLOYER, M. A. (1962). *Recenti. Progr. Med.*, **32**, 473.
FODOR, J., and LOJDA, Z. (1956). *Physiol. bohemoslov.*, **5**, 275.
FODOR, J., FÁBRY, P., and LOJDA, Z. (1958a). *Experientia (Basel)*, **14**, 184.

REFERENCES

FODOR, J., ZEMPLÉNYI, T., LOJDA, Z., and FELT, V. (1958b) *Čas Lék. čes.*, **97**, 316.
FODOR, J., FÁBRY, P., and LOJDA, Z. (1960). *Virchows Arch. path. Anat.*, **333**, 582.
FONNESU, A., and SEVERI, C. (1956). *J. biophys. biochem. Cytol.* **2**, 293.
FONTAINE, R., MANDEL, P., PANTESCO, V., and KEMPF, E. (1960). *Strasbourg méd.*, **9**, 605.
FOUQUET, J. P. (1961). *Ann. Histochim.*, **6**, 153.
FOX, H. (1933). In *Arteriosclerosis, Survey of the Problem*, p. 153, ed. E. V. Cowdry. New York: Macmillan Co.
FRENCH, J. E. (1964). In *Biological Aspects of Vascular Disease*, p. 24, eds. D. G. Chalmers and G. A. Gresham. London: Cambridge Univ. Press.
FRENCH, J. E. (1966). *Int. Rev. exp. Path.*, **5**, 253.
FRENCH, J. E., and JENNINGS, M. A. (1965). In *Comparative Atherosclerosis*, p. 25, eds. J. C. Roberts, Jr., and R. Straus. New York: Harper and Row.
FRENCH, J. E. JENNINGS, M.A., POOLE, J. C. F., ROBINSON, D. S., and FLOREY, H. W. (1963). *Proc. roy. soc. B.*, **158**, 24.
FRIED, G. H., and ZWEIFACH, B. W. (1955). *Anat. Rec.*, **121**, 97.
FRIED, G. H., GREENBERG, N., and ANTOPOL, W. (1961). *Proc. Soc. exp. Biol. (N.Y.)*, **107**, 523.
FRIEDEN, C. (1963). In *The Enzymes*, Vol. 7, p. 3, eds. P. D. Boyer, H. Lardy and K. Myrbäck. New York: Academic Press.
FRIEDEN, E. H. (1964). In *Actions of Hormones and Molecular Processes*, p. 509, eds. G. Litwack and D. Kritchevsky. New York: J. Wiley & Sons.
FRIEDMAN, M., and BYERS, S. O. (1951). *Proc. Soc. exp. Biol. (N.Y.)*, **78**, 528.
FRIEDMAN, M., and BYERS, S. O. (1954). *Amer. J. Physiol.*, **179**, 201.
FRIEDMAN, M., and BYERS, S. O. (1955). *Proc. Soc. exp. Biol. (N.Y.)*, **90**, 496.
FRIEDMAN, M., BYERS, S. O., and ROSEMAN, R. H. (1957). *Proc. Soc. exp. Biol. (N.Y.)*, **95**, 586.
FRIEDMANN, H. C. (1963). In *Methods of Enzymatic Analysis*, pp. 596–601, ed. H. U. Bergmeyer. New York: Academic Press.
FRUTON, J. S. (1960). In *The Enzymes*, Vol. 4, p. 233, eds. P. D. Boyer, H. Lardy and K. Myrbäck. New York: Academic Press.
FRUTON, J. S., and SIMMONDS, S. (1958). *Biochemistry*, 2nd edit. New York: J. Wiley & Sons.
FULTON, G. P., MAYNARD, F. L., RILEY, J. F., and WEST, G. B. (1957). *Physiol. Rev.*, **37**, 221.
FURMAN, R. H., HOWARD, R. P., NORCIA, L. N., and KEATY, E. C. (1958). *Amer. J. Med.*, **24**, 80.
GAARDER, A., JONSEN, J., LALAND, S., HELLEM, A., and OWREN, P. A. (1961). *Nature (Lond.)*, **192**, 531.
GALBRAITH, P. A., PERRY, W. F., and BEAMISH, R. E. (1959). *Lancet*, **1**, 222.
GALLAGHER, C. H., and REES, K. R. (1960). *Nature (Lond.)*, **187**, 148.
GALLAGHER, C. H., JUDAH, J. D., and REES, K. R. (1956). *J. Path. Bact.*, **72**, 247.
GALLAI-HATCHARD, J. J., and THOMPSON, R. H. S. (1965). *Biochim. biophys. Acta (Amst.)*, **98**, 128.
GARDNER, D. L. and LAING, C. P. (1965). *J. Path. Bact.*, **90**, 399.
GARVIN, J. E. (1961). *J. exp. Med.*, **114**, 51.
GEER, J. C., MCGILL, H. C., and STRONG, J. P. (1961). *Amer. J. Path.*, **38**, 263.
GEIRINGER, E. (1951). *J. Path. Bact.*, **6**, 201.
GEREBTZOFF, M. A. (1953). *Acta anat. (Basel)*, **19**, 366.
GERÖ, S., GERGELY, J., DÉVÉNYI, T., JAKAB, L., SZÉKELY, T., and VIRÁG, S. (1961). *J. Atheroscler. Res.*, **1**, 67.
GERÖ, S., GERGELY, J., DÉVÉNYI, T., VIRÁG, S., SZÉKELY, J., and JAKAB, L. (1962a). In *Metabolismus Parietis Vasorum*, p. 34, eds. B. Prusík, Z. Reiniš and O. Riedl. Prague: State Medical Publ. House.
GERÖ, S., GERGELY, J., FARKAS, K., DÉVÉNYI, T., KOSCÁR, L., JAKAB, L., SZÉKELY, J., and VIRÁG, S. (1962b). *J. Atheroscler. Res.*, **2** 276.
GERSCHENSON, L., MALINOW, M. R., LACUARA, J. L., and MOGUILEVSKY, H. C. (1962). *J. Atheroscler. Res.*, **2**, 365.
GERTLER, M. M., HUDSON, P. B., and JOST, H. (1953). *Geriatrics*, **8**, 500.
GETTY, R. (1965). In *Comparative Atherosclerosis*, p. 11, eds. J. C. Roberts, Jr. and R. Straus. New York: Harper and Row.
GHERONDACHE, C. N. (1963). *J. clin. Endocr.*, **23**, 1024.
GIERTSEN, J. C. (1965). *Acta path. microbiol. scand.*, **65**, 245.

GILFILLAN, R. F., SBARRA, A. J., and BARDAWIL, W. A. (1960). *Fed. Proc.*, **19**, 144.
GILLMAN, T. (1957). *Lancet*, 2, 1117.
GILLMAN, T. (1958). *Lancet*, 2, 901.
GILLMAN, T. (1964). In *Biological Aspects of Occlusive Vascular Disease*, p. 3, eds. D. G. Chalmers and G. A. Gresham. London: Cambridge Univ. Press.
GILLMAN, T., and GILBERT, C. (1957). *Exp. Med. Surg.*, **15**, 181.
GILLMAN, T. and HATHORN, M. (1959). *Nature (Lond.)*, **183**, 1139.
GILLMAN, T., HATHORN, M., and PENN, J. (1957). In *Connective Tissue*, p. 120, eds. R. E. Turnbridge and D. A. Hall. Oxford: Blackwell.
GILLMAN, T., GRANT, R. A. and HATHORN, M. (1960). *Brit. J. exp. Path.*, **41**, 1.
GLAGOV, S., ROWLEY, D. A., and KOHOUT, R. I. (1961). *Arch. Path.*, **72**, 558.
GLENNER, G. G., COHEN, L. A., and FOLK, J. E. (1965). *J. Histochem. Cytochem.*, **7**, 416.
GLIMCHER, M. J. (1959). In *Connective Tissue, Thrombosis and Atherosclerosis*, p. 140, ed. I. H. Page. New York: Academic Press
GLOCK, G. E., and MCLEAN, P. (1958). *Proc. roy. Soc. B*, **149**, 354.
GLOMSET, J. A. (1962). *Biochim. biophys. Acta (Amst.)*, **65**, 128.
GLOMSET, J. A., PARKER, F., TJADEN, M., and WILLIAMS, R. H. (1962). *Biochim. biophys. Acta (Amst.)*, **58**, 398.
GOMORI, G. (1939). *Proc. Soc. exp. Biol. (N.Y.)*, **42**, 23.
GOMORI, G. (1946). *Arch. Path.*, **41**, 121.
GOMORI, G. (1952). *Microscopic Histochemistry. Principles and Practice*. Urbana: Univ. Chicago Press.
GONZALES, I. E. (1963). In *Evolution of the Atherosclerotic Plaque*, p. 151, ed. R. J. Jones. Urbana: Univ. Chicago Press.
GONZALES, I. E., and FURMAN, R. H. (1965). In *Comparative Atherosclerosis*, p. 329, eds. J. C. Roberts, Jr., and R. Straus. New York: Harper and Row.
GORE, I., and LARKEY, B. J. (1960). *J. Lab. clin. Med.*, **56**, 839.
GOTTLIEB, M., and LALICH, J. J. (1954). *Amer. J. Path.*, **30**, 851.
GOULD, B. S. (1960). *Vitam. and Horm.*, **18**, 89.
GRAFNETTER, D., and ZEMPLÉNYI, T. (1958). *Čs. Fysiol.*, **7**, 457.
GRAFNETTER, D. and ZEMPLÉNYI, T. (1959). *Hoppe-Seylers Z. physiol. Chem.*, **316**, 218.
GRAFNETTER, D., and ZEMPLÉNYI, T. (1961). *Cor et vasa (Praha)*, **3**, 63.
GRAFNETTER, D. and ZEMPLÉNYI, T. (1962a). In *Metabolismus Parietis Vasorum*, p. 185, eds. B. Prusík, Z. Reiniš and O. Riedl. Prague: State Medical Publ. House.
GRAFNETTER, D., and ZEMPLÉNYI, T. (1962b). *Experientia (Basel)*, **18**, 85.
GRANT, R. A. (1967). *J. Atheroscler. Res.*, **7**, 463.
GREEN, M. N., KWAN CHUNG TSOU, BRESSLER, R., and SELIGMAN, A. M. (1955). *Arch. Biochem.*, **57**, 458.
GRESHAM, G. A., and HOWARD, A. N. (1960). *Brit. J. exp. Path.*, **41**, 395.
GRESHAM, G. A., and Howard, A. N. (1963). *J. Atheroscler. Res.*, **3**, 161.
GRISOLIA, S. (1962). In *Methods in Enzymology*, Vol. 5, p. 236, eds. S. P. Colowick and N. O. Kaplan. New York: Academic Press.
GROLLMAN, A., ASHWORTH, C., and SUKI, W. (1963). *Arch. Path.*, **75**, 618.
GROSS, L., EPSTEIN, E. Z., and KUGEL, M. A. (1934). *Amer. J. Path.*, **10**, 253.
GRUNBAUM, B. W., GEARY, J. R., Jr., GRANDE, F., ANDERSON, J. T., and GLICK, D. (1957). *Proc. Soc. exp. Biol. (N.Y.)*, **94**, 613.
GRÜNBERG, W. (1964). *Bull. Soc. Roy. Zool. d'Anvers*, No. 34, 21.
GRYDER, R. M., and POGELL, B. M. (1960). *J. biol. Chem.*, **235**, 558.
GUHA, S., and WEGMANN, R. (1960). *Bull. Soc. Chim. biol. (Paris)*, **42**, 115.
GUTSTEIN, W. H., LAZZARINI-ROBERTSON, A., and LA TAILLADE, J. N. (1963). *Amer. J. Path.*, **42**, 61.
HAHN, P. F. (1943). *Science*, **98**, 19.
HAIMOVICI, H., and MAIER, N. (1964). *Arch. Surg.*, **89**, 961.
HAIMOVICI, H., and MAIER, N. (1966). *J. Atheroscler. Res.*, **6**, 62.
HAIMOVICI, H., MAIER, N., and STRAUSS, L. (1958). *Arch. Surg.* **76**, 282.
HALL, D. A. (1953). *Biochem. J.*, **55**, 35P.
HALL, D. A. (1955). *Biochem. J.*, **59**, 459.
HALL, D. A. (1958). *Biochem. J.*, **70**, 5P.
HALL, D. A. (1961a). *J. Atheroscler. Res.*, **1**, 173.

REFERENCES

HALL, D. A. (1961b). *Biochem. J.*, **78**, 491.
HALL, D. A. (1964a). *Elastolysis and Ageing*, Springfield, Ill.: C. C. Thomas.
HALL, D. A. (1964b). Paper read at the Internat. Conference on Enzymology and Immunology, October 24, Bordeaux.
HALL, D. A., and CZERKAWSKI, J. W. (1961a). *Biochem. J.*, **80**, 128.
HALL, D. A., and CZERKAWSKI, J. W. (1961b). *Biochem. J.*, **80**, 134.
HALL, D. A., and WILKINSON, J. E. (1963). *Nature (Lond.)*, **197**, 454.
HALL, D. A., REED, R., and TURNBRIDGE, R. E. (1952). *Nature (Lond.)*, **170**, 264.
HAMOIR, G., GODFROID, A., and MOREAU-COLLINET, C. M. (1961). *Arch. Internat. Physiol. Biochim.*, **69**, 595.
HARKNESS, M. L. R., HARKNESS, R. D., and MCDONALD, D. A. (1957). *Proc. roy. Soc. B.*, **146**, 541.
HARRIS, P. M., and ROBINSON, D. S. (1961). *Biochem. J.*, **80**, 253.
HARTROFT, W. S., and THOMAS, W. A. (1957). *J. Amer. med. Ass.*, **164**, 1899.
HARTROFT, W. S., and THOMAS, W. A. (1963). In *Atherosclerosis and Its Origin*, p. 439, eds. M. Sandler and G. H. Bourne. New York: Academic Press.
HARTROFT, W. S., RIDOUT, J. H., SELLERS, E. A., and BEST, C. H. (1952). *Proc. Soc. exp. Biol. (N.Y.)*, **81**, 384.
HARUKI, F., and KIRK, J. E. (1965). *Proc. Soc. exp. Biol. (N.Y.)*, **118**, 479.
HASHIMOTO, Y., and KOBERNICK, S. D. (1964). *Proc. Soc. exp. Biol. (N.Y.)*, **115**, 212.
HASLAM, R. J. (1964). *Nature (Lond.)*, **202**, 765.
HASS, G. M. (1954). In *Symposium on Atherosclerosis* p. 24, ed. I. H. Page. Washington, D.C.: National Academy of Sciences, National Research Council, Pub. No. 338.
HATEFI, Y. (1963). In *The Enzymes*, Vol. 7, p. 495, eds. P. D. Boyer, H. Lardy and K. Myrbäck. New York: Academic Press.
HATEFI, Y., HAAVIK, A. G., and IURTSHUK, P. (1961). *Biochim. biophys. Acta (Amst.)*, **52**, 106.
HAUST, M. D., and MORE, R. H. (1963). In *Evolution of the Atherosclerotic Plaque*, p. 51, ed. R. J. Jones. Urbana: Univ. Chicago Press.
HAVEL, R. J. (1958). *Amer. J. Clin. Nutr.*, **6**, 662.
HELLEM, A. J. (1960). *The Adhesiveness of Human Blood Platelets in vitro*. Oslo: Oslo Univ. Press.
HELLEM, A. J. (1964). In *Biological Aspects of Occlusive Vascular Disease*, p. 220, eds. D. G. Chalmers and G. A. Gresham. London: Cambridge Univ. Press.
HEPTINSTALL, R. H., and PORTER, K. A. (1957). *Brit. J. exp. Path.*, **38**, 55.
HEPTINSTALL, R. H., BARKLEY, H., and PORTER, K. A. (1958). *Angiology*, **9**, 84.
HERNANDEZ, H. H., and CHAIKOFF, I. L. (1957). *J. Biol. Chem.*, **228**, 447.
HESS, B. (1963). *Enzymes in Blood Plasma*. New York: Academic Press.
HESS, R., and STÄUBLI, W. (1963). *Amer. J. Path.*, **43**, 301.
HESS, B., and WALTER, S. I. (1960). *Klin. Wschr.*, **38**, 1080.
HESS, B., and WALTER, S. I. (1961). *Ann. N.Y. Acad. Sci.*, **94**, 890.
HESS, R., SCARPELLI, D. G., and PEARSE, A. G. E. (1958). *J. biophys. biochem. Cytol.*, **4**, 753.
HIGGINBOTHAM, F. H., and HIGGINBOTHAM, A. C. (1967). *J. Atheroscler. Res.*, **7**, 89.
HILL, M. (1957). *Nature (Lond.)*, **180**, 654.
HILZ, H. (1962). *J. Atheroscler. Res.*, **2**, 252.
HIRSCH, S. (1952). *Cardiologia (Basel)*, **20**, 27.
HIRSCH, P. F., LOSSOW, W. J., and CHAIKOFF, I. L. (1956). *J. biol. Chem.*, **221**, 509.
HJELMAN, G. (1954). *Soc. Sci. Fennica*, Comment. Biol., 15.
HOFFMAN, J. F. (1962). *Circulation*, **26**, 1201.
HOFFMAN-OSTENHOF, O. (1954). *Enzymologie*. Wien: Springer-Verlag.
HOFSTEE, B. M. J. (1960). In *The Enzymes*, Vol. 4, p. 485, eds. P. D. Boyer, H. Lardy and K. Myrbäck. New York: Academic Press.
HOLLENBERG, C. H. (1959). *Amer. J. Physiol.*, **197**, 667.
HOLLENBERG, C. H. (1960). *J. clin. Invest.*, **39**, 1282.
HOLLETT, C., and NESTEL, P. J. (1960). *Amer. J. Physiol.*, **199**, 803.
HOLMAN, R. L., MCGILL, H. C., Jr., STRONG, J. P., and GEER, J. C. (1958). *Amer. J. Path.*, **34**, 209.
HOLMAN, R. L., MCGILL, H. C., Jr., STRONG, J. P., GEER, J. C., and GUIDRY, M. A. (1959). In *Hormones and Atherosclerosis*, p. 123, ed. G. Pincus. New York: Academic Press.
HOMOLKA, J., and ANGEROVÁ, M. (1963). *Čas. Lék. čes.*, **102**, 1304.

HOMOLKA, J., KŘIŠŤANOVÁ, D., and VEČEREK, B. (1966). *Clin. chim. Acta*, **13**, 125.
HONOUR, A. J., and MITCHELL, J. R. A. (1963). *Nature (Lond.)*, **197**, 1019.
HOŘEJŠÍ, J. (1948). *Fundamentals of Chemical Determinations in Internal Medicine.* Prague: Czech Graphical Union (In Czech.)
HOŘEJŠÍ, J., MAŠEK, K., AND PUDLAK, P. (1963). *Fundamentals of Clinical Biochemistry in Internal Medicine.* Prague: State Medical Publ. House. (In Czech.)
HORŇÁČEK, J., SÁGNEROVÁ, H., TOMÁNKOVÁ, M., and TRČKA, V. (1962). *Čs. Fysiol.*, **11**, 512.
HORŇÁČEK, J., TRČKA, V., and VEJDĚLEK, Z. (1966). *Čs. Fysiol.* **15**, 33.
HOVIG, T. (1963a). *Thrombos. Diathes. haemorrh. (Stuttg.)*, **9**, 248.
HOVIG, T. (1963b). *Thrombos. Diathes. haemorrh. (Stuttg.)*, **9**, 264.
HOWARD, A. N. and GRESHAM, G. A. (1964). In *Biological Aspects of Occlusive Vascular Disease*, p. 347, esd. D. G. Chalmers and G. A. Gresham. London: Cambridge Univ. Press.
HOWARD, A. N., GRESHAM, G. A., JONES, D. and JENNINGS, J. W. (1965). *J. Atheroscler. Res.*, **5**, 330.
HOWARD, A. N., GRESHAM, G. A., BOWYER, D. E., and LINDGREN, F. T. (1966). Paper read at the International Symposium on Recent Advances in Atherosclerosis, Athens.
HUDSON, R. E. B. (1965). *Cardiovascular Pathology.* London: Arnold.
HUEPER, W. C. (1944). *Arch. Path.*, **38**, 162, 245, 350.
HUEPER, W. C. (1945). *Arch. Path.*, **39**, 51, 117, 187.
HUGUES, J. (1960). *C. R. Soc. Biol. (Paris)*, **154**, 866.
HUMMEL, K. P., and BARNES, L. L. (1938). *Amer. J. Path.*, **14**, 121.
HUMPRREYS, E. M. (1957). *Quart, J. exp. Physiol.*, **42**, 96.
IBRAHIM, S. A., SANDERS, H., and THOMPSON, R. H. S. (1964). *Biochem. J.*, **93**, 588.
IGNATOWSKI, A. (1908). *Ber. mil. mediz. Akad.* (Petersburg), **16**, 174. (Cited by Anitschkow, 1933.)
JACOBSON, W. (1964). In *Ciba Foundation Symposium on Cellular Injury*, p. 136, eds. A. V. S. de Reuck and J. Knight. London: Churchill.
JAENICKE, L., and LYNEN, F. (1960). In *The Enzymes*, Vol. 3, p. 3. eds. P. D. Boyer, H. Lardy and K. Myrbäck. New York: Academic Press.
JANOUŠEK, Št. (1959). *Čas. Lék. čes.*, **97**, 206.
JAQUES, L. B. (1961). *Canad. J. Biochem.*, **39**, 643.
JELÍNKOVÁ-TENOROVÁ, M., and HRŮZA, Z. (1963). *Gerontologia (Basel)*, **7**, 168.
JENNINGS, M. A., FLOREY, H. W., and ROBINSON, D. S. (1960). *Quart. J. exp. Physiol.*, **45**, 298.
JENNINGS, M. A., FLOREY, H. W., STEHBENS, W. E., and FRENCH, J. E. (1961). *J. Path. Bact.*, **81**, 49.
JONES, H. B., GOFMAN, J. W., LINDGREN, F. T., LYON, T. P., GRAHAM, D. M., STRISOWER, B., and NICHOLS, A. V. (1951). *Amer. J. Med.*, **11**, 358.
JUDAH, J. D., and AHMED, K. (1964). *J. cell. comp. Physiol.*, **64**, 355.
JUDAH, J. D., AHMED, K., and MCLEAN, A. E. M. (1964). In *Ciba Foundation Symposium on Cellular Injury*, p. 187, eds. A. V. S. de Reuck and J. Knight. London: Churchill.
KAGAN, A., DAWBER, T. R., KANNEL, W. B., and REVOTSKIE, N. (1962). *Fed. Proc.*, **21**, Suppl. 11, 52.
KAHN, S. G. and SLOCUM, A. (1967a). *Amer. J. Physiol.*, **213**, 367.
KAHN, S. G. and SLOCUM, A. (1967b). *Amer. J. Physiol.*, **213**, 373.
KALCKAR, H. M. (1953). *Biochem. biophys. Acta (Amst.)*, **12**, 250.
KANNEL, W. B., BARRY, P., and DAWBER, T. R. (1962). *Proc. 4th World Cardiol. Congr.*, Mexico City.
KAPLAN, N. O. (1960). In *The Enzymes*, Vol. 3, p. 106, eds. P. D. Boyer, H. Lardy and K. Myrbäck. New York: Academic Press.
KAPLAN, N. O. (1963). *Bact. Rev.*, **27**, 155.
KAPLAN, N. O. (1964). *Brookhaven Sympos. Biol.*, **17**, 131.
KAPLAN, N. O., and CAHN, R. D. (1962). *Proc. nat. Acad. Sci. (Wash.)*, **48**, 2123.
KAPLAN, N. O., and LIPMANN, F. (1948). *J. biol. Chem.*, **174**, 37.
KAPLAN, A., and NARAHARA, A. (1953a). *J. Lab. clin. Med.*, **41**, 819.
KAPLAN, A., and NARAHARA, A. (1953b). *J. Lab. clin. Med.*, **41**, 825.
KAPLAN, N. O., CIOTI, M. M., HAMOLSKY, M., and BIEBER, R. E. (1960). *Science*, **131**, 392.
KARMEN, A. (1955). *J. clin. Invest.*, **34**, 131.

KATZ, L. N., and PICK, R. (1963). In *Evolution of the Atheroslerotic Plaque*, p. 251, ed. J. Jones. Urbana: Univ. Chicago Press.
KATZ, L. N., and STAMLER, J. (1953). *Experimental Atherosclerosis*. Springfield. Ill.: C. C. Thomas.
KAWAMURA, R. (1927). *Neue Beiträge zur Morphologie und Physiologie der Cholesterinsteatose*. Jena: Fischer.
KAYAHAN, S. (1960). *Lancet*, 2, 667.
KEARNEY, E. B. (1960). *J. biol. Chem.*, 235, 865.
KEILIN, D., and HARTREE, E. F. (1939). *Proc. roy. Soc B*, 127, 167.
KEMPF, É., and MANDEL, P. (1961). *C. R. Acad. Sci. (Paris)*, 253, 2155.
KENNEDY, E. P. (1957). *Ann. Rev. Biochem.*, 26, 119.
KENNEDY, E. P. (1958). In *Chemistry of Lipids as Related to Atherosclerosis*, p. 325, ed. J. H. Page. Springfield, Ill.: C. C. Thomas.
KENNEDY, E. P. (1960). In *The Enzymes*, Vol. 2, p. 63, ed. P. D. Boyer. New York: Academic Press.
KENNY, G. E., and FINK, B. R. (1966). *Fed. Proc.*, 25, 297.
KIM, J. J., PAOLETTI, R., and VERTUA, R. (1960). *Atompraxis*, 6, 55.
KING, E. J., and ALLOTT, E. N. (1947). In *Recent Advances in Clinical Pathology*, p. 222, ed. S. C. Dyke. London: Churchill.
KIRK, J. E. (1959*a*). *J. Geront.*, 14, 288.
KIRK, J. E. (1959*b*). *J. Geront.*, 14, 181.
KIRK, J. E. (1959*c*). *J. Geront.*, 14, 447.
KIRK, J. E. (1959*d*). *Ann. N.Y. Acad. Sci.*, 72, 1006.
KIRK, J. E. (1960*a*). *J. Geront.*, 15, 139.
KIRK, J. E. (1960*b*). *J. Geront.*, 15, 136.
KIRK, J. E. (1960*c*). *J. Geront.*, 15, 262.
KIRK, J. E. (1961*a*). *J. Geront.*, 16, 25.
KIRK, J. E. (1961*b*). *J. Geront.*, 16, 243.
KIRK, J. E. (1962*a*). *J. Geront.*, 17, 276.
KIRK, J. E. (1962*b*). *J. Geront.*, 17, 154.
KIRK, J. E. (1962*c*). *J. Geront.*, 17, 369.
KIRK, J. E. (1962*d*). *J. Geront.*, 17, 158.
KIRK, J. E. (1963*a*). In *Atherosclerosis and Its Origin*, p. 67, eds. M. Sandler and G. H. Bourne. New York: Academic Press.
KIRK, J. E. (1963*b*). *Clin. Chem.*, 9, 776.
KIRK, J. E. (1964*a*). *Clin. Chem.*, 10, 184.
KIRK, J. E. (1964*b*). *Clin. Chem.*, 10, 306.
KIRK, J. E. (1965*a*). *J. Geront.*, 20, 357.
KIRK, J. E. (1965*b*). *Lab. Invest.*, 14, 573.
KIRK, J. E. (1966*a*). *J. Geront.*, 21, 420.
KIRK, J. E. (1966*b*). In *Perspectives in Experimental Gerontology*, p. 182, ed. Shock. Springfield, Ill.: C. C. Thomas.
KIRK, J. E. (1966*c*). *J. Lab. clin. Med.*, 68, 888.
KIRK, J. E., and DYRBYE, M. (1956). *J. Geront.*, 11, 129.
KIRK, J. E., and HANSEN, P. (1953). *J. Geront.*, 8, 150.
KIRK, J. E., and KIRK, T. E. (1959). *Fed. Proc.*, 18, 261.
KIRK, J. E., and LAURSEN, T. J. S. (1955). *J. Geront.*, 10, 18.
KIRK, J. E., and PRAETORIUS E. (1950). *Science*, 111, 334.
KIRK, J. E., and RITZ, E. (1966). *Proc. Soc. exp. Biol. (N.Y.)*, 122, 1201.
KIRK, J. E., and SANWALD, R. (1966). *J. Atheroscler. Res.*, 6, 440.
KIRK, J. E., and SØRENSEN, L. B. (1956). *J. Geront.*, 11, 373.
KIRK, J. E., EFFERSØE, P. G., and CHIANG, S. P. (1954*a*). *J. Geront.*, 9, 10.
KIRK, J. E., HANSEN, P. F., EFFERSØE, P. G., and IVERSEN, K. (1954*b*). *J. biol. Chem.*, 208, 17.
KIRK, J. E., LAURSEN, T. J. S., and SCHAUS, R. (1955). *J. Geront.*, 10, 178.
KIRK, J. E., MATZKE, J. R., BRANDSTRUP, N., and WANG, I. (1958). *J. Geront.*, 13, 24.
KIRK, J. E., WANG, I., and BRANDSTRUP, N. (1959). *J. Geront.*, 14, 25.
KITTINGER, G. W., WEXLER, B. C., and MILLER, B. P. (1960). *Proc. Soc. exp. Biol. (N.Y.)*, 104, 616.

Kittinger, G. W., Wexler, B. C., and Miller, B. F. (1962). In *Metabolismus Parietis Vasorum*, p. 60, eds. B. Prusík, Z. Reiniš and O. Riedl. Prague: State Medical Publ. House.
Kjaerheim, A., and Hovig, T. (1962). *Thrombos. Diathes. haemorrh. (Stuttg.)*, **7**, 1.
Kleinzeller, A. (1942). *Biochem. J.*, **36**, 729.
Klembovskij, A. I. (1963). *Kardiologija, Moskva*, **3**, 31.
Klingenberg, M. (1963). In *Methods of Enzymatic Analysis*, p. 528, ed. H. U. Bergmeyer. New York: Academic Press.
Klinkenberg, M., and Bücher, Th. (1960). *Ann. Rev. Biochem.*, **29**, 669.
Klinkenberg, M., and Schollmeyer, P. (1961). *Biochem. Z.*, **333**, 338.
Klotzsch, H., and Bergmeyer, H. U. (1963). In *Methods of Enzymatic Analysis*, p. 363, ed. H. U. Bergmeyer. New York: Academic Press.
Knox, W. E. (1960). In *The Enzymes*, Vol. 2, p. 254, eds. P. D. Boyer, H. Lardy and K. Myrbäck. New York: Academic Press.
Kochakian, C. D. (1945). *J. biol. Chem.*, **161**, 115.
Kochakian, C. D. (1947). *Recent Progr. Hormone Res.*, **1**, 177.
Korn, E. D. (1955). *J. biol. Chem.*, **215**, 1.
Korn, E. D. (1958). In *Chemistry of Lipides as Related to Atherosclerosis. A Symposium*, p. 169, ed. I. H. Page. Springfield, Ill.: C. C. Thomas.
Korn, E. D., and Quigley, Th. W., Jr. (1955). *Biochem. biophys. Acta (Amst.)*, **18**, 143.
Koštíř, J. (1960). *General Biochemistry*. Prague: State Technical Publ. House. (In Czech.)
Krantz, J. C., Jr., Carr, C. J., and Bryant, H. H. (1951). *J. Pharmacol. exp. Ther.*, **102**, 16.
Krčílek, A., Janoušek, V., and Šerák, L. (1962). In *Metabolismus Parietis Vasorum*, p. 50, eds. B. Prusík, Z. Reiniš and O. Riedl. Prague: State Medical Publ. House.
Krebs, E. G., and Fischer, E. H. (1963). *Vitam. and Horm.*, **22**, 399.
Krebs, E. G., Graves, D. J., and Fischer, E. H. (1959). *J. biol. Chem.*, **234**, 2867.
Kritchevsky, D., Whitehouse, M. W., and Staple, E. (1960). *J. Lipid Res.*, **1**, 154.
Kritsmann, M. G., and Bavina, M. V. (1955). In *Atheroscleroz*, p. 127, ed. N. N. Anichkov. Moscow: Medgiz.
Krompecher, S. (1960). *Nova Acta Leopoldina*, **22**, 146.
Kubie, G., Reichl, D., and Felt, V. (1956). *Nature (Lond.)*, **178**, 210.
Kubowitz, F., and Ott, P. (1943). *Biochem. Z.*, **314**, 94.
Kuby, S. A. and Noltmann, E. A. (1962). In *The Enzymes*, Vol. 6, p. 515, eds. P. D. Boyer, H. Lardy and K. Myrbäck. New York: Academic Press.
Kuff, E. L., and Hogeboom, G. H. (1956). In *Enzymes, Units of Biological Structure and Function*, p. 235. New York: Academic Press.
Kuhn, E., Páv, J., and Placer, Z. (1959). *Rev. Czech. Med.*, **5**, 261.
Labella, F. S. (1957). *Nature (Lond.)*, **180**, 1360.
Lacuara, J. L., Gerschenson, L., Moguilevsky, H. C., and Malinow, M. R. (1962). *J. Atheroscler. Res.*, **2**, 496.
Lande, K. E., and Sperry, W. M. (1936). *Arch. Path.*, **22**, 301.
Lands, W. E. M. (1960). *J. biol. Chem.*, **235**, 2233.
Lansing, A. I., Alex, M., and Rosenthal, T. B. (1950). *J. Geront.*, **5**, 112, 314.
Lansing, A. I., Rosenthal, T. B., Alex, M., and Dempsey, E. W. (1952). *Anat. Rec.* **114**, 55
Lapin, B. A., and Yakovleva, L. A. (1963). *Comparative Pathology in Monkeys*, p. 144. Springfield, Ill.: C. C. Thomas.
Lardy, H. (1962). In *The Enzymes*, Vol. 6, p. 67, eds. P. D. Boyer, H. Lardy and K. Myrbäck. New York: Academic Press.
Lardy, H. A., and Wellman, H. (1952). *J. biol. Chem.*, **195**, 215.
Lathe, G. H., and Walker, M. (1958). *Biochem. J.*, **70**, 705.
Laursen, T. J. S., and Kirk, J. E. (1955a). *J Geront.*, **10**, 26.
Laursen, T. J. S., and Kirk, J. E. (1955b). *J. Geront.*, **10**, 18.
Lazovskaya, L. N. (1943). *Biokhimiya*, **8**, 171.
Lazzarini-Robertson, A., Jr. (1963). In *Fundamentals of Vascular Grafting*, p. 79, eds. S. A. Weselowski and C. Dennis. New York: McGraw-Hill Co.
Lazzarini-Robertson, A., Jr. (1966). Paper read at the International Symposium on Atherosclerosis, Athens, June, 1966.
Le Breton, E., and Pantaléon, J. (1947). *Arch. Sci. physiol.*, **1**, 63.
Ledvina, M. (1958). *Mucoproteins of the Organism*. Prague: State Medical Publ. House. (In Czech.)

LEHNINGER, A. L. (1959). In *The Arterial Wall*, p. 220, ed. A. L. Lansing. Baltimore: Williams and Wilkins.
LEHNINGER, A. L. (1962). *Physiol. Rev.*, **42**, 467.
LEHNINGER, A. L. (1965). *The Mitochondrion Molecular Basis of Structure and Function*. New York: Benjamin.
LEITES, F. L. (1963a). *Pat. Fiziol. éksp. Ter. (Moskva)*, **7**(2), 50.
LEITES, F. L. (1963b). *Vestn. Akad. med. Nauk SSSR*, **18**(7), 37.
LEITES, F. L. (1963c). *Probl. Endokr. Gormonoter.* **6**, 29.
LEITES, F. L. (1963d). *Vestnik AMN SSSR*, **18**(7), 37.
LEITES, F. L. (1963e). *Arkh. Pat*, **25**(6), 37.
LEITES, F. L. (1963f). *Vop. Pitan.*, **22**(6), 37.
LEITES, F. L. (1965a). *Probl. Endokr. Gormonoter*, **9**(6), 29 and **11**(3), 83.
LEITES, F. L. (1965b). *Dokl. Akad. Nauk SSSR, Otd. Biokh.*, **165**(5), 1175.
LEITES, F. L. (1965c). *Pat. Fiziol. éksp. Ter.* **9**(2), 12.
LEITES, F. L., and FEDOSEEV, A. N. (1964). *Archif Patologii (Moskva)*, **26**(9), 15.
LEITES, F. L., and FUKS, B. B. (1966). *Bjul. Exp. Biol. Med.*, **31**(5), 46.
LEITES, F. L., and GOLOSOVSKAYA, M. A. (1964). *Dokl. Akad. Nauk SSSR, Otd. Biokh.*, **156**(4), 91
LEITES, F. L., and GOLOSOVSKAYA, M. A. (1966). *Arch. Anat*, **49**(7), 61.
LEITES, F. L., and LEMPERT, B. L. (1964). *Dokl. Akad. Nauk SSSR, Otd. Biokh.*, **157**(3), 672.
LEITES, F. L., and LEMPERT, B. L. (1965). *Byull. eksp. Biol. Med.*, **60**(7), 123.
LEITES, F. L., and LEMPERT, B. L. (1967). *Cor et vasa (Praha)*. (In Press.)
LEITES, S. M., and CHOW-SU (1962a). *Klinitscheskaja Medicina (Moskva)*, **40**(7), 17.
LEITES, S. M., and CHOW-SU (1962b). *Vop. med. Khim.*, **8**(3), 289.
LELOIR, L. F., and CARDINI, C. E. (1953). *Biochim. biophys. Acta (Amst.)*, **12**, 15.
LELOIR, L. F., and CARDINI, C. E. (1960). In *The Enzymes*, Vol. 2, p. 39, eds. P. D. Boyer, H. Lardy and K. Myrbäck. New York: Academic Press.
LEMPERT, B. L. (1965). Thesis to obtain the degree of "Candidate of Science", Moscow. (In Russian.)
LEMPERT, B. L., and LEITES, F. L. (1963). *Byull. eksp. Biol. Med. (Moskva)*, **56**(10), 25.
LEMPERT, N., STEIN, A., and DOYLE, J. (1961). *Arch. Path.*, **72**, 386.
LESTER, R. L., and SMITH, A. L. (1961). *Biochim. biophys. Acta (Amst.)*, **47**, 475.
LEVINTOW, L., and MEISTER, A. (1954). *J. biol. Chem.*, **209**, 265.
LEVONEN, E., RAEKALLIO, J., and UOTILA, U. (1960). *Nature (Lond.)*, **188**, 677.
LEVVY, G. A., and CONCHIE, J. (1964). *Progr. Biophys.*, **14**, 105.
LEVVY, G. A., and MARSH, C. A. (1960). In *The Enzymes*, Vol. 4, p. 397, eds. P. D. Boyer, H. Lardy and K. Myrbäck. New York: Academic Press.
LEVY, G. A., KERR, L. M. H., and CAMPBELL, J. G. (1948). *Biochem. J.*, **42**, 462.
LEVY, S. W. (1958). *Rev. canad. Biol.*, **17**, 1.
LEWIS, L. A., GREEN, A. A., and PAGE, I. H. (1952). *Amer. J. Physiol.*, **171**, 391.
LINDBERG, O., LOW, H., CONOVER, T. E., and ERNSTER, L. (1961). In *Biological Structure and Function*, Vol. 2, p. 3, eds. T. W. Goodwin and O. Lindberg. New York: Academic Press.
LINDSAY, S., and CHAIKOFF, I. L. (1963). In *Atherosclerosis and Its Origin*, p. 350, eds. M. Sandler and G. H. Bourne. New York: Academic Press.
LINDSAY, S., and CHAIKOFF, I. L. (1966). *J. Atheroscler. Res.*, **6**, 36.
LINDSAY, S., NICHOLS, C. W., Jr., and CHAIKOFF, I. L. (1955). *Arch. Path.*, **59**, 173.
LITWACK, G., and KRITCHEVSKY, D. (1964). *Actions of Hormones on Molecular Processes*. New York: J. Wiley & Sons.
LOEVEN, W. A. (1963a). *Acta physiol. pharmacol. neerl.*, **12**, 497.
LOEVEN, W. A. (1963b). *Internat. Rev. Connect. Tissue Res.*, **1**, 183.
LOEVEN, W. A. (1963c). *Acta physiol. pharmacol. neerl.*, **12**, 57.
LOEVEN, W. A. (1964). Paper read at the Internat. Conf. on Enzymology and Immunology, October, 24, Bordeaux.
LOEVEN, W. A. (1967). *Europ. J. Pharmacol.*, **1**, 254.
LOFLAND, H. B., Jr., MOURY, D. M., HOFFMAN, C. W., and CLARKSON, T. B. (1965). *J. Lipid Res.*, **6**, 112.
LOHMANN, K. (1934). *Biochem. Z.*, **271**, 264.
LOJDA, Z. (1958a). *Acta histochem. (Jena)*, **5**, 236.

LOJDA, Z. (1958b). *Azocoupling Reactions in Histochemical Detection of Enzymes*. (In Czech.) Prague: State Medical Publ. House.
LOJDA, Z. (1961a). *Čs. Morfol.*, **9**, 179.
LOJDA, Z. (1961b). Paper delivered at the IVth Internat. Congress of Angiology, Prague, 4–9, Sept. See *Čs. Morfologie* (1962). **10**, 46.
LOJDA, Z. (1962). In *Metabolismus Parietis Vasorum*, p. 232, eds. B. Prusík, K. Reiniš and O. Riedl. Prague: State Medical Publ. House.
LOJDA, Z. (1965a). *Folia morph. (Prague)*, **13**, 84.
LOJDA, Z. (1965b). *Int. Symp. Morphology, Histochemistry, Vascular Wall*, Part II, p. 364. Fribourg.
LOJDA, Z., and FELT, V. (1960). *Experientia (Basal)*, **16**, 514.
LOJDA, Z., and FRIČ, P. (1966a). *Giorn. dell' Arterioscler.*, **4**, 28.
LOJDA, Z., and FRIČ, P. (1966b). *J. Atheroscler. Res.*, **6**, 264.
LOJDA, Z., and ZEMPLÉNYI, T. (1958a). *Čs. Fysiol.*, **7**, 503.
LOJDA, Z., and ZEMPLÉNYI, T. (1958b). Paper delivered at the First Congr. of the Italian Society for Histochemistry, Messina, 15 Oct., 1958. See *Monit. Zool. Ital.*, **67**, Suppl. 291, 1959.
LOJDA, Z., and ZEMPLÉNYI, T. (1960). In *Modern Problems of Cardiology*, p. 261, ed. I. T. Speranskij. Moscow. (In Russian).
LOJDA, Z., and ZEMPLÉNYI, T. (1961). *J. Atheroscler. Res.*, **1**, 101.
LOOMEIJER, F. J. (1961). *J. Atheroscler. Res.*, **1**, 62.
LOOMEIJER, F. J., and OSTENDORF, J. P. (1959). *Circulat. Res.*, **7**, 466.
LOWRY, O. H., ROSEBROUGH, N. J., FAAR, A. L., and RANDALL, R. J. (1951). *J. biol. Chem.*, **193**, 265.
LOWRY, O. H., ROBERTS, N. R., WU, M., HIXON, W. S. M., and CRAWFORD, E. J. (1954). *J. biol. Chem.*, **207**, 19.
LUGINBÜHL, H., and JONES, J. E. T. (1965). In *Comparative Atherosclerosis*, p. 3, eds. J. C. Roberts, Jr., and R. Straus. New York: Harper and Row.
LYNEN, F., and WIELAND, O. (1955). In *Methods in Enzymology*, Vol. 1, p. 566, eds. S. P. Colwick and N. O. Kaplan. New York: Academic Press.
LYNEN, F., KNAPPE, J., LORCH, E., JÜTTIN, G., and RINGELMANN, E. (1959). *Angew. Chem.*, **71**, 481.
LYNEN, F., MATSUHASHI, M., NUMA, S., and SCHWEIZER, E. (1963). In *The Control of Lipid Metabolism*, p. 43, ed. J. K. Grant. New York: Academic Press.
MAGAREY, F. R. (1957). *J. Path. Bact.*, **73**, 125.
MAIER, N., and HAIMOVICI, H. (1957). *Proc. Soc. exp. Biol. (N.Y.)*, **95**, 425.
MAIER, N., and HAIMOVICI, H. (1958). *Amer. J. Physiol.*, **195**, 476.
MAIER, N., and HAIMOVICI, H. (1965a). *Proc. Soc. exp. Biol. (N.Y.)*, 118, 258.
MAIER, N., and HAIMOVICI, H. (1965b). *Circulat. Res.*, **16**, 65.
MALINOW, M. R. (1960). *Circulat. Res.*, **8**, 506.
MALINOW, M. R. (1962). In *Metabolismus Parietis Vasorum*, p. 285, eds. B. Prusík, Z. Reiniš and O. Riedl. Prague: State Medical Publ. House.
MALINOW, M. R., and MOGUILEVSKY, J. A. (1961). *J. Atheroscler. Res.*, **1**, 417.
MALINOW, M. R., HOJMAN, D., and PELLEGRINO, A. (1954). *Acta cardiol. (Brux.)*, **5**, 480.
MALINOW, M. R., FERNANDEZ, M. A., GIMENO, A. L., and BUR, G. E. (1959). *Nature (Lond.)*, **183**, 1262.
MALINOW, R. M., MOGUILEVSKY, J. A., and BUMASHNY, E. (1961). *J. Atheroscler. Res.*, **1**, 128.
MALINOW, M. R., MOGUILEVSKY, J. A., and LACUARA, J. L. (1962). *Circulat. Res.*, **10**, 624.
MALLOV, S. (1964). *Circulat. Res.*, **14**, 357.
MALMSTRÖM, B. G. (1961). In *The Enzymes*, Vol. 5, p. 471, eds. P. D. Boyer, H. Lardy and K. Myrbäck. New York: Academic Press.
MANDEL, P. (1956). *Expos. ann. Biochim. med.*, **18**, 187.
MANDEL, P. (1962). In *Metabolismus Parietis Vasorum*, p. 25, eds. B. Prusík, Z. Reiniš and O. Riedl. Prague: State Medical Publ. House.
MANDEL, P., and KEMPF, E. (1960). *C. R. Soc. Biol. (Paris)*, **144**, 791.
MANDEL, P., and KEMPF, E. (1961). *Biochem. biophys. Acta (Amst.)*, **51**, 184.
MANDEL, P., and KEMPF, E. (1963). *J. Atheroscler. Res.*, **3**, 233.
MANDEL, P., PANTESCO, V., KEMPF, E., and FONTAINE, R. (1959). *C. R. Soc. Biol. (Paris)*, **153**, 343.

REFERENCES

MANDEL, P., POIREL, G., and SIMARD-DUQUESNE, N. (1966). *J. Atheroscler. Res.*, **6**, 463.
MANN, G. V., and ANDRUS, S. B. (1956). *J. Lab. clin. Med.*, **48**, 533.
MARKERT, C. L. (1963). *Science*, **140**, 1329.
MARMORSTON, J., MAGIDSON, O., LEWIS, J. J., MEHL, J., MOORE, F. J., and BERNSTEIN, J. (1958). *New Engl. J. Med.*, **258**, 583.
MARTIN, D. B., and VAGELOS, P. R. (1962). *J. biol. Chem.*, **237**, 1787.
MASORO, E. J. (1962). *J. Lipid Res.*, **3**, 149.
MASSEY, V. (1960). *Biochim. biophys. Acta (Amst.)*, **37**, 314.
MASSEY, V. (1963). In *The Enzymes*, Vol. 7, p. 275, eds. P. D. Boyer, H. Lardy and K. Myrbäck. New York: Academic Press.
MATZKE, J. R., KIRK, J. E., and WANG, I. (1957). *J. Geront.*, **12**, 279.
MAY, P. (1955). *C. R. Acad. Sci. (Paris)*, **241**, 1347.
MCDONALD, D. A. (1960). *Blood Flow in Arteries*. London: E. Arnold.
MCGILL, H. C., Jr. (1965). In *Comparative Atherosclerosis*, p. 354, eds. J. C. Roberts, Jr., and R. Straus. New York: Harper and Row.
MCGILL, H. C., Jr. (1966). Paper read at the Conf. on Atherosclerosis, November 21–23, New York. (In press.)
MCGILL, H. C., and GEER, J. C. (1963). In *Evolution of the Atherosclerotic Plaque*, p. 65, ed. R. J. Jones. Urbana: Univ. Chicago Press.
MCGILL, H. C., Jr., STRONG, J. P., HOLMAN, R. L., and WERTHESSEN, N. T. (1960). *Circulat. Res.*, **8**, 670.
MCGILL, H. C., FRANK, M. H., and GEER, J. C. (1961). *Arch. Path.*, **71**, 96.
MCGILL, H. C., Jr., GEER, J. C., and STRONG, J. P. (1963). In *Atherosclerosis and Its Origin*, p. 39, eds. M. Sandler and J. C. Bourne. New York: Academic Press.
MCLEAN, A. E. M. (1960). *Nature (Lond.)*, **185**, 191.
MCMILLAN, G. C. (1965). In *Comparative Atherosclerosis*, p. 360, eds. J. C. Roberts, Jr. and R. Straus, New York: Harper and Row.
MCMILLAN, G. C., KLATZO, I., and DUFF, G. L. (1954). *Lab. Invest.*, **3**, 451.
MEISTER, A. J. (1950). *J. biol. Chem.*, **199**, 373.
MENDEL, B., and RUDNEY, H. (1943). *Biochem. J.*, **37**, 59.
MENTL, S. (1966). Paper read at the meeting of the Czechoslovak Cardiol. Society, Brno, November, 1966.
MEYER, K. (1958). *Fed. Proc.*, **17**, 1075.
MICHAL, G. and BERGMEYER, H. V. (1963). In *Methods of Enzymatic Analysis*, p. 512, ed. H. U. Bergmeyer. New York: Academic Press.
MIDDLETON, C. C., CLARKSON, T. B., LOFLAND, H. D., and PRICHARD, R. W. (1964). *Arch. Path.*, **78**, 16.
MILLER, B. F., AIBA, T., KEYES, F. P., KURRERI, P. W., and BRANWOOD, A. W. (1966). *J. Atheroscler. Res.*, **6**, 352.
MILLER, O. N., HAMILTON, J. G., and GOLDSMITH, G. A. (1958). *Circulation*, **18**, 489.
MILLER, O. N., HAMILTON, J. G., and GOLDSMITH, G. A. (1962). *Amer. J. Clin. Nutr.*, **10**, 285.
MISTRY, S. P., and DAKSHINAMURTI (1964). *Vitam. and Horm.*, **22**, 1.
MITCHELL, J. R. A., and SCHWARTZ, C. J. (1965). *Arterial Disease*. Oxford: Blackwell.
MITCHELL, J. R. A., SCHWARTZ, C. J., and ZINGER, A. (1964). *Brit. med. J.*, **I**, 205.
MOMMAERTS, W. F. H. M., and GREEN, I. (1954). *J. biol. Chem.*, **208**, 833.
MORRISON, J. F. (1954). *Biochem. J.*, **58**, 685.
MORRISON, J. F., and PETERS, R. A. (1954). *Biochem. J.*, **58**, 473.
MOSES, C. (1954). *Circulat. Res.*, **2**, 243.
MOSKOWITZ, M. S. (1967). In *Lipid Metabolism in Tissue Culture Cells*, p. 49, eds. G. H. Rothblat and D. Kritchevsky. Philadelphia: Wistar Inst. Press.
MOSKOWITZ, M. S., and MOSKOWITZ, A. A. (1965). *Science*, **149**, 72.
MOVAT, H. Z., MORE, R. H., and HAUST, M. D. (1958). *Amer. J. Path.*, **34**, 1023.
MRHOVÁ, O., and ZEMPLÉNYI, T. (1965). *Quart. J. exp. Physiol.*, **50**, 289.
MRHOVÁ, O., ZEMPLÉNYI, T., and LOJDA, Z. (1963*a*). *Quart. J. exp. Physiol.*, **48**, 61.
MRHOVÁ, O., ZEMPLÉNYI, T., and LOJDA, Z. (1963*b*). *J. Atherocler. Res.*, **3**, 44.
MUKHERJEE, S., KUNITAKE, G., and ALFIN-SLATER, R. B. (1958). *J. biol. Chem.*, **230**, 9.
MÜLLER, E., and NEUMANN, W. (1959). *Frankfurt. Z. Path.*, **70**, 174.
MUNCH-PETERSEN, A. (1955). *Acta chem. scand.*, **9**, 1523.

MUNRO, A. F., and RIFKIND, B. M. (1964). In *Biological Aspects of Occlusive Vascular Disease*, p. 50, eds. D. G. Chalmers and G. A. Gresham. London: Cambridge Univ. Press.
MUNRO, A. F., RIFKIND, B. M., LIEBENSCHEUTZ, H. J., CAMPBELL, R. S. F., and HOWARD, B. R. (1961). *J. Atheroscler. Res.*, **1**, 296.
MUSILOVÁ, H., JELÍNEK, J., and ALBRECHT, J. (1966). *Physiol. bohemoslov.* **15**, 525.
MUSTARD, J. F. (1966). Paper read at the Conference on Atherosclerosis: Recent Advances, Nov. 21, New York. (In press.)
MUSTARD, J. F., MURPHY, E. A., ROWSELL, H. C., and DOWNIE, H. G. (1964). *J. Atheroscler. Res.*, **4**, 1.
MYASNIKOV, A. L. (1960). *Ateroskleroz.* Moskva: Medgiz.
MYERS, D. K. (1960). In *The Enzymes*, Vol. 4, p. 475, eds. P. D. Boyer, H. Lardy and K. Myrbäck. New York: Academic Press.
NACHLAS, M. M., and SELIGMAN, A. M. (1949a). *J. nat. Cancer Inst.*, **9**, 415.
NACHLAS, M. M., and SELIGMAN, A. M. (1949b). *Anat. Rec.*, **105**, 677.
NACHLAS, M. M., CRAWFORD, D. T., and SELIGMAN, A. M. (1957a). *J. Histochem. Cytochem.*, **5**, 264.
NACHLAS, M. M., TSOU, K. C., DE SOUZA, E., CHENG, C. S., and SELIGMAN, A. M. (1957b). *J. Histochem. Cytochem.*, **5**, 420.
NACHLAS, M. M., CRAWFORD, D. T., GOLDSTEIN, T. P., and SELIGMAN, A. M. (1958). *J. Histochem. Cytochem.*, **6**, 445.
NACHLAS, M. M., MARGULIES, S. I., and SELIGMAN, A. M. (1960a). *J. biol. Chem.*, **235**, 2739.
NACHLAS, M. M., MARGUILES, S. I., GOLDBERG, J. D., and SELIGMAN, A. M. (1960b). *Analyt. Biochem.*, **1**, 317.
NAIMI, S., GOLDSTEIN, R., NOTHMAN, M. M., WILGRAM, G. F., and PROGER, S. (1962). *J. clin. Invest.*, **41**, 1708.
NARPOZZI, A. (1957). *Boll. Soc. ital. Biol. sper.*, **33**, 467.
NAUGHTON, M. A., and SANGER, F. (1961). *Biochem. J.*, **78**, 156.
NAVARATNAM, V., and PALKAMA, A. (1965). *Acta anat. (Basel)*, **60**, 445.
NEILANDS, J. B. (1952). *J. biol. Chem.*, **199**, 373.
NEILANDS, J. B., and STUMPF, P. K. (1958). *Outlines of Enzyme Chemistry*, 2nd edit. New York: J. Wiley & Sons.
NELSON, B. D. (1966). *Proc. Soc. exp. Biol. (N.Y.)*, **121**, 998.
NEWMAN, W., FEIGIN, I., WOLF, A., and KABAT, E. A. (1950). *Amer. J. Path.*, **26**, 257.
NEWSHOLME, E. A., RANDLE, P. J., and MANCHESTER, K. L. (1962). *Nature (Lond.)*, **193**, 270.
NIEFT, M. L. (1949). *J. biol. Chem.*, **177**, 151.
NIEWIAROWSKI, I., POPLAWSKI, A., and PROKOPOWICZ, J. (1963). *Thrombs. Diathes. haemorrh. (Stuttg.)*, **9**, 126.
NIKKILÄ, E. (1953). *Scand. J. Lab. clin. Invest.*, **5**, Suppl. 8.
NISSELBAUM, J. S., and BODANSKY, O. (1959). *J. biol. Chem.*, **234**, 3276.
NISSELBAUM, J. S., and BODANSKY, O. (1963). *Ann. N.Y. Acad. Sci.*, **103**, 930.
NOLTMANN, E. A., and KUBY, S. A. (1963). In *The Enzymes*, Vol. 7, p. 223, eds. P. D. Boyer, H. Lardy and K. Myrbäck. New York: Academic Press.
NOVELLI, G. D. (1957). In *Methods in Enzymology*, Vol. 3, p. 913, eds. S. P. Colowick and N. O. Kaplan. New York: Academic Press.
NOVIKOFF, A. B. (1959). *J. Histochem. Cytochem.*, **7**, 301.
NOVIKOFF, A. B. (1960a). In *Developing Cell Systems and Their Control*, p. 167, ed. D. Rudnick. New York: Ronald Press.
NOVIKOFF, A. B. (1960b). *J. Histochem. Cytochem.*, **8**, 345.
NOVIKOFF, A. B. (1961a). In *Analytical Cytology*, p. 69, ed. R. E. Mellors. New York: McGraw-Hill.
NOVIKOFF, A. B. (1961b). In *The Cell*, Vol. II, p. 423, eds. J. Brachet and A. E. Mirsky. New York: Academic Press.
NUMA, S., MATSUHASHI, M., and LYNEN, F. (1961). *Biochem. Z.*, **334**, 203.
NUZUM, Fr., SEEGAL, B., GARLAND, R., and OSBORN, E. M. (1926). *Arch. intern. Med.*, **37**, 733.
O'BRIEN, J. R. (1963). *J. Atheroscler. Res.*, **3**, 262.
OCHOA, S. (1955). In *Methods in Enzymology*, Vol. 1, p. 685, eds. S. P. Colowick and N. O. Kaplan. New York: Academic Press.
ÖDEGAARD, A. E., SKÅLHEGG, B. A., and HELLEM, A. J. (1964). *Thrombos. Diathes. haemorrh. (Stuttg.)*, **11**, 317.

REFERENCES

OESTER, Y. T., DAVIS, O. F., and FRIEDMAN, B. (1955). *Amer. J. Path.*, **31**, 717.
OLIVER, M. F., and BOYD, G. S. (1959). *Lancet*, 2, 690.
OLSON, J. A., and ANFINSEN, C. B. (1952). *J. biol. Chem.*, **197**, 67.
PADYKULA, H. A., and HERMAN, E. (1955a). *J. Histochem. Cytochem.*, 3, 161.
PAGE, I. H., and BROWN, H. B. (1952). *Circulation*, 6, 681.
PAGE, I. H., GREEN, J. G., and LAZARINI-ROBERTSON, A., Jr. (1966). *Ann. intern. Med.*, **64**, 189.
PALAY, S. L., and KARLIN, L. J. (1959). *J. biophys. biochem. Cytol.*, **5**, 373.
PAOLETTI, P., TESSARI, L., and VERTUA, R. (1959). *Ric. Sci.*, **29**, 2382.
PARISH, W. E. (1964). In *Biological Aspects of Occlusive Vascular Disease*, p. 84, eds. D. G. Chalmers and G. A. Gresham. London: Cambridge Univ. Press.
PARKER, F. (1960). *Amer. J. Path.*, 36, 19.
PARTRIDGE, S. M., DAVIS, H. F., and ADAIR, G. S. (1955). *Biochem. J.*, **61**, 11.
PATELSKI, J. (1964). *Esteraza Cholesterolowa tetnicy głównėj*. Varšava: Państwowy zaklad wydawnictw lekarskich.
PATELSKI, J., and SZENDZIKOWSKI, St. (1961). Paper delivered at the IVth Internat. Congress of Angiology, Prague, Sept. 4–9 (In press.)
PATELSKI, J., and SZENZIKOWSKI, S. (1962a). In *Metabolismus Parietis Vasorum*, p. 55, eds. B. Prusík, Z. Reiniš and O. Riedl. Prague: State Medical Publ. House.
PATELSKI, J., and SZENDIKOWSKI, S. (1962b). *Bull. Soc. Amis Sci. Poznan, Ser. C.*, **11**, 37.
PATELSKI, J., ROŻYNKOVA, D., and PALUSZAK, J. (1962). *Acta med. pol.*, 3, 417.
PATELSKI, J., WALIGÓRA, Z., SZULE, S., BOWYER, D. E., HOWARD, A. N., and GRESHAM, G. A. (1966). Paper read to the International Symposium on Recent Advances in Atherosclerosis, Athens, May 30–June 2. (In press.)
PATERSON, J. C., and MILLS, J. (1958). *Arch. Path.*, **66**, 335.
PATERSON, J. C., MILLS, J., and MOFFATT, M. (1957). *Arch. Path.*, **64**, 129.
PÁV, J., and WENKEOVÁ, J. (1960). *Nature (Lond.)*, **185**, 926.
PÁV, J., WENKEOVÁ, J., and KUHN, E. (1961). *Čas. Lék. čes.*, **100**, 273.
PEARSE, A. G. E. (1960). *Histochemistry, Theoretical and Applied*. London: J. and A. Churchill.
PETROFF, J. P. (1922). *Beitr. Path. Anat.*, **71**, 115.
PFLEIDERER, E. (1932). *Arch. Path. Anat.*, **284**, 154.
PFLEIDERER, G., and JECKEL, D. (1957). *Biochem. Z.*, **329**, 370.
PHILLIPS, T. H., BURCH, G. E., and HIBBS, R. G. (1960). *Circulat. Res.*, 8, 692.
PICK, R., and KATZ, L. N. (1965). In *Comparative Atherosclerosis*, p. 77, eds. J. C. Roberts, Jr., and R. Straus. New York: Harper and Row.
PILGERAM, L. O., and GREENBERG, D. M. (1954). *Science*, **120**, 760.
PIZER, L. I. (1962). In *The Enzymes*, Vol. 6, p. 179, eds. P. D. Boyer, H. Lardy and K. Myrbäck. New York: Academic Press.
PLAGEMANN, P. G. W., GREGORY, K. F., and WRÓBLEWSKI, F. (1960a). *J. biol. Chem.*, **235**, 2282.
PLAGEMANN, P. G. W., GREGORY, K. F., and WRÓBLEWSKI, F. (1960). *J. biol. Chem.*, **235**, 2288.
PLAUT, G. W. E. (1963). In *The Enzymes*, p. 7, eds. P. Boyer, H. Lardy and K. Myrbäck. New York: Academic Press.
POGELL, B. M., and GRYDER, R. M. (1957). *J. biol. Chem.*, **228**, 701.
POLLAK, O. J. (1947). *Arch. Path.*, **43**, 387.
POLLAK, O. J. (1956). *Circulation*, **14**, 503.
POLLAK, O. J. (1957). *Circulation*, **16**, 1084.
POMERANCE, A. (1958). *J. Path. Bact.*, **76**, 55.
POOLE, J. C. F., and FRENCH, J. E. (1961). *J. Atheroscler. Res.*, 1, 251.
POSTNOV, U. V. (1959). *Sci. Progr. Med. Inst. Riasansk*, 7, 26.
POSTNOV, U. V. (1965). Paper read at the annual Scientific session of the Institute of Therapy of the Medical Academy, January 28, Moscow.
POUCHLEV, A., YOUROUKOVA, Z., and KIPROV, D. (1966). *J. Atheroscler. Res.*, **6**, 342.
PRESS, E. M., PORTER, R. R., and CEBRA, J. (1960). *Biochem. J.*, **74**, 501.
PRESSMAN, B. C., and LARDY, H. A. (1956). *Biochem. biophys. Acta (Amst.)*, **21**, 458.
PRIOR, J. T., and HARTMANN, W. H. (1956). *Amer. J. Path.*, **32**, 417.
PRIOR, J. T., KURTZ, D. M., and ZIEGLER, D. D. (1961). *Arch. Path.*, **71**, 672.
RAABO, E. (1963). *Scand. J. clin. Lab. Invest.*, **15**, 405.
RACKER, E. (1952). *Biochim. biophys. Acta (Amst.)*, **4**, 211.

RACKER, E. (1955). *J. biol. Chem.*, **217**, 855.
RANDLE, P. J., GARLAND, P. B., HALES, C. N., and NEWSHOLME, E. A. (1963). *Lancet*, **1**, 785.
RAPAPORT, S. M. (1964). *Medizinische Biochemie*, 2nd edit. Berlin: Volk und Gesundheit.
REES, K. R. (1964). In *Ciba Foundation Symposium on Cellular Injury*, p. 53, eds. A. V. S. de Reuck and J. Knight. London: Churchill.
REES, K. R., SINHA, K. P., and SPECTOR, W. G. (1961). *J. Path. Bact.*, **81**, 107.
REINIŠ, Z., PUCHMAYER, V., PROCHÁZKA, R., DUBEN, Z., VANĚČEK, R., and KUBÁT, K. (1961*a*). *Čas Lék. čes.*, **100**, 669.
REINIŠ, Z., PUCHMAYER, V., VANĚČEK, R., KUBÁT K., and DUBEN, Z. (1961*b*) *Cor et vasa (Praha)*, **3**, 178.
REIS, J. L. (1951). *Biochem. J.*, **48**, 548.
REITMAN, S., and FRANKEL, S. (1957). *Amer. J. clin. Path.*, **28**, 56.
RICHTER, G. (1962). *Biochim. biophys. Acta (Amst.)*, **61**, 144.
RICHTERICH, R. (1952*a*). *Acta anat. (Basel)*, **14**, 263.
RICHTERICH, R. (1952*b*). *Acta anat. (Basel)*, **14**, 342.
RIFKIND, B. M., and MUNRO, A. F. (1963). *J. Atheroscler. Res.*, **3**, 268.
RIGG, K. J., FINLAYSON, R., SYMONS, C., HILL, K. R., and FIENNES, R. N. T. W. (1960). *Proc. Zool. Soc. Lond.*, **135**, 157.
RILEY, T. F. (1959). *The Mast Cells*. Edinburgh: E. and S. Livingstone.
RILEY, J. F., and WEST, G. B. (1953). *J. Physiol. (Lond.)*, **120**, 528.
RILEY, M., and PARDEE, A. B. (1962). *Amer. Rev. Microbiol.*, **16**, 1.
RINDFLEISCH, E. (1872). *A Manual of Pathological Histology*. Quoted by Mitchell and Schwartz (1965). In *Arterial Disease*. Oxford: Blackwell.
RINEHART, J. F., and GREENBERG, L. D. (1949). *Amer. J. Path.*, **25**, 481.
RINGLER, R. L., MINAKAMI, S., and SINGER, T. P. (1960). *Biochem. biophys. Res. Commun.* **3**, 417.
RITZ, E., and KIRK, J. E. (1967). *Experientia (Basel)*, **23**, 16.
RIVIN, A. V., and DIMITROFF, S. P., (1954). *Circulation*, **9**, 533.
ROBERT, L., and SAMUEL, P. (1957). *Experientia (Basal)*, **13**, 167.
ROBERTS, J. C., MOSES, C., and WILKINS, R. H. (1959). *Circulation*, **20**, 511.
ROBERTSON, A. L., Jr. (1965). *Cleveland Clin. Quart.*, **32**, 99.
ROBERTSON, A. L., Jr. (1967). In *Lipid Metabolism in Tissue Culture Cells*, p. 115, eds. G. H. Rothblat and D. Kritchevsky. Philadelphia: Wistar Inst. Press.
ROBERTSON, W. B., GREER, J. C., STRONG, J. P., and McGILL, H. C., Jr. (1963). *J. exp. mol. Pathol.*, Suppl. **1**, 28.
ROBINSON, D. S. (1960). *J. Lipid Res.*, **1**, 332.
ROBINSON, D. S. (1963). *Adv. Lipid Res.*, **1**, 134.
ROBINSON, D. S., and HARRIS, P. M. (1959). *Quart. J. exp. Physiol.*, **44**, 80.
ROBINSON, R. W., HIGANO, N., and COHEN, W. D. (1959). *Arch. intern. Med.*, **104**, 908.
RODBARD, S. (1959). *Ann. intern. Med.*, **50**, 1339.
RODBARD, S. (1962). *Circulat. Res.*, **11**, 664.
ROE, J. H. (1934). *J. biol. Chem.*, **107**, 15.
ROMANUL, F. C. A., and BANNISTER, R. G. (1962). *J. cell. Biol.*, **15**, 73.
ROSALKI, S. B., and WILKINSON, J. H. (1960). *Nature (Lond.)*, **188**, 1110.
ROSENBERG, P., and DETTBARN, W. D., (1965). *Life Sciences*, **4**, 567–572.
ROTHBLAT, G. H., HARTZELL, R., Jr., MIALHE, H., and KRITCHEVSKY, D. (1967). In *Lipid Metabolism in Tissue Culture Cells*, p. 129, eds. G. H. Rothblat and D. Kritchevsky. Philadelphia: Wistar Inst. Press.
ROUGHTON, F. J. W., and BOOTH, V. H. (1938). *Biochem. J.*, **32**, 2049.
ROUGHTON, F. J. W., and BOOTH, V. H. (1946). *Biochem. J.*, **40**, 309.
ROWEN, J. W., and KORNBERG, A. (1951). *J. biol. Chem.*, **193**, 497.
ROWEN, R. (1964). *Biochim. biophys. Acta (Amst.)*, **84**, 761.
ROWEN, R., and MARTIN, J. (1963). *Biochim. biophys. Acta (Amst.)*, **70**, 396.
ROWLEY, D. A., and BENDITT, E. P. (1956). *J. exp. Med.*, **103**, 399.
RUBINSTEIN, L. J. (1966). Paper read at the Conference on Atherosclerosis, November 21–23, New York. (In press.)
RUDZIT, K. (1959). *Heparinocytes (basophil leucocytes and mast cells) in the Clinique and experiment*. Acad. of Sciences, Riga. (In Russian.)
RUSS, E. M., EDER, H. A., and BARR, D. P. (1951). *Amer. J. Med.*, **11**, 468.

REFERENCES

RYU, S. (1959). *Jap. Circulat. J. (En.)*, 23, 1047.
SACHAR, L. A., WINTER, K. K., SICHER, N., and FRANKEL, S. (1955). *Pro. Soc. exp. Biol. (N.Y.)*, 90, 323.
SACHS, N. (1954). *S. Afr. J. med. Sci.*, 19, 113.
SANADI, D. R. (1963). In *The Enzymes*, Vol. 7, p. 307, eds. P. D. Boyer, H. Lardy and K. Myrback. New York: Academic Press.
SANDLER, M. (1960). *Anat. Rec.*, 136, 271.
SANDLER, M., and BOURNE, G. H. (1960a). *Circulat. Res.*, 8, 1274.
SANDLER, M., and BOURNE, G. H. (1960b). *J. Geront.*, 15, 32.
SANDLER, M., and BOURNE, G. H. (1962a). *J. Amer. med. Ass.*, 179, 43.
SANDLER, M., and BOURNE, G. H. (1962b). *Nature (Lond.)*, 193, 4821.
SANDLER, M., and BOURNE, G. H., eds. (1963). In *Atherosclerosis and Its Origin*, p. 515. New York: Academic Press.
SANWALD, R., and KIRK, J. E. (1965a). *J. Atheroscler. Res.*, 5, 497.
SANWALD, R., and KIRK, J. E. (1965b). *Proc. Soc. exp. Biol. (N.Y.)*, 118, 1088.
SANWALD, R., and KIRK, J. E. (1965c). *Klin. Wschr.*, 43, 940.
SANWALD, R., and KIRK, J. E. (1966). *Nature (Lond.)*, 209, 912.
SAPPINGTON, S. W., and COOK, H. S. (1936). *Amer. J. med. Sci.*, 192, 822.
SARDA, L., and DESNUELLE, P. (1958). *Biochim. biophys. Acta (Amst.)*, 30, 513.
SAUDEK, C. C., ADAMS, C. W. M., and BAYLISS, O. B. (1966). *J. Path. Bact*, 92, 265.
SAXL, H. (1957). *Gerontologia (Basel)*, 1, 142.
SBARRA, A. J., SHIRLEY, W., GILFILLAN, R. F., and BARDAVIL, W. A. (1961). 30, 162.
SCARPELLI, D. G., HESS, R., and PEARSE, A. G. E. (1958). *J. biophys. biochem. Cytol.*, 4, 747.
SCHAUS, R., KIRK, J. E., and LAURSEN, T. J. S. (1955). *J. Geront.*, 10, 170.
SCHLIEF, H., SCHMIDT, C. G., and HILLENBRAND, H. J. (1954). *Z. ges. exp. Med.*, 122, 497.
SCHMIDT, C. G., and HILLENBRAND, H. J. (1953). *Z. ges. exp. Med.*, 120, 685.
SCHNEIDER, W. C. (1945). *J. biol. Chem.*, 161, 293.
SCHÖNHEIMER, R. (1924). *Biochem. Z.*, 147, 258.
SCHÖNHEIMER, R. (1926). *Z. Physiol. Chem.*, 160, 61.
SCHWARTZ, C. J., and MITCHELL, J. R. A. (1962a). *Circulat. Res.*, 11, 63.
SCHWARTZ, C. J., and MITCHELL, J. R. A. (1962b). *Postgrad. med. J.*, 38, 25.
SCHWERT, G. W., and WINER, A. D. (1963). In *The Enzymes*, Vol. 7, p. 127, eds. P. D. Boyer, H. Lardy and K. Myrbäck. New York: Academic Press.
SELIGMAN, A. M. (1963). In *Methods in Enzymology*, Vol. 6, p. 889, eds. S. P. Colowick and N. O. Kaplan. New York: Academic Press.
SELIGMAN, A. M., NACHLAS, M. M., and MOLLOMO, M. C. (1949). *Amer. J. Physiol.*, 159, 337.
ŠERÁK, L., KRČÍLEK, A., JANOUŠEK, V. (1962). In *Metabolismus Parietis Vasorum*, p. 483, eds. B. Prusík, Z. Reiniš and O. Riedl. Prague: State Medical Publ. House.
ŠEVELA, M., and TOVÁREK, J. (1964). *Čas. Lék. čes.*, 103, 419.
SHAH, S. N., LOSSOW, J., and CHAIKOFF, I. L. (1964). *Biochem. biophys. Acta (Amst.)*, 84, 176.
SHIMAMOTO, T. (1963). *J. Atheroscler. Res.*, 3, 87.
SHORE, M. L., ZILVERSMIT, D. B., and ACKERMAN, R. F. (1955). *Amer. J. Physiol.*, 181, 527.
SHORE, P. A., and ALPERS, H. S. (1963). *Fed. Proc.*, 22, 504.
SHYAMALA, G., NICHOLS, C. W., Jr., and CHAIKOFF, I. L. (1966). *Life Science*, 5, 1191.
SILLER, G. W. (1965). In *Comparative Atherosclerosis*, p. 66, eds. J. C. Roberts and R. Straus. New York: Harper and Row.
SINGER, T. P., and KEARNEY, E. B. (1954). *Biochem. biophys. Acta (Amst.)*, 15, 151.
SINGER, T. P., and KEARNEY, E. G. (1963). In *The Enzymes*, Vol. 7, p. 383, eds. P. D. Boyer, H. Lardy and K. Myrbäck. New York: Academic Press.
SINGER, T. P., and LUSTY, C. J. (1963). In *Methods of Enzymatic Analysis*, p. 346, ed. E. H. Bergmeyer. New York: Academic Press.
SJÖVALL, H., and WIHMAN, G. (1934). *Acta path. microbiol. scand.*, Suppl. 20, 1934.
SKOLD, B. H., and GETTY, R. (1961). *J. Amer. vet. med. Ass.*, 139, 655.
SKOŘEPA, J. (1965). *Sborn. lek.*, 67, 285. (In Czech.)
SKOU, J. C. (1964). *Progr. Biophys. and molec. Biol.*, 14, 131.
SLATER, T. F., GREENBAUM, A. L., and WANG, D. Y. (1963). In *Ciba Foundation Symposium on Lysosomes*, p. 311. London: Churchill.
SLEIN, M. W. (1955). In *Methods in Enzymology*, Vol. 1, p. 304, eds. S. P. Colowick and N. O. Kaplan. New York: Academic Press.

SMITH, D. E. (1958). *Amer. J. Physiol.*, **193**, 573.
SMITH, D. E. (1963). *Ann. N.Y. Acad. Sci.*, **103**, 40.
SMITH, E. B. (1965). *J. Atheroscler. Res.*, **5**, 224.
SMITH, E. B., and MILLS, G. T. (1953). *Biochem. J.*, **54**, 164.
SMITH, E. L., and HILL, R. L. (1960). In *The Enzymes*, Vol. 3, p. 37, eds. P. D. Boyer, H. Lardy and K. Myrbäck. New York: Academic Press.
SMITH, T. C., KOCHAKIAN, C. D., and FONDALE, E. (1953). *Amer. J. Physiol.*, **174**, 247.
SMUCKLER, E. A., ISERI, O. A., and BENDITT, E. P. (1961). *Biochem. biophys. Res. Commun.*, **5**, 270.
SMUCKLER, E. A., ISERI, O. A., and BENDITT, E. P. (1962). *J. exp. Med.*, **116**, 55.
SOBEL, A. E. (1955). *Ann. N.Y. Acad. Sci.*, **60**, 713.
SØRENSEN, L. B., and KIRK, J. E. (1956). *J. Geront.*, **11**, 28.
SPAIN, D. M., and ARISTIZABAL, N. (1962). *Arch. Path.*, **73**, 82.
SPERRY, W. M., and STOYANOFF, V. A. (1938). *J. biol. Chem.*, **126**, 77.
SPINKS, A. (1952). *J. Physiol. (Lond.)*, **117**, 35P.
SPINKS, A., and BURN, J. H. (1952). *Brit. J. Pharmacol.*, **7**, 93.
SSOLOWJEW, A. (1929). *Z. ges. exp. Med.*, **69**, 94.
STADTMAN, E. R. (1963). *Bact. Rev.*, **27**, 170.
STAMLER, J. (1963). In *Atherosclerosis and Its Origin*, p. 231, eds. M. Sandler and G. H. Bourne. New York: Academic Press.
STARE, F. J., ANDRUS, S. B., and PORTMAN, O. W. (1963). Primates in Medical Research with Special Reference to New World Monkeys. In *Proceedings of Conference on Research with Primates* p. 59, ed. D. E. Pickering.
STAROKODAMSKY, L. M., and SSOBOLEW, L. W. (1909). *Frankfurt. Z. Path.*, **3**, 912.
STEHBENS, W. E. (1960). *Amer. J. Path.*, **36**, 289.
STEIN, A. A., and HARRIS, J. (1964). *Surgery*, **56**, 413.
STEIN, A. A., ROSENBLUM, J., and LEATHER, R. (1966). *Arch. Path.*, **81**, 548.
STEIN, Y., and STEIN, O. (1962). *J. Atheroscler. Res.*, **2**, 400.
STEIN, Y., and STEIN, O. (1966). *Biochim. biophys. Acta (Amst.)*, **116**, 95.
STEIN, Y., STEIN, O., and SHAPIRO, B. (1963). *Biochem. biophys. Acta (Amst.)*, **70**, 1963.
STEIN, Y., EISENBERG, S., and STEIN, O. (1966). Paper read at the International Symposium on Recent Advances in Atherosclerosis, Athens, May 30–June 2. (In press.)
STEINER, J. W., and BAGLIO, C. M. (1963). *Lab. Invest.*, **12**, 765.
STERN, J. R., SHAPIRO, B., STADTMAN, E. R., and OCHOA, S. (1951). *J. biol. Chem.*, **193**, 703.
STEWART, J. A., and PAPACONSTANTINOU, J. (1966). *Biochem. biophys. Acta (Amst.)*, **121**, 69.
STOLK, A. (1959a). *Naturwissenschaften*, **46**, 361.
STOLK, A. (1959b). *Naturwissenschaften*, **46**, 409.
STRAUB, F. B. (1939). *Biochem. J.*, **33**, 787.
STRAUS, R., and ROBERTS, J. C., Jr. (1965). In *Comparative Atherosclerosis*, p. 365, eds. J. C. Roberts, Jr., and R. Straus. New York: Harper and Row.
STREHLER, B. L., and TOTTER, J. R. (1952). *Arch. Biochem.*, **40**, 28.
STROMINGER, J. L., KALCKAR, H. M., AXELROD, J., and MAXWELL, E. S. (1954). *J. Amer. chem. Sec.*, **76**, 6411.
STRONG, J. P., and MCGILL, H. C., Jr. (1962). *Amer. J. Path.*, **40**, 37.
STRONG, J. P., KRITCHEVSKY, D., and MARTINEZ, R. D. (1966). *Arch. Path.*, **81**, 544.
STUCKEY, N. W. (1912). *Zbl. allg. Path. path. Anat.*, **23**, 910.
SÜDHOF, H. (1950). *Pflugers Arch. ges. Physiol.*, **252**, 551.
SUNDBERG, M. (1955). *Acta path. microbiol. scand.* Suppl. 107.
SUTHERLAND, E. W., and RALL, T. W. (1960). *Pharmacol. Rev.*, **12**, 265.
SWELL, L., and TREADWELL, C. R. (1955). *J. biol. Chem.*, **212**, 141.
SZABÓ, I. K., and CSEH, G. (1962). *Naturwissenschaften*, **49**, 260.
SZABÓ, R., TÉNYI, M., and VARGA, L. (1962). *Kisérl. Orvostud.*, **14**, 507.
SZENDZIKOWSKI, S. (1963). *Folia Histochem. Cytochem.*, **1**, Suppl. I, p. 149.
SZENDZIKOWSKI, S. (1963). *Folia Histochm. Cytochem.*, **1**, Suppl.I, p. 149.
SZENDZIKOWSKI, S., and PATELSKI, J. (1960). Paper read at the 1st Internat. Congr. of Cytochemistry, Paris, and Histochemistry, 1960; see (1962). *Ann. Histochem.*, **4**, 377.
SZENDZIKOWSKI, S., and PATELSKI, J. (1962). In *Metabolismus Parietis Vasorum*, p. 202, eds. B. Prusík, Z. Reiniš and O. Riedl. Prague: State Medical Publ. House.

SZENDZIKOWSKI, S., PATELSKI, J., and PEARSE, A. G. E. (1961–1962). *Enzymol. Biol. Clin.* (*Basel*), **1**, 125.
SZNAJDERMAN, M., and OLIVER, M. F. (1963). *Lancet*, 2, 962.
TAKEUCHI, T. (1958). *J. Histochem. Cytochem.*, **6**, 208.
TALALAY, P., HUGGINS, C., and FISHMAN, W. H. (1946). *J. biol. Chem.*, **166**, 757.
TANZER, M. L., and GILVARG, C. (1959). *J. biol. Chem.*, **234**, 3201.
TAPPEL, A. L., SAWANT, P. L., and SHIBKO, S. (1963). In *Lysosomes* (Ciba Foundation Symposium), p. 284, eds. A. V. S. de Reuck and M. P. Cameron. London: Churchill.
TAYLOR, C. B. (1954). In *Symposium on Atherosclerosis*, p. 91, ed. I. H. Page, Washington: National Academy of Sciences, National Research Council, Publ. 338.
TAYLOR, C. B. (1965). In *Comparative Atherosclerosis*, p. 215, eds. J. C. Roberts and R. Straus. New York: Harper and Row.
TAYLOR, C. B., BALDWIN, D., and HASS, G. M. (1950). *Arch. Path.*, **49**, 623.
TAYLOR, C. B., COX, G. E., HALL-TAYLOR, B. J., and NELSON, L. G. (1954). *Circulation*, **10**, 613.
TAYLOR, C. B., COX, G. E., MANALLO-ESTRELLA, P., and SOUTHWORTH, J. (1962). *Arch. Path.*, **74**, 16.
TAYLOR, C. B., TRUEHEART, R. E., and COX, G. E. (1963). *Arch. Path.*, **76**, 14.
TEITELBAUM, J. I., COOPERBERG, A. A., and KALANT, N. (1962). *Canad. med. Ass. J.*, **87**, 1001.
TEXON, M. (1963). In *Atherosclerosis and Its Origin*, p. 167, eds. M. Sandler and G. H. Bourne. New York: Academic Press.
THOMAS, W. A., and HARTROFT, W. S. (1959a). *J. Nutr.*, **69**, 4.
THOMAS, W. A., and HARTROFT, W. S. (1959b). *Circulation*, **19**, 65.
THOMAS, W. A., HARTROFT, W. S., and O'NEAL, R. M. (1959). *J. Nutr.*, **69**, 325.
THOMPSON, R. H. S. (1963). In *Metabolism and Physiological Significance of Lipids*, p. 541, eds, R. M. C. Dawson and D. N. Rhodes. New York: J. Wiley & Sons.
THOMPSON, R. H. S., and TICKNER, A. (1951). *J. Physiol. (Lond.)*, **115**, 34.
THOMPSON, R. H. S., and TICKNER, A. (1953). *J. Physiol. (Lond.)*, **121**, 623.
TISCHENDORF, F., and CURRI, S. B. (1959). *Acta histochem. (Jena)*, **8**, 158.
TOMKINS, G. M., and YIELDING, K. L. (1964). In *Actions of Hormones on Molecular Processes*, p. 209, eds. G. Litwack and D. Kritchevsky. New York: J. Wiley & Sons.
TOMKINS, G. M., YIELDING, K. L., and CURRAN, J. (1961). *Proc. nat. Acad. Sci. (Wash.)*, **47**, 270.
TRČKA, V., HORŇÁČEK, J., and VEJDĚLEK, Z. (1967). *Arzneimittelforsch.* (In press.)
TRUMP, B. F., and ERICSSON, J. L. (1965). In *The Inflammatory Process*, eds. B. W. Zweifach, L. Grant and R. T. McCluskey. New York: Academic Press.
TUROLLA, E. (1962). *G. Geront.*, **10**, 271.
UTTER, M. F. (1960). In *The Enzymes*, Vol. 2, p. 75, eds. P. D. Boyer, H. Lardy and K. Myrbäck. New York: Academic Press.
UVNÄS, B. (1958). *J. Pharm. Pharmacol.*, **10**, 1.
VANĚČEK, M., and TRČKA, V. (1959). *Čs. Fysiol.*, **8**, 462.
VASTESAEGER, M. M., and DELCOURT, R. (1961). *Nutr. et. Dieta (Basel)*, **3**, 174.
VASTESAEGER, M. M., and DELCOURT, R. (1962). *Circulation*, **26**, 841.
VEČEREK, B., and KŘIŠŤANOVÁ, D. (1967). *Vnitrni Lek.*, **13**, 60.
VEČEREK, B., ŠTĚPÁN, J., and HYNIE, J. (1967). *Collection Czechosl. Chem. Comm.* (In press.)
VEEGER, C., and MASSEY, V. (1960). *Biochim. biophys. Acta (Amst.)*, **37**, 181.
VERNON, H. M. (1913-14). *J. Physiol. (Lond.)*, **47**, 320.
VESELL, E. S. (1966a). In *Progress in Medical Genetics*, p. 128, eds. H. G. Steinberg and H. G. Bearn.
VESELL, E. S. (1966b). *Nature (Lond.)*, **210**, 421.
VESELL, E. S., and BEARN, A. G. (1957). *Proc. Soc. exp. Biol. (N.Y.)*, **94**, 96.
VESELL, E. S., and BEARN, A. G. (1961). *J. clin. Invest.*, **40**, 586.
VESELL, E. S., and BEARN, A. G. (1962). *J. gen. Physiol.*, **45**, 553.
VESELL, E. S., and POOL, P. E. (1966). *Proc. nat. Acad. Sci. (Wash.)*, **55**, 756.
VILLEE, C. A., and HAGERMAN, D. D. (1958). *J. biol. Chem.*, **233**, 583.
VIRCHOW, R. (1856). In *Gesammelte Abh. zur wiss. Mediz.* Frankfurt: Staatsdruckerei.
WACHSMUTH, E. D., and PFLEIDERER, G. (1963). *Biochem. Z.*, **336**, 545.
WACHSTEIN, M., and MEISEL, E. (1957). *Amer. J. clin. Path.*, **27**, 13.

WAGENVOORT, C. A. (1954). *Acta anat. (Basel)*, **21**, 70.
WAITE, B. M., and WAKIL, S. J. (1962). *J. biol. Chem.*, **237**, 2750.
WAKERLIN, G. E., MOSS, W. G., NEVILLE, J. B., and BOURQUE, J. E. (1951). In *Trans. 5th Congr. on Factors Regulating Blood Pressure*, p. 193, eds. B. W. Zweifach and E. Shorr. New York: Josiah Macy Jr., Found.
WAKERLIN, G. E., MOSS, W. G., and KIELY, M. S. (1957). *Circulat. Res.*, **5**, 426.
WAKIL, S. J., and GIBSON, D. M. (1960). *Biochim. biophys. Acta (Amst.)*, **41**, 122.
WAKIL, S. J., and HÜBSCHER, G. (1960). *J. biol. Chem.*, **235**, 1554.
WAKIL, S. J., TITCHENER, E. B., and GIBSON, D. M. (1958). *Biochim. biophys. Acta (Amst.)*, **29**, 225.
WALIGORA, Z. (1966). *Pozn. Towarzy. Przyjac. Nauk, Wydz. lek., Prace Kom. Med. doświad.*, **34**, 317. (In Polish.)
WANG, I., and KIRK, J. E. (1959). *J. Geront.*, **14**, 444.
WANG, I., and KIRK, J. E. (1960). *J. Geront.*, **15**, 35.
WARBURG, O., and CHRISTIAN, W. (1933). *Biochem. Z.*, **266**, 377.
WARBURG, O., and CHRISTIAN, W. (1936). *Biochem. Z.*, **286**, 81.
WARTMAN, W. B. (1933). *Amer. J. med. Sci.*, **186**, 27.
WATERS, L. L. (1954). In *Symposium on Atherosclerosis*, p. 112. Washington: Nat. Acad. Sci., Nat. Res. Counc.
WATERS, L. L. (1956). *Yale J. Biol. Med.*, **29**, 9.
WATERS, L. L. (1965). In *Comparative Atherosclerosis*, p. 196, eds. J. C. Roberts, Jr. and R. Straus. New York: Harper and Row.
WATSON, W. C. (1958). *Brit. J. exp. Path.*, **39**, 540.
WEBSTER, G. R. (1962). *Biochim. biophys. Acta (Amst.)*, **64**, 573.
WEBSTER, G. R., and ALPERN, R. J. (1964). *Biochem. J.*, **90**, 35.
WEGMANN, R., and FOUQUET, J. P. (1961). *Ann. Histochem.*, **6**, 61.
WEISS, H. S. (1959). *J. Geront.*, **14**, 19.
WEISS, H. S., and FISHER, H. (1959). *Amer. J. Physiol.*, **197**, 1219.
WERTHEIMER, H. E., and BEN-TOR, V. (1960). *Arch. Kreisl.-Forsch.*, **33**, 25.
WERTHEIMER, H. E., and BEN-TOR, V. (1961). *Circulat. Res.*, **9**, 23.
WEXLER, B. C. (1964). *J. Atheroscler. Res.*, **4**, 57.
WEXLER, B. C., and JUDD, J. T. (1966). *Nature (Lond.)*, **209**, 383.
WEXLER, B. C., and MILLER, B. F. (1958). *Science*, **127**, 590.
WEXLER, B. C., and MILLER, B. F. (1959). *Proc. Soc. exp. Biol. (N.Y.)*, **100**, 573.
WEXLER, B. C., BROWN, T. E., and MILLER, B. F. (1960). *Circulat. Res.*, **8**, 278.
WHEREAT, A. F. (1961a). *Circulation*, **9**, 571.
WHEREAT, A. F. (1961b). *Circulation*, **24**, 1070.
WHEREAT, A. F. (1964). *J. Atheroscler. Res.*, **4**, 272.
WHITE, N. K., EDWARDS, J. E., and DRY, T. J. (1950). *Circulation*, **1**, 645.
WIELAND, T., and PFLEIDERER, G. (1957). *Biochem. Z.*, **329**, 112.
WIELAND, T., and PFLEIDERER, G. (1961). *Ann. N.Y. Acad. Sci.*, **94**, 691.
WIELAND, T., and PFLEIDERER, G. (1962). *Angew. Chemie*, Internat. Ed. **1**, 169.
WIELAND, T., PFLEIDERER, G., and ORTANDERL, F. (1959). *Biochem. Z.*, **331**, 103.
WILENS, S. L. (1943). *Amer. J. Path.*, **19**, 293.
WILENS, S. L. (1947). *Arch. internat. Med.*, **79**, 129.
WILENS, S. L. (1951a). *Amer. J. Path.*, **27**, 825.
WILENS, S. L. (1951b). *Science*, **114**, 389.
WILGRAM, G. F. (1958). *Proc. Soc. exp. Biol. (N.Y.)*, **99**, 496.
WILKINS, R. H., ROBERTS, J. C., and MOSES, C. (1959). *Circulation*, **20**, 527.
WILKINSON, J. H., COOKE, K. B., ELLIOTT, B. A., and PLUMMER, D. T. (1961). *Biochem. J.*, **80**, 29P.
WILLIAMS-ASHMAN, H. G., and LIAO, S. (1964). In *Actions of Hormones on Molecular Processes*, p. 482, eds. G. Litwack and D. Kritchevsky. New York: J. Wiley & Sons.
WILLS, E. D. (1965). *Advanc. Lipid Res.*, **3**, 197.
WILSON, C., and BYROM, F. B. (1939). *Lancet*, **1**, 136.
WILSON, J. D. (1962). *J. clin. Invest.*, **41**, 153.
WINTERNITZ, M. C., THOMAS, R. M., and LE COMPTE, P. M. (1938). *The Biology of Atherosclerosis*. Springfield, Ill.: C. C. Thomas.
WINZLER, R. J., DEVOR, A. W., MEHL, J. W., and SMITH, I. M. (1948). *J. clin. Invest.*, **27**, 609.

WISSLER, R. W. (1965). In *Comparative Atherosclerosis*, p. 342, eds. J. C. Roberts, Jr. and R. Straus. New York: Harper and Row.
WISSLER, R. W., EILERT, M. L., SCHROEDER, M. A., and COHEN, L. (1954). *Arch. Path.*, **57**, 333.
WOLFFE, J. B., DIGLIO, V. A., DALE, A. D., MCGINNIS, G. E., DONNELLY, D. J., PLUNGIAN, M. B., SPROWLS, J., JAMES, F., EINHORN, C., and WERKHEISER, G. (1949). *Amer. Heart J.*, **38**, 467.
WOLINSKY, H., and GLAGOV, S. (1967). *Circulat. Res.*, **20**, 409.
WOLKOFF, K. (1924). *Virchows Arch. path. Anat.*, **252**, 208.
WOLKOVA, K. G. (1929). *Beitr. path. Anat.*, **82**, 555.
WOLKOVA, K. G. (1930). *Beitr. path. Anat.*, **85**, 386.
WOLKOVA, K. G. (1953). In *Ateroskleroz*, p. 53. Moscow: Medgiz. (In Russian.)
WOLLEMAN, M., and KOCSAR, L. (1964). *J. Atheroscler. Res.*, **4**, 367.
WORLD HEALTH ORGANIZATION. (1958). *Wld. Hlth. Org. techn. Rep. Ser.*, No. 143.
WOYDA, W. C., BERKAS, E. M., and FERGUSON, D. J. (1960). *Surg. Forum*, **11**, 174.
WRÓBLEWSKI, F., and LADUE, J. S. (1955). *Proc. Soc. exp. biol. (N.Y.)*, **90**, 210.
WUEST, J. H., Jr., DRY, T. J., and EDWARDS, J. E. (1953). *Circulation*, **7**, 801.
YAMAGIWA, K., and ADACHI, O. (1914–1916). *Verh. d. Japan path. Ges.*, **4–6**, 55. (Quoted by Fox, 1933.)
YONETANI, T. (1963). In *The Enzymes*, Vol. 8. p. 41, eds. P. D. Boyer, H. Lardy and K. Myrbäck, New York: Academic Press.
YOUNG, W., GOFMAN, J. W., TANDY, R., MALAMUD, N., and WATERS, E. (1960). *Amer. J. Cardiol.*, **6**, 294.
YUNIS, A. A., FISCHER, E. H., and KREBS, E. G. (1962). *J. biol. Chem.*, **237**, 2809.
ZÁHOŘ, Z. (1963). *Čas. Lék. čes.*, **102**, 764.
ZÁHOŘ, Z., and CZABANOVÁ, V. (1965). *J. Atheroscler. Res.*, **5**, 338.
ZEMPLÉNYI, T. (1956). Paper read at the Congr. of the Czech. Med. and Cardiological Soc., published in *Arteriosclerosis*, p. 26, Prague: State Medical Publ. House. (In Czech.)
ZEMPLÉNYI, T. (1958a). *Rev. Czech. Med.*, **4**, 189.
ZEMPLÉNYI, T. (1958b). *Čas. Lék. čes.*, **97**, 1230.
ZEMPLÉNYI, T. (1960). *Rev. Czech. Med.*, **6**, 43.
ZEMPLÉNYI, T. (1962). *J. Atheroscler. Res.*, **2**, 2.
ZEMPLÉNYI, T. (1964a). *Advanc. Lipid Res.*, **2**, 235.
ZEMPLÉNYI, T. (1964b). *G. dell'Arterioscler.*, **2**, 169.
ZEMPLÉNYI, T. (1966). *Studies of Local Metabolic Factors in the Pathogenesis of Atherosclerosis. Dissertation*. Prague: Faculty of Medicine. (In Czech.)
ZEMPLÉNYI, T., and GRAFNETTER, D. (1958a). *Brit. J. exp. Path.*, **39**, 99.
ZEMPLÉNYI, T., and GRAFNETTER, D. (1958b). *Čas. Lék. čes.*, **97**, 638.
ZEMPLÉNYI, T., and GRAFNETTER, D. (1958c). *Čs. Fysiol.*, **7**, 589.
ZEMPLÉNYI, T., and GRAFNETTER, D. (1959a). *Brit. J. exp. Path.*, **40**, 312.
ZEMPLÉNYI, T., and GRAFNETTER, D. (1959b). *Arch. int. Pharmacodyn.*, **122**, 57.
ZEMPLÉNYI, T., and GRAFNETTER, D. (1959c). *Gerontologia (Basel)*, **3**, 55.
ZEMPLÉNYI, T., and GRAFNETTER, D. (1959d). *Čas. Lék. čes.*, **98**, 97.
ZEMPLÉNYI, T., and GRAFNETTER, D. (1960a). *Acta Tertii Europaei de Cordis Scientia Conventus*, Roma, p. 1043.
ZEMPLÉNYI, T., and GRAFNETTER, D. (1960b). *Čs. Fysiol.*, **9**, 486.
ZEMPLÉNYI, T., and MRHOVÁ, O. (1962). *Čs. Fysiol.*, **11**, 226.
ZEMPLÉNYI, T., and MRHOVÁ, O. (1963). *Brit. J. exp. Path.*, **44**, 278.
ZEMPLÉNYI, T., and MRHOVÁ, O. (1964a). Paper read at Symposium on Atherosclerosis, Leipzig: Published in *Gefäszwand und Blutplasma* II, p. 207, eds. R. Emmrich and E. Perlick. Jena: G. Fischer.
ZEMPLÉNYI, T., and MRHOVÁ, O. (1964b). Paper read at the Conference on Enzymes and Immunology, Bordeaux, October 10. *Arch. Mal. Coeur* (1966) Suppl. 3, p. 145.
ZEMPLÉNYI, T., and MRHOVÁ, O. (1965a). *J. Atheroscler. Res.*, **5**, 548.
ZEMPLÉNYI, T., and MRHOVÁ, O. (1965b). Paper read at the 2nd Internat. Symposium on Drugs affecting Lipid Metabl. Milano, Sept. 9. Published in *Progr. Biochem. Pharmacol.*, **2**, 141.
ZEMPLÉNYI, T., and MRHOVÁ, O. (1966). Paper read at the Internat. Symposium on Problems and Prospectives in Medical Therapy of Atherosclerosis, Parma, Nov. 5., Italy. Published in *G. dell'Arterioscler.* (In press.)

ZEMPLÉNYI, T., LOJDA, Z., and GRAFNETTER, D. (1958). *Čs. Fysiol.*, **7**, 355.
ZEMPLÉNYI, T., LOJDA, Z., and GRAFNETTER, D. (1959*a*). *Circulat. Res.*, **7**, 286.
ZEMPLÉNYI, T., FODOR, J., and LOJDA, Z. (1959*b*). *Čs. Fysiol.*, **8**, 263.
ZEMPLÉNYI, T., FODOR, J., and LOJDA, Z. (1960*a*). *Quart. J. exp. Physiol.*, **45**, 50.
ZEMPLÉNYI, T., GRAFNETTER, D., and LOJDA, Z. (1960*b*). 5th Internat. Conf. Biochem. Probl. Lipids, Marseilles. In *Enzymes of Lipid Metabolism*, p. 203, ed. P. Desnuelle. Oxford: Pergamon Press.
ZEMPLÉNYI, T., GRAFNETTER, D., LOJDA, Z., and MRHOVÁ, O. (1962*a*). *Rev. Czech. Med.*, **8**, 124.
ZEMPLÉNYI, T., MRHOVÁ, O., LOJDA, Z., and GRAFNETTER, D. (1962*b*). In *Metabolismus Parietis Vasorum*, p. 63, eds. B. Prusík, Z. Reiniš and O. Riedl. Prague: State Medical Publ. House.
ZEMPLÉNYI, T., MRHOVÁ, O., and LOJDA, Z. (1963*a*). *J. Atheroscler. Res.*, **3**, 50.
ZEMPLÉNYI, T., KNÍŽKOVÁ, I., LOJDA, Z., and MRHOVÁ, O. (1963*b*). *Cor et vasa (Praha)*, **5**, 107.
ZEMPLÉNYI, T., LOJDA, Z., and MRHOVÁ, O. (1963*c*). In *Atherosclerosis and Its Origin*, p. 459, eds. M. Sandler and G. H. Bourne. New York: Academic Press.
ZEMPLÉNYI, T., HLADOVEC, J., and MRHOVÁ, O. (1965*a*). *J. Atheroscler. Res.*, **5**, 540.
ZEMPLÉNYI, T., MRHOVÁ, O., and GRAFNETTER, D. (1965*b*). *Bull. Soc. Roy. Zool. d'Anvers*, **37**, 55.
ZEMPLÉNYI, T., MRHOVÁ, O., URBANOVÁ, D., and LOJDA, Z. (1965)*c*. Paper read at a Symposium on Comparative Aspects of Atherosclerosis, June 1965, Antwerp: Published in *Acta Zool. Pathol. Antverp.*, No. 39, **45**, 1966.
ZEMPLÉNYI, T., MRHOVÁ, O., and GRAFNETTER, D. (1966*a*). In *Lipid Research*, eds. J. Suva, F. Musil and P. Sobotka. *Plzeňský lék. Sborn.*, Suppl., **16**, p. 123.
ZEMPLÉNYI, T., MRHOVÁ, O., URBANOVÁ, D., KRUML, J., and SOPH, A. (1966*b*). *G. dell Arterioscler.*, **4**, 12.
ZEMPLÉNYI, T., MRHOVÁ, O., URBANOVÁ, D., and KOHOUT, M. (1966*c*). Paper read at the Internat. Symposium on Recent Advances in Atherosclerosis, May 31, Athens, Greece. Published in *Progr. Biochem. Pharmacol.* (In press.)
ZEMPLÉNYI, T., MRHOVÁ, O., URBANOVÁ, D., and KOHOUT, M. (1966*d*). Paper read at the Conference on Atherosclerosis, November 21–23, New York. Published in *Ann. N.Y. Acad. Sci.* (In press.)
ZEMPLÉNYI, T., MRHOVÁ, O., and HLADOVEC, J. (1966*e*). Paper read at the Conference on Dietetic Aspects of Infancy and Ageing, September 24, Rimini, Italy. Published in *Atti del V Convegno Internazionale sugli Aspetti Dietetici*, p. 418, Roma: Societa Editrice Universo.
ZEMPLÉNYI, T., NENOV, D., VRANEŠIČ, M., MRHOVÁ, O., and URBANOVÁ, D. (1967). *Cor et vasa (Praha)*. (In press.)
ZILVERSMIT, D. B., and MCCANDLESS, E. L. (1959). *J. Lipid Res.*, **1**, 118.
ZILVERSMIT, D. B., SWEELEY, C. C., and NEWMAN, H. A. I. (1961*a*). *Circulat. Res.*, **9**, 235.
ZILVERSMIT, D. B., MCCANDLESS, E. L., JORDAN, P. H., HENLEY, W. S., and ACKERMAN, R. F. (1961*b*). *Circulation*, **23**, 370.
ZOLLINGER, H. U. (1948). *Amer. J. Path.*, **24**, 569.
ZSOLDOS, S. J., and HEINEMANN, H. O. (1964). *Amer. J. Physiol.*, **206**, 615.
ZUCKER, M. B., and BORRELLI, J. (1962). *Proc. Soc. exp. Biol. (N.Y.)*, **109**, 779.

INDEX

INDEX

Acid phosphatase, 60
   in aortic layers, 165
   assay methods, 99, 107
   and "atherogenic diet", 139
   biological significance, 127
   in bovine aortas, 62
   in calciferol-fed rats, 149
   and calcium phosphate deposits, 151
   in calf's ascending aortas v. pulmonary arteries, 194
   in chicken's ascending v. abdominal aortas, 183
   in chicken's large arteries (histochem.), 185
   distribution, 61
   in DOCA + salt hypertension, 154, 157
   in duck's ascending v. abdominal aortas, 183
   in duck's large arteries, 185
   in experimental rabbit atherosclerosis, 122, 125
   in human aortic segments (ageing effects), 178
   in human arteries, 62
   in human ascending v. abdominal aortas, 174
   in human brachial v. femoral arteries, 175
   in human large arteries (histochem.), 179
   in human pulmonary arteries v. aortas, 173
   in human atherosclerosis, 62
   and injury, 127, 151, 196
   and injury by hypertension, 157
   and interspecies differences, 89
   and lysosomal activation, 166
   lysosomal and cytoplasmic form, 62
   and lysosomal damage, 159
   in macaque's ascending v. abdominal aorta, 187
   in macaque's large arteries (histochem.), 188
   in pig's ascending v. abdominal aortas, 192
   at pig's brachiocephalic artery orifice, 192

Acid phosphatase—*cont.*
   in orchidectomized rats, 113
   in ovariectomized rats, 113
   and phagocytosis, 127
   and photosynthesis, 61
   properties, 61
   and renal artery stenosis, 157
   and semisynthetic diet, 139
   and spermatozoal metabolism, 61
   in susceptible v. resistant arteries, 196
   in thromboangiitis, 62
   and "thrombogenic diet", 139
Acid phosphomonoesterase, *see* Acid phosphatase
Acid ribonuclease, and lysosomal damage, 159
Acidosis
   and atherogenesis, 13, 166
   and vascular injury, 166, 199
Acetyl-CoA carboxylase, 82
Acetylcholinesterase, 59
   assay methods, 60
   biological significance, 59
   distribution, 60
   inhibition characteristics, 60
   specificity, 59
Aconitate hydratase, 29
   assay methods, 29
   distribution, 29
   in human arteries, 29
   in human pulmonary arteries v. aortas, 176
   in human atherosclerosis, 29
   inhibition by fluoroacetate, 142
   properties, 29
Aconitase, *see* Aconitate hydratase
Acyl-CoA synthetase, 79
Active CO_2
Active glucose, *see* UDPglucose
Active glycerol, *see* Glycerol 3-phosphate
ACTH, and arteriosclerosis in rats, 147
Adenosine diphosphate, *see* ADP
Adenylate kinase, 50
   in bovine arteries, 50
   distribution, 50
   properties, 50

Adenylpyrophosphatase, *see* ATP pyrophosphohydrolase and ATPase

ADP
as coenzyme, 80
and "factor R", 141
and injured tissue, 142
and platelet adhesiveness and aggregation, 141
release from platelets of, 142
release from red cells of, 142
substrate-linked phosphorylation of, 165

Adenosinetriphosphatase, *see* ATPase

Aerobic glycolysis, physiological significance of, 11
as survival mechanism, 11

Alanine aminotransferase, 57
assay methods, 106
optical test for, 106
in experimental rabbit atherosclerosis, 133

Alcohol dehydrogenase, 7

Aldolase
assay methods, 20
in bovine arteries, 20
in human arteries, 20
properties, 20

Aliesterases (B-Esterases), 204

Alkaline phosphatase, 60
and active transport, 126
in adventitia, 62
and androgens, 126
assay methods, 98
and "atherogenic diet", 139
biological significance of, 126
and bone formation, 61
in bovine aortas, 62
in calciferol-fed rats, 149
and calcification, 61, 126, 151
in capillaries, 151
in chicken's ascending v. abdominal aorta, 183
and collagen synthesis, 126
distribution, 61
in DOCA + salt hypertension, 157
in endothelial cells, 63
in experimental rabbit atherosclerosis, 122, 125
and fibre formation, 114
in fibrous lesions, 62
and fibrous proteins, 61, 126, 151
and ground substance, 126

Alkaline phosphatase—*cont.*
in growing tissues, 61
histochemical methods for, 107
interspecies differences, 87
and intimal vascularization, 63, 127
and oestradiol, 114
in orchidectomized rats, 113
in ovariectomized rats, 113
properties, 61
and semisynthetic diet, 139
sex differences, 110
species differences, 62
and "thrombogenic diet", 138
and tissue proliferation, 159

Alkaline phosphomonoesterase, *see* Alkaline phosphatase

Allosteric sites, and sex hormones, 116

Amino-acyl-t-RNA synthetases, 4

Aminotransferases, *see* Transaminases

Animal vessels, preparation for enzyme studies, 95

Anticodons, 4

Aorta
structure in the chicken, 183
structure in the duck, 183
structure in the rhesus macaque, 187
structure of segments, 182

Aorta, O_2 uptake, 8
in chickens, 9
and cholesterol feeding, 9
and gonadectomy, 109
and heparin, 8
and hormones, 8
interspecies differences, 9
polarographic assay for, 9
in rabbits, 8
in rats, 8
and renal hypertension, 8
and sex hormones, 109
topochemical aspects, 8
Warburg technique, 9

Aorta, RQ, 10

Apyrase, 48

Arteriosclerosis in rats, 50, 147

Artery
acidification effects, 13
glycogen content of, 17
interspecies structural differences, 121
Krebs cycle activity in, 12
metabolic problems of, 8
nourishment of, 10
oxydative phosphorylation in, 13

INDEX

Artery—*cont.*
  pentose phosphate pathway in, 13
Arylsulphatase, 64
  assay methods, 64
  biological significance, 64
  distribution, 64
  in human arteries, 64
  in human atherosclerosis, 64
Aspartate aminotransferase, 57
  assay methods, 57, 104
  in bovine aortas, 57
  in experimental rabbit atherosclerosis, 133
  optical test for, 105
  in ovariectomized rats, 113
  in "thrombogenic diet", 138
Atherocytes, and foam cells, 171
Atherophils, and smooth muscle cells, 171
Atherosclerosis
  in aorta homografts, 197
  combination theory of, 146
  differences in susceptibility, 168
  and embryological development, 168
  early stages, 195
  and elastase, 209
  and elastic tissue damage, 210
  and erect posture, 168
  filtration theory of, 153
  grading of, 108
  and haemodynamics, 168
  and hypertension, 151
  local vascular factors, 195
  and mast cells, 207
  morphological sequence of lesions, 197
  nutritional-metabolic theory of, 108
  and ovariectomy, 109
  production in rats, 147
  in pulmonary artery autografts, 197
  sex differences, 108
  and smooth muscle cells, 146
  and stilboestrol treatment, 109
  susceptibility of arterial segments, 181
Atherosclerosis, Anseriformes
  histology, 183
  occurrence, 183
Atherosclerosis, Apes, 186
Atherosclerosis, avian, 181
Atherosclerosis, bovine, 193
Atherosclerosis, comparative, 181
Atherosclerosis, coronary
  electron microscopy of, 171

Atherosclerosis, chicken
  histology, 182
  natural history, 182
Atherosclerosis, experimental
  and excess choline, 135
  grading of, 107
  and hypertension, 152
  and low choline diet, 135
  and mast depletion, 208
  and smooth muscle cells, 146
Atherosclerosis, experimental, in chicken
  distribution of lesions, 182
  morphology, 182
Atherosclerosis, experimental, in rabbit, 119
  alfalfa content of diet, 122
  complicated lesions, 121
  distribution of lesions, 121
  and endocrine glands, 121
  and intermittent lipaemia, 121
  morphology, 119
  serum cholesterol, 122
  serum glucosamine, 122
  serum mucoprotein, 122
Atherosclerosis, experimental, in rats, 147
Atherosclerosis, experimental, in rhesus macaque, 186
Atherosclerosis, human
  in brachial artery, 168
  in diaphragmatic artery, 168
  distribution of lesions, 121
  in femoral artery, 168
  in internal mammary artery, 168
  main types of lesion, 169
  natural history of, 169
  in pulmonary artery, 168
  in renal artery, 168
  in thoracic and abdominal aorta, 168
Atherosclerosis, pig
  in ascending v. abdominal aorta, 192
  distribution of lesions, 189
  natural history of, 189
Atherosclerosis, rhesus macaque
  distribution of lesions, 187
  occurrence, 186
  natural history of, 187
Atherosclerosis, subhuman primates
  natural history of, 185
  occurrence, 185
Atherosclerosis, squirrel monkey, 186

Atherosclerosis, susceptibility, to
and acid phosphatase, 176
and enzyme activities, 181
and intimal thickening, 198
and medial muscle-cell density, 197
and TCA enzymes, 176
and vascular injury, 196
Atherosclerosis, turkey, 182
Atmungsferment, 43
ATPase, 46
and age, 49
in aortic layers, 161, 165
in aortic segments, 49
assay methods, 48, 100
and "atherogenic diet", 139
in bovine arteries, 48
in calciferol-fed rats, 149
in canine arteries, 48
in chicken's ascending v. abdominal aortas, 183
distribution, 47
in DOCA + salt hypertension, 154
in duck's ascending v. abdominal aortas, 183
in experimental atherosclerosis, 49, 123, 125, 127
and fatty acids, 143
histochemical "gradient", 49
histochemical methods for, 107
in human arteries, 48
in human atherosclerosis, 48
in human brachial v. femoral arteries, 175
in human large arteries (histochem.), 175
in human pulmonary artery v. aorta, 173
interspecies differences, 175
in macaque's ascending v. abdominal aorta, 187
in macaque's large arteries (histochem.), 188
and myosin, 47
in orchidectomized rats, 113
in ovariectomized rats, 113
in pig's ascending v. abdominal aorta, 192
"preatherosclerotic" changes, 50
properties, 47
in rat aortas, 49
and sarcosomes, 47
sex differences, 110

ATPase—*cont.*
species differences, 49
and thrombogenesis, 143
and "thrombogenic diet", 139
and vasodilators, 49
ATP diphosphohydrolase, 48
ATP phosphohydrolase (*see* ATPase), 47
ATP production
during glycolysis, 21
and removal of sclerogenic lipids, 199
and vascular enzyme synthesis, 199
and vascular phospholipid synthesis, 199
ATP pyrophosphohydrolase, 47, 50
Autolysis, in experimental atherosclerosis, 69
Arginine phosphate, 45
Arylesterases, (A-Esterases), 204
Avidin, inhibition of carboxylases by, 82

Baranowski enzyme, *see* Glycerol-3-phosphate dehydrogenase
Biotin, 82
and age, 83
assay methods, 83
distribution, 82
in human arteries, 83
in human ascending v. abdominal aorta, 83, 180
in human atherosclerosis, 83
in human venous tissue, 83
metabolic significance, 82
sex differences, 83
Biotin-enzymes, 82
Blood clotting, and fatty acids, 144

Calciferol-lesions, morphology, 149
Carbonate dehydratase, properties, 70
assay methods, 71
biological significance, 70
distribution, 70
function in the artery, 71
in human aortas, 71
Carbonic anhydrase, *see* Carbonate dehydratase
Carboxybiotin-enzymes, 82
Carboxylases, 82
Carboxytransferase, 82
Cathepsins, 68
assay methods, 69
biological significance, 69
classification, 68

INDEX

Cathepsins—*cont.*
  distribution, 69
  lysosomal characteristics of, 69
  in human arteries, 69
  in human atherosclerosis, 69
Cell membrane, permeability of, 6
Cholesterol
  in aorta, 176
  assay method, 122
  overloading of artery by, 131
  sclerogenic activity, 203
  and vascular injury, 203
Cholesterolaemia
  and bile acids, 135
  sex differences, 114
  and thyroid function, 135
Cholesterolesterase, 226
  assay methods, 227
  distribution, 226
  inhibition-activation characteristics, 226
  in pig's aortas, 227
  properties, 226
Cholesterolesterase, vascular
  and age, 228
  and "atherogenic diets", 227
  in chickens, 228
  in experimental rabbit atherosclerosis, 227
  properties, 227
  sex differences, 228
  in thoracic v. abdominal aortas, 228
Cholesterol esterification
  in arteries, 228
  and lecithin, 226
  mechanisms of, 226
  in plasma, 226
  and sclerogenic activity, 228
  in tissue culture cells, 228
Cholesterol esters, selective hydrolysis, 288
Chromosomes, 3
Citrate, and malonyl-CoA pathway, 82
Cholinesterase, 59
  assay methods, 60
  and carboxylesterase, 60
  in different arteries, 60
  distribution, 60
  in DOCA + salt hypertension, 157
  in human arteries, 60
  inhibition characteristics, 60
  interference with esterolytic activity, 225

E.B.—18

Cholinesterase—*cont.*
  interference with lipolytic activity, 225
  species differences, 60
  specificity, 60
Citrate synthase, 27
  assay method, 28
  distribution, 27
  properties, 27
  in human arteries, 28
  in human atherosclerosis, 28
Citric acid cycle, *see* Krebs cycle or TCA cycle,
Clearing factor
  activity in aorta, 219
  and lipoprotein lipase, 205
Clearing reaction
  and free fatty acids, 205
  methods of study, 205
Collagen
  in large arteries, 175
  and platelet aggregation, 141
Condensing enzyme, *see* Citrate synthase
Coding system, deciphering of, 4
Codon, 4
Coenzyme A, 78
  in aortic segments, 79
  assay methods, 78
  distribution, 78
  enzymatic reactions of, 78
  free energy of hydrolysis of, 78
  in human arteries, 79
  in human atherosclerosis, 79
  in human thoracic v. abdominal aorta, 180
  mechanism of action, 78
  species differences, 79
  structure, 78
Coenzymes, 72
Cofactors, 72
  and age, 87
  classification, 72
  in decarboxylation reactions, 72
  in isomerisation reactions, 72
  in oxido-reduction reactions, 72
  in transport reactions, 72
  and vitamins, 72
Creatine kinase, 45
  and age, 46
  assay methods, 46
  distribution, 46
  in human arteries, 46
  in human atherosclerosis, 46
  properties, 46

INDEX

Creatine phosphate, 45
Cytidine nucleotides, and phospholipids, 80
   in bovine aortas, 81
Cytidine triphosphate, in aortic mid zone, 161
Cytochrome oxidase, 43
   in aortic segments, 44
   assay methods, 44
   in experimental atherosclerosis, 44
   and gonadectomy, 114
   properties, 43
   species differences, 44
   in vascular tissue, 44
Cytochrome system, 27

DEAE-Sephadex, for LDH isozyme assay, 7, 103
Dehydrogenases (*see also* individual enzymes)
   histochemical methods for, 106
   NT-PMS method for, 102
   tetrazolium methods for, 97, 102
Dehydrogenase systems, 98
   in calciferol-fed rats, 150
   NT methods for, 102
Deoxyribonucleic acid, *see* DNA
Derepression, 5
Diabetes, and liver citrate, 82
Diaphorase, 131
   assay method, 43
   in human arteries, 43
   in human atherosclerosis, 43
   (*see also* Lipoamide dehydrogenase)
Diet
   "atherogenic", 137
   Larsen's, 138
   "thrombogenic", 136
2,6 Dichlorophenol indophenol, as hydrogen acceptor, 97
DNA
   assay method, 106
   and genes, 3
   nucleotide sequence of, 3
   template, 3

Ediol, composition of 214
Ediol, activated
   preparation of, 214
Elastase
   and aortic lipolytic activity, 213
   and arterial injury, 213
   assay methods, 209

Elastase—*cont.*
   and atherosclerosis, 209
   in atherosclerosis treatment, 210
   and elastomucase, 210
   and elastoproteinase, 209
   fractionation, 209, 212
   in human aorta, 209, 213, 219
   and lipolysis, 212
   and liproprotein lipase, 209–211
   mucopolysaccharide release by, 210
   in pancreas, 210
   and production of clearing factor, 211
   properties, 209
   proteolytic action, 210
   site of production, 212
   systemic function of, 212
Elastase inhibitor, and atherosclerosis, 210
Elastin
   in large arteries, 175
   and lipoprotein, 210
Elastolipoproteinase
   relation to elastase, 211
   and experimental rabbit atherosclerosis, 212
   and arterial lipid infiltration, 213
Elastomucase
   and arterial lipid infiltration, 213
   and clearing factor, 211
   esterolytic activity of, 211
   and lipase activity, 211
   and lipoprotein lipase, 212
Electron transport particles, 42
Emboli, and pulmonary atherosclerosis, 168
Endopeptidases, 67
Endoplasmic reticulum, 4
Energy-rich compounds, and hydrolytic products, 47
Enolase
   assay methods, 22
   distribution, 22
   in human arteries, 22
   properties, 22
Enzymes
   allosteric sites, 7
   arbitrary units, 6
   assay methods, 95 (*see also* individual enzymes)
   biosynthesis of, 3
   general properties, 3
   localization, 6
   specific activity, 6
   of TCA cycle, 27

Enzyme activities (*see also* individual enzymes)
   and age, 87–89, 196
   and androgens, 115
   in bovine arteries, 193
   calculation of, 95
   in children v. adult aortas, 196
   comparison of biochemical and histochemical findings, 95, 126, 131, 151, 179, 189
   and gonadectomy, 109, 113, 115
   hormonal regulation of, 113
   in human aorta v. inferior cava, 89, 94
   in human atherosclerotic aortas, 87, 90, 91 (*see also* individual enzymes)
   in human coronary arteries, 87, 90
   in injury by calciferol feeding, 147, 151
   in injury by hypertension, 157
   interspecies differences, 87, 93
   and lipid catabolism, 166
   and lipid transport, 166
   in macaque's arteries, 185
   modifying factors, 6
   in normal rabbit aortas, 133
   pattern in aortic segments, 195
   in pig's large arteries (histochem.), 193
   in reproductive organs, 117
   sex differences, 108–118
   and smooth muscle cells, 180, 197
   standard unit of, 100
   and vascular injury, 147
   in venous tissue, 89
Enzyme nomenclature, 15
Enzyme structure, and sex hormones, 115
Enzymes, vascular
   in the chicken, 181
   in the duck, 181
   in the pig, 189
Esterases
   properties, 204
   and organophosphorus compounds, 204
Esterolytic activity, vascular
   adaptation to excess lipid, 225
   and adrenalectomy, 219
   and alloxan diabetes, 221
   and calciferol feeding, 218
   and cortisone, 219

Esterolytic activity—*cont.*
   in experimental rabbit atherosclerosis, 216
   and gonadectomy, 114, 219
   and noradrenaline, 219
   sex differences, 219
   species differences, 218
   and stress, 219
   and sympathetic nervous system, 219
Exopeptidases, 67

Factor R, and platelet adhesiveness, 141
FAD, structure of, 76
Fatty acids
   incorporation in aortic lipids, 229
   and platelet aggregation, 143
   and sclerogenic activity of lipids, 203
Fatty streaks
   in bovine aortas, 193
   characteristics, 169
   distribution, 169
   relation to age, 170
   transformation into fibrous plaques, 172, 181, 185, 197, 198
   in sequence of atherosclerotic lesions, 197
   sex differences, 108
   in subhuman primate atherosclerosis, 185
Feedback inhibition, 6
Fibrogenesis, and local hypoxia, 166
Fibrous plaques
   in bovine aortas, 193
   and haemodynamic vascular stress, 199
   in sequence of atherosclerotic lesions, 197
   sex differences, 108
Flavin-adenine dinucleotide (FAD), 76
Flavin mononucleotide (FMN), 76
Flavin nucleotides, 76
   assay methods, 76
   constituents of, 76
   distribution, 76
   in dog aortas, 77
   and fluorescence, 76
   in human aortas, 77
   oxidation-reduction reactions of, 76
Free energy change, 11, 39, 47

INDEX

Free nucleotides
   and age, 81
   assay methods, 81
   in bovine aortas, 76, 81
   and mucopolysaccharide synthesis, 82
   and protein synthesis, 82
Fructose diphosphate aldolase, *see* Aldolase
Fructokinase, 18
Fumarase
   and age, 30
   distribution, 29
   in human arteries, 30
   in human atherosclerosis, 30
   in human pulmonary arteries v. aortas, 176
   properties, 29
Fumarate hydratase, *see* Fumarase

Galactokinase, 18
Genetic code, 4
Genetic information, 3, 5
Glomset enzyme, *see* Lecithin: cholesterol fatty acid transferase
Glucokinase, 18
Glucosamine, in experimental rabbit atherosclerosis, 133
Glucose, utilization in arteries, 18
Glucose fatty acid cycle, 25
Glucosephosphate isomerase, 19
   and age, 20
   assay method, 19, 106
   in experimental rabbit atherosclerosis, 133
   distribution, 19
   in human arteries, 19
   in rabbit arteries, 20
Glucose 6-phosphate dehydrogenase, 37
   and age, 38
   assay method, 37
   distribution, 37
   in DOCA hypertension, 38
   in human arteries, 38
   in human atherosclerosis, 38
   in optical tests, 37
   properties, 37
   in rabbit aortas, 38
   in rat aortas, 38
   sex differences, 115
Glucose-1-phosphate uridylyltransferase, 25
   in rat aortas, 26

β**-Glucuronidase,** 65
   and age, 66
   assay methods, 65, 104
   in atherosclerosis-resistant arteries, 66
   biological significance, 65, 133
   distribution, 65
   in DOCA + salt hypertension, 154
   in experimental atherosclerosis, 66, 133
   histochemical method for, 107
   in human arteries, 65
   in human atherosclerosis, 66
   lysosomal character of, 65
   and lysosomal damage, 159
   in macaque's ascending v. abdominal aortas, 187
   and mucopolysaccharides, 65
   and neoplastic growth, 65
   properties, 65
   sex differences, 110
   in spontaneous rat arteriosclerosis, 66
   and tissue injury, 65, 133
Glutamate, ammonia-carrier function of, 56
Glutamate dehydrogenase, 7, 53
   and age, 54
   assay method, 54
   biological significance, 53
   distribution, 53
   hormone actions on, 54
   in human arteries, 54
   in human atherosclerosis, 54
   properties, 54
   and steroid hormones, 116
   and sex hormones, 116
Glutamine-fructose-6-phosphate aminotransferase, *see* Hexosephosphate aminotransferase
Glutathione, 77
   assay methods, 77
   and *cis-trans* isomerases, 77
   constituents of, 77
   distribution, 77
   and formaldehyde dehydrogenase, 77
   and glyoxalase system, 77
   and hormonal influences, 77
   in human arteries, 78
   in human atherosclerosis, 78
Glutathione: dehydroascorbate oxidoreductase, 77
Glutathione oxidized (GSSG), 77
Glutathione reduced (GSH), 77

INDEX

Glutathione reductase, 55
  assay method, 56
  distribution, 56
  in human arteries, 56
  in human atherosclerosis, 56
  properties, 56
  and tumours, 56
Glyceraldehydephosphate dehydrogenase, in human arteries, 26
Glycerol 3-phosphate
  and hypoxia, 41
  and phospholipid synthesis, 230
  and triglyceride synthesis, 230
Glycerol 3-phosphate cycle, 41
Glycerol-3-phosphate dehydrogenase
  in atherosclerosis, 26, 41
  and $NADH_2$ transfer, 41
  in rabbit atherosclerosis, 42, 128
Glycogen breakdown, 15
Glycogen pathway, 25
Glycogen phosphorylase, 15
  and age, 17
  assay methods, 17
  properties, 15
  in human arterial tissue, 17
  and smooth muscle atrophy, 17
Glycolysis, aerobic, 161
  in arterial tissue, 11
Glycolysis, arterial
  and age, 167
  in atherosclerotic plaques, 167
Glycolytic enzymes, in human pulmonary artery v. aorta, 176
Glyoxalase I, see Lactoyl-glutathione lyase
Glyoxalase II, see Hydroxyacylglutathione hydrolase
Glyoxalase system, 51
GOT, see Aspartate aminotransferase
GPT, see Alanine aminotransferase
Ground substance, relation to aortic oxygen tension, 166
Guanosine nucleotides
  distribution, 81
  metabolic functions of, 80
  in protein biosynthesis, 81
Guanosine phosphorylase, 58
Guanosine triphosphate (GTP) 4

Haemostasis, 141
Hb-splitting enzyme, 69
Heparin-clearing reaction, 205

Hexokinase, 17
  and age, 18
  assay methods, 18
  distribution, 18
  in human arteries, 18
  properties, 18
Hexosamine, assay method, 122
Hexosephosphate aminotransferase, 58
  assay method, 58
  in aortic segments, 58
  distribution, 58
  in human arteries, 58
  in human atherosclerosis, 58
  in human ascending v. abdominal aorta, 58, 180
  properties, 58
Histochemical methods, 106
Human vessels
  preparation for enzyme studies, 95
Hydrolases, 59
3-Hydroxyacyl-CoA dehydrogenase, 52
  assay method, 52
  distribution, 52
  in human arteries, 52, 203
  in human atherosclerosis, 53
  in human thoracic v. abdominal aorta, 53, 180
Hydroxyacylglutathione hydrolase
  in human arteries, 51
  in human atherosclerosis, 51
  properties, 51
α-Hydroxybutyrate dehydrogenase
  assay method, 167
  in atherosclerotic arteries, 167
  in human arteries, 167
  and LDH isozymes, 166
Hypertension
  and atherosclerosis, 151
  and ischaemic heart disease, 151
  and vascular permeability, 153
Hypertension, DOCA + salt, ultrastructural changes, 153
Hypoxia, arterial
  adaptive enzyme changes, 161
  and atherosclerosis, 161
  and lipid film, 161
Hypoxia, local
  and collagen production, 198
  and fibrogenic cell function, 166
  and mucopolysaccharide production, 166, 198

Inducer, 5
Induction, of enzyme synthesis, 5
Injury
  by cholesterol, 131
  connective tissue reaction to, 134
Injury, cellular
  and intracellular ATP, 158
  and leakage of enzymes, 159
  and lysosomes, 159
  and membrane calcium, 158
  mitochondrial changes in, 158
  and mitochondrial swelling, 159
  and sodium transport, 158
Injury, hepatic
  and endoplasmic reticulum, 158
  and fatty changes, 158
  and leakage of inorganic ions, 157
  and leakage of macromolecules, 157
  and permeability changes, 157
  uncoupling of oxidative
    phosphorylation by, 158
Injury, hypoxic
  in aortic mid zone, 165
  and glycolysis, 199
  and lipid film, 131
Injury, vascular
  agents causing, 145
  and atherogenesis, 145, 147, 160
  and energy production, 166, 199
  and enzymes, 145
  and experimental atherosclerosis, 145
  and experimental hypertension, 151
  and haemodynamic stress, 197
  and human atherosclerosis, 145
  and hyperlipaemia, 135
  and hypertension, 153
  inflammatory reaction to, 145
  and intimal thickening, 146
  and lateral pressure, 197
  and lipid accumulation, 145
  and mast cells, 208
  mesenchymal reaction to, 146
  morphology, 146
  and mucopolysaccharides, 146
  and oxydative phosphorylation, 199
  by platelets, 142
  and pulsatile flow, 197
  regenerative processes, 146
  repair processes, 146, 171
  and shearing strain, 197
  and smooth muscle cells, 146
  and suction pressure, 197

Injury, vascular—*cont.*
  and susceptibility to atherosclerosis, 196
  and turbulence, 197
  and wall tension, 197
Inorganic pyrophosphatase, in human arteries, 49
Inosine nucleotides, metabolic functions of, 80
Intimal thickening
  and arterial blood supply, 161
  in coronary arteries, 109
  distribution in pig arteries, 194
  electron microscopy, 171, 190
  and lipid accumulation, 171, 190
  and lipid transport, 198
  and metabolic injury, 198
  and morphology of atherosclerosis, 121, 198
  and mural thrombi, 198
  sex differences, 109
  and susceptibility to atherosclerosis, 198
  and vascular haemodynamic stress, 198
Intimomedial junction, lipid accumulation in, 171
Intracellular oxidation
  localization, 41
  and oxidative phosphorylation, 39
Ischaemic heart disease
  and ovariectomy, 109
  sex differences, 108
Isocitrate dehydrogenase, 30
  assay method, 30
  distribution, 30
  in human arteries, 31
  in human atherosclerosis, 31
  in human pulmonary arteries v. aortas, 176
  properties, 30
  sex differences, 115
Isomerases, 71
Isozymes, 7
  biological significance of, 7
  isolation of, 7
  molecular basis of, 7
  subunits of, 7
  (*see also* Lactate dehydrogenase isozymes)

Ketokinase, 18
Krebs cycle enzymes, *see* TCA enzymes

Lactate
 and atherogenesis, 166
 and hypoxic vascular injury, 199
Lactate dehydrogenase, 23
 and age, 24
 in aortic mid zone, 161
 assay methods, 23, 101
 and "atherogenic diet", 139
 biological significance of, 42
 in bovine aortas, 24
 in calciferol-fed rats, 149
 in chicken's ascending v. abdominal aortas, 183
 in chicken's large arteries (histochem.), 185
 in consecutive aortic layers, 24, 164
 distribution, 23
 in DOCA + salt hypertension, 154
 in duck's ascending v. abdominal aortas, 183
 in duck's large arteries (histochem.), 185
 in experimental rabbit atherosclerosis, 128
 in human arteries, 23
 in human large arteries (histochem.), 178
 in human ascending v. abdominal aortas, 173
 in macaque's ascending v. abdominal aortas, 187
 in macaque's large arteries (histochem.), 188
 in pig's ascending v. abdominal aortas, 192
 properties, 23
 in rabbit aortas, 24
 in rat aortas, 24
 regulation of $NADH_2/NAD$ ratio by, 165
 sex differences, 109
Lactate dehydrogenase isozymes, 8, 23, 161–165
 adaptive changes of, 165
 adsorption on DEAE-Sephadex, 164
 and age, 166
 in aorta, 162
 assay methods, 102
 and atherosclerosis, 161, 166
 and castration, 166
 in consecutive aortic layers, 165
 electrophoretic fractionation of, 102, 162, 164

Lactate dehydrogenase isozymes—*cont.*
 functional role of, 162
 inhibition by pyruvate, 162
 inhibition by sulphite, 162
 in metabolically different tissues, 163
 molecular hybrids, 163
 and oxygen tension, 163
 parent forms of, 163
 sex differences, 164
 species differences, 166
 subunits, 162, 164
 tetramers, 162
 and thyroid function, 166
 in tissues, 162
Lactoyl-glutathione lyase
 in human arteries, 51
 in human atherosclerosis, 51
 properties, 51
Lecithin
 and cholesterol esterification, 199
 conversion to lysolecithin, 227
Lecithin: cholesterol fatty acid transferase, 203
 in arterial tissue, 228, 231
 and catecholamines, 227
 in plasma, 227
 and sex hormones, 227
Leucine aminopeptidase, 67
 and age, 68
 assay method, 67
 distribution, 67
 in experimental atherosclerosis, 68
 in human arteries, 68
 in human organs, 68
 isozymes, 67
 properties, 67
 in rabbit aortas, 68
 in rat aortas, 68
Ligases, ATP as cofactor of, 72
Lipase
 in aorta, 218
 definitions of, 204
 characteristics of, 204
Lipids
 in aorta and age, 176
 and inflammatory reaction, 171
 and sex hormones, 109
 and smooth endoplasmic reticulum, 158
 species differences in biosynthesis, 79
 and vascular injury, 203
Lipoamide dehydrogenase, 42
 properties, 42
 (*see also* Diaphorase)

Lipolytic activity, vascular, 203
  adaptation to excess lipid, 225
  and ACTH, 221
  and age, 214
  assay method, 214
  and calciferol feeding, 218
  and carbon disulphide, 221
  in cholesterol-fed rats, 219
  and dietary lipids, 221
  in experimental rabbit atherosclerosis, 216, 224
  and heparin, 208
  hydrolases involved, 216
  and hypertension, 221
  inhibition-activation characteristics, 216
  inhibition by Ca^{++}, 151, 219
  inhibition by mucopolysaccharides, 151, 219, 221
  interference with cholinesterase, 225
  sex differences, 114, 219
  significance, 231
  species differences, 214
  and supply of fatty acids, 231
  and vascular defence mechanisms, 199, 225, 231

Lipoprotein lipase
  biological function, 205
  in cholesterol-fed rabbits (histochem.), 224
  distribution, 205, 224
  and elastase, 209
  and heparin, 205
  in human atherosclerosis (histochem.), 224
  histochemical method for, 224
  localization, 205
  and nutrition state, 206
  properties, 205
  and "Tween esterase", 212
  in venous tissue (histochem.), 224

Lohman reaction, 45
Lyases, 70
Lysolecithin
  acylation in arterial tissue, 230
Lysomes
  and cell injury, 62, 159
  functions, 62
  and hydrocortisone, 62
Lysomal enzymes, 61, 159

Malate dehydrogenase
  assay method, 33, 101
  in aortic mid zone, 162

Malate dehydrogenase—*cont.*
  and "atherogenic diet", 139
  in bovine aortas, 33
  in calciferol-fed rats, 149
  in calf's ascending v. abdominal aortas, 194
  in consecutive aortic layers, 165
  in chicken's ascending v. abdominal aortas, 183
  in chicken's large arteries (histochem.), 185
  distribution, 33
  in DOCA + salt hypertension, 154
  in duck's ascending v. abdominal aortas, 183
  in duck's large arteries (histochem.), 185
  in experimental rabbit atherosclerosis, 128
  in human ascending v. abdominal aortas, 173
  in human aortic segments (and age), 178
  in human arteries, 33
  in human atherosclerosis, 33
  in human brachial v. femoral arteries, 175
  in human large arteries (histochem.), 178
  in human pulmonary arteries v. aortas, 173, 176
  isozymes, 33
  in macaque's ascending v. abdominal aortas, 187
  in macaque's large arteries (histochem.), 187
  optical test for, 101
  in orchidectomized rats, 113
  in ovariectomized rats, 113
  in pig's ascending v. abdominal aortas, 192
  properties, 33
  and renal artery stenosis, 157
  and "thrombogenic diet", 138
Malate dehydrogenase (decarboxylating), *see* Malic enzyme
Malic enzyme, 33
  assay method, 34
  distribution, 34
  in human arteries, 34
  in human atherosclerosis, 34
  properties, 34
  sex differences, 110, 115

Malonyl-CoA pathway, 82
Mannosephosphate isomerase, 71
    assay method, 71
    distribution, 71
    in human arteries, 71
    in human atherosclerosis, 71
    properties, 71
Mast cells
    and atherosclerosis, 207
    and coronary thrombosis, 208
    and degenerative vascular changes, 208
    and fat feeding, 207
    functions of, 207
    and histamine-liberators, 207
    and histamine production, 207
    and hyaluronic acid production, 207
    morphology, 207
    and serotonin production, 207
    and vascular injury, 208
    and vascular repair processes, 208
Mediocalcinosis in rats
    sex differences, 114
    and excess vitamin D, 147
Messenger RNA (mRNA), 4, 115
Metamorphosis, viscous, 141
Meyerhof-Green enzyme, 41
Mitochondria
    criterion of intactness, 10, 159
    and injury, 158
    swelling-contraction cycle of, 6, 159
Mitochondrial swelling, and loss of enzymes, 158
Monoamine oxidase, 54
    assay methods, 55
    biological functions, 55
    distribution, 54
    in guinea-pig arteries, 55
    and hyperthyroidism, 55
    properties, 55
    in rabbit arteries, 55
Mucopolysaccharides, and local hypoxia, 166
    and injury, 146
Mucoproteins
    assay method, 122
    and binding of lipids, 149
    and metastatic calcification, 148
Muscular contraction, 45
Muscular dystrophy, and lysosomal enzymes, 159
Myokinase, *see* Adenylate kinase

NAD
    biosynthesis, 59, 63
    in bovine aortas, 76
    metabolic functions, 73
    in rat tissues, 73
NADH$_2$-cytochrome c reductase, 42
    and age, 43
    assay method, 43
    in human arteries, 43
    in human atherosclerosis, 43
NADH$_2$-dehydrogenase, 42
    properties, 42
NADH$_2$-tetrazolium reductase, 24
    in aortic mid zone, 161
NADI reaction, 43
NADP
    biosynthesis, 59
    metabolic functions, 73
    in rat tissues, 73
    and reductive biosynthetic processes, 73
Nicotinamide nucleotide coenzymes, 73
    assay methods, 73, 75
    in human aortas, 75
    mechanism of action, 73
    (*see also* NAD and NAPD)
Non-specific carboxylesterase
    and age, 223
    and alloxan diabetes (histochem.), 223
    in arterial tissue, 218
    assay methods, 100
    in calciferol-fed rats, 149, 151
    in experimental canine atherosclerosis (histochem.), 223
    in chicken's ascending v. abdominal aortas, 183
    in cholesterol granulomas (histochem.), 223
    and cholinesterase (histochem.), 222
    in DOCA + salt hypertension, 157, 221
    and dietary fat (histochem.), 223
    in duck's ascending v. abdominal aortas, 183
    effect of actinomycin D, 225
    in experimental atherosclerosis in rats (histochem.), 223
    in experimental chicken atherosclerosis (histochem.), 223
    in experimental rabbit atherosclerosis, 127, 223
    histochemical methods for, 107

Non-specific carboxylesterase—*cont.*
    in human ascending v. abdominal aortas, 224
    in human atherosclerosis, 224
    in human large arteries (histochem.), 179
    in human pulmonary arteries v. aortas, 173
    and hypothyroidism (histochem.), 223
    and immobilization stress (histochem.), 223
    inhibition characteristics (histochem.), 222
    and injury by hypertension, 157
    interference with cholinesterase, 218
    in macaque's ascending v. abdominal aortas, 187
    in orchidectomized rats, 110
    in ovariectomized rats, 113
    at pig's brachiocephalic artery ostia, 192
    and renal artery stenosis, 157
    and semisynthetic diet, 139
    species differences (histochem.), 222
    and "thrombogenic diet", 139
    and vascular injury, 196
Nucleoside 5'-diphosphates, as phosphate carriers, 80
Nucleoside-diphosphate kinase, 80
Nucleoside triphosphates, 46
5'-Nucleotidase, 63
    and age, 64
    assay method, 99
    in aortic mid zone, 161
    biological significance of, 63
    in bovine arteries, 64
    and calcification, 126, 151
    in calciferol-fed rats, 149
    in calf's ascending aortas v. pulmonary arteries, 194
    in consecutive aortic layers, 165
    distribution, 63
    in experimental rabbit atherosclerosis, 123, 125, 127
    in guinea-pig arteries, 64
    histochemical method for, 107
    in human arteries, 63
    in human ascending v. abdominal aortas, 174
    in human atherosclerosis, 49, 64
    in human brachial v. femoral arteries, 176

5'-Nucleotidase—*cont.*
    in human large arteries (histochem.), 179
    in human pulmonary arteries v. aortas, 173
    interspecies differences, 89
    in macaque's ascending v. abdominal aortas, 187
    in macaque's large arteries (histochem.), 189
    in orchidectomized rats, 113
    in ovariectomized rats, 113
    in pig's ascending v. abdominal aortas, 192
    properties, 63
    in rabbit arteries, 64
    and renal artery stenosis, 157
    sex differences, 110, 115

Oestradiol dehydrogenases, 115
Operators, 5
Operons, 5
Optical tests
    auxiliary reactions for, 96
    and changes of nicotinamide ring, 73
    general principles, 96
    indicator reactions for, 96
    primary reactions for, 96
Oxidative decarboxylation, 42
Oxidative phosphorylation, 39, 199
Oxidoreductases, 52

Pasteur effect, 11, 25
Pentose phosphate pathway
    biological functions, 35
    in tumours, 35
Pentosyltransferases, 58
Peripheral vascular disease, sex differences, 108
Phenazine alkylsulphates, as hydrogen acceptors, 97
Phenazine methosulphate
    as oxidation-reduction acceptor, 31
    and tetrazolium methods, 97, 98
Phosphagen, 45
Phosphate acetyltransferase, 79
Phosphofructokinase reaction, 21, 25
    in human arteries, 26
Phosphoglycerate kinase, 20
    distribution, 21
    in human ascending v. abdominal aortas, 21
    properties, 21

Phosphoglucomutase, 18
  and age, 19
  assay methods, 19
  distribution, 19
  in human arteries, 19
  properties, 18
  in rat aortas, 19
Phosphogluconate dehydrogenase, 37
  assay method, 37
  distribution, 37
  in human arteries, 38
  in rabbit aortas, 38
  in rat aortas, 38
  sex differences, 115
Phosphoglycerate kinase, 21
  assay method, 21
  in human arteries, 21
  properties, 21
  in venous tissue, 21
Phosphoglycerate phosphomutase, 21
Phosphoglyceromutase, 21
  assay methods, 22
  distribution, 21
  in human arteries, 22
  properties, 21
Phospholipids
  and lipid solubilization, 199
  lyso-derivatives of and acylation
    systems, 229
  and sclerogenic activity of lipids, 199,
    203
  synthesis and aortic lysolecithin, 230
Phosphomonoesterase I, *see* Alkaline
    phosphatase
Phosphomonoesterase II, *see* Acid
    phosphatase
Phosphopyruvate hydratase, *see* Enolase
Phosphorylase kinase, 16
Phosphorylase phosphatase, 16
Plaques
  in bovine aorta, 193
  distribution and age, 170
  distribution and blood pressure, 170
  distribution and sex, 170
  in subhuman primate atherosclerosis,
    185
Plaques, complicated
  in Caucasians and Negroes, 170
  characteristics, 169
  in sequence of atherosclerotic
    lesions, 197
Plaques, fibrous
  and age, 170

Plaques, fibrous—*cont.*
  distribution, 169
  in Caucasians and Negroes, 170
  characteristics, 169
  in aortic segments, 170
Platelet aggregation
  and adrenaline-ATP system, 143
  inhibition by monoiodoacetate, 142
  and saturated fatty acids, 143
  and thrombogenesis, 141
  and unsaturated fatty acids, 143
  and vascular injury, 142
Phospholipase A, 229
  in arterial tissue, 230
  assay methods, 229
  distribution, 229
  inhibition-activation characteristics,
    229
  production of lyso-derivatives, 229
  properties, 229
Phospholipase A, vascular
  and "atherogenic diets", 230
  inhibition-activation characteristics,
    230
  in pig's aortas, 230
  species differences, 230
Polypeptides, aminoacid sequence of, 3
Polysomes, 4
PR enzyme, *see* Phosphorylase
    phosphatase
Prosthetic groups, and coenzymes, 72
Proteinases, 68
Proteins
  assay method, 106
  synthesis, 3
Proteolytic activity, in experimental
    atherosclerosis, 69
Proteosynthesis, in rabbit aortas, 69
Pseudocholinesterase, *see* Cholinesterase
Purine bases, 3
Purine nucleoside phosphorylase, 58
  assay method, 59
  distribution, 59
  in human arteries, 59
  in human atherosclerosis, 59
  properties, 58
  sex differences, 115
Purine and pyrimidine nucleotides, 79
  assay methods, 81
  and transphosphorylation, 80
Pyridoxal phosphate, and glycogen
    phosphorylase, 17
Pyrimidine bases, 3

Redox potential
   and free energy change, 39
   and NADH$_2$/NAD system, 39
   and respiratory chain, 39
   and succinate/fumarate system, 39
Regulator genes, 5
Repressors, 5
Respiratory chain, 39
   inhibitors of, 41
Respiratory control, 9
   and ADP, 10
   and ATP, 10
Ribitol, and flavin nucleotides, 76
Riboflavin, 76
   in dog aortas, 77
   in human aortas, 77
Ribonucleic acid, *see* RNA
Ribosephosphate isomerase, 38
   assay method, 38
   distribution, 38
   in human arteries, 38
   in human atherosclerosis, 38
Ribosomes, 4
RNA
   base sequence of, 3
   and injury, 158
RNA nucleotidyltransferase, 3, 115
   and Actinomycin D, 225
RNA-polymerase, and sex hormones, 115

Scar tissue, vascular, 146
Sclerosis, Mönckeberg's, 135
Sex differences, 108
Smooth muscle cells
   affinity for lipids, 147
   and atherogenesis, 146, 171, 189, 197
   and enzyme activities, 180, 185, 189
   and foam cells, 120, 171, 191
   in human large arteries, 178
   in human thoracic v. abdominal aorta, 175
   and lipid accumulation, 171, 191
   and multipotential cells, 187
   and vascular injury, 146
   vulnerability of, 201
Sorbitol dehydrogenase, in human arteries, 26
Starvation, and liver citrate, 82
Succinate dehydrogenase
   assay methods, 32, 102
   in calciferol-fed rats, 151

Succinate dehydrogenase—*cont.*
   in DOCA + salt hypertension, 157
   in duck's large arteries (histochem.), 184
   and gonadectomy, 114
   histochemical method for, 106
   in human arteries, 32
   in human ascending v. abdominal aortas, 174
   in human large arteries (histochem.), 178
   in human pulmonary arteries v. aortas, 173
   in macaque's ascending v. abdominal aortas, 187
   in macaque's large arteries (histochem.), 188
   properties, 31
Succinate dehydrogenase system
   in experimental rabbit atherosclerosis, 128, 131
   in calciferol-fed rats, 149
Succinoxidase system, 31
   in aortic segments, 32
   in human arteries, 32
   species differences, 32
Synthases, 70

TCA cycle, 27
   energy balance of, 39
TCA enzymes
   and injury by hypertension, 157
   in susceptible v. resistant arteries, 196
   and tissue injury, 151
   and vascular injury, 196
   and vascular lipid metabolism, 203
Tetrazolium methods
   for dehydrogenases, 102
   and phenazine methosulphate, 97
   and redox potential, 97
   site of reduction in, 97
   and succinoxidase system, 97
Thrombin production, 141
Thrombogenesis, and platelet aggregation 141
   (*see also* Platelet aggregation)
Thrombus formation, 141
   and inhibition of TCA cycle, 142
   and vascular injury, 142
Thymidine phosphorylase, 58
Tissue respiration, 10
   and mitochondrial ATPase, 10

Tissue respiration—*cont.*
  rate limiting factors, 9
  uncoupling from phosphorylation, 10
Transaminases, 56
  assay methods, 104
  in clinical diagnosis, 57
  distribution, 57
  in experimental atherosclerosis, 57
  in human arteries, 57
  in human atherosclerosis, 57
  and pyridoxal phosphate, 56
  in rabbit aorta, 57
  and urea formation, 56
Transhydrogenases, 35
  and oestrogens, 115
Transport ATPase, 47
Tricarboxylic acid cycle, *see* TCA cycle
Tricarboxylic acid cycle enzymes, *see* TCA enzymes
Transfer RNA (tRNA), 4
Tween esterase
  in arteries (histochem.), 222
  in atherosclerotic rabbits (histochem.), 222

UDPglucose (UDPG), 19
  and glycogen synthesis, 26, 80
  and mucopolysaccharides, 26, 80
  synthesis of, 80
UDPG-dehydrogenase, 26
UDPG-glycogen glucosyltransferase, 25
  assay method, 26
Uridine phosphorylase, 58
Uridine 5'-diphosphate
  and transfer of glycosyl group, 80
Uridine diphosphate glucose, *see* UDPglucose
Uridine nucleotides, distribution, 80
Useful energy, 12
  and chemical energy, 39
  in poorly vascularized tissue, 13

Vascular defence mechanisms
  and lipolysis, 199
  and phospholipid synthesis, 199
Vasa vasorum, and arterial blood supply, 161

Zwischenferment, *see* Glucose 6-phosphate dehydrogenase

DATE DUE			
NOV 2 3 1986			
NOV 1 2 1986			
			PRINTED IN U.S.A.